Yan Meng and Yaochu Jin (Eds.)

Bio-Inspired Self-Organizing Robotic Systems

Studies in Computational Intelligence, Volume 355

Editor-in-Chief
Prof. Janusz Kacprzyk
Systems Research Institute
Polish Academy of Sciences
ul. Newelska 6
01-447 Warsaw
Poland
E-mail: kacprzyk@ibspan.waw.pl

Further volumes of this series can be found on our homepage: springer.com

Vol. 332. Jianguo Zhang, Ling Shao, Lei Zhang, and Graeme A. Jones (Eds.)
Intelligent Video Event Analysis and Understanding, 2011
ISBN 978-3-642-17553-4

Vol. 333. Fedja Hadzic, Henry Tan, and Tharam S. Dillon
Mining of Data with Complex Structures, 2011
ISBN 978-3-642-17556-5

Vol. 334. Álvaro Herrero and Emilio Corchado (Eds.)
Mobile Hybrid Intrusion Detection, 2011
ISBN 978-3-642-18298-3

Vol. 335. Radomir S. Stankovic and Radomir S. Stankovic
From Boolean Logic to Switching Circuits and Automata, 2011
ISBN 978-3-642-11681-0

Vol. 336. Paolo Remagnino, Dorothy N. Monekosso, and Lakhmi C. Jain (Eds.)
Innovations in Defence Support Systems – 3, 2011
ISBN 978-3-642-18277-8

Vol. 337. Sheryl Brahnam and Lakhmi C. Jain (Eds.)
Advanced Computational Intelligence Paradigms in Healthcare 6, 2011
ISBN 978-3-642-17823-8

Vol. 338. Lakhmi C. Jain, Eugene V. Aidman, and Canicious Abeynayake (Eds.)
Innovations in Defence Support Systems – 2, 2011
ISBN 978-3-642-17763-7

Vol. 339. Halina Kwasnicka, Lakhmi C. Jain (Eds.)
Innovations in Intelligent Image Analysis, 2010
ISBN 978-3-642-17933-4

Vol. 340. Heinrich Hussmann, Gerrit Meixner, and Detlef Zuehlke (Eds.)
Model-Driven Development of Advanced User Interfaces, 2011
ISBN 978-3-642-14561-2

Vol. 341. Stéphane Doncieux, Nicolas Bredeche, and Jean-Baptiste Mouret(Eds.)
New Horizons in Evolutionary Robotics, 2011
ISBN 978-3-642-18271-6

Vol. 342. Federico Montesino Pouzols, Diego R. Lopez, and Angel Barriga Barros
Mining and Control of Network Traffic by Computational Intelligence, 2011
ISBN 978-3-642-18083-5

Vol. 343. Kurosh Madani, António Dourado Correia, Agostinho Rosa, and Joaquim Filipe (Eds.)
Computational Intelligence, 2011
ISBN 978-3-642-20205-6

Vol. 344. Atilla Elçi, Mamadou Tadiou Koné, and Mehmet A. Orgun (Eds.)
Semantic Agent Systems, 2011
ISBN 978-3-642-18307-2

Vol. 345. Shi Yu, Léon-Charles Tranchevent, Bart De Moor, and Yves Moreau
Kernel-based Data Fusion for Machine Learning, 2011
ISBN 978-3-642-19405-4

Vol. 346. Weisi Lin, Dacheng Tao, Janusz Kacprzyk, Zhu Li, Ebroul Izquierdo, and Haohong Wang (Eds.)
Multimedia Analysis, Processing and Communications, 2011
ISBN 978-3-642-19550-1

Vol. 347. Sven Helmer, Alexandra Poulovassilis, and Fatos Xhafa
Reasoning in Event-Based Distributed Systems, 2011
ISBN 978-3-642-19723-9

Vol. 348. Beniamino Murgante, Giuseppe Borruso, and Alessandra Lapucci (Eds.)
Geocomputation, Sustainability and Environmental Planning, 2011
ISBN 978-3-642-19732-1

Vol. 349. Vitor R. Carvalho
Modeling Intention in Email, 2011
ISBN 978-3-642-19955-4

Vol. 350. Thanasis Daradoumis, Santi Caballé, Angel A. Juan, and Fatos Xhafa (Eds.)
Technology-Enhanced Systems and Tools for Collaborative Learning Scaffolding, 2011
ISBN 978-3-642-19813-7

Vol. 351. Ngoc Thanh Nguyen, Bogdan Trawiński, and Jason J. Jung (Eds.)
New Challenges for Intelligent Information and Database Systems, 2011
ISBN 978-3-642-19952-3

Vol. 352. Nik Bessis and Fatos Xhafa (Eds.)
Next Generation Data Technologies for Collective Computational Intelligence, 2011
ISBN 978-3-642-20343-5

Vol. 353. Igor Aizenberg
Complex-Valued Neural Networks with Multi-Valued Neurons, 2011
ISBN 978-3-642-20352-7

Vol. 354. Ljupco Kocarev and Shiguo Lian (Eds.)
Chaos-Based Cryptography, 2011
ISBN 978-3-642-20541-5

Vol. 355. Yan Meng and Yaochu Jin (Eds.)
Bio-Inspired Self-Organizing Robotic Systems, 2011
ISBN 978-3-642-20759-4

Yan Meng and Yaochu Jin (Eds.)

Bio-Inspired Self-Organizing Robotic Systems

Springer

Yan Meng
Department of Electrical and Computer Engineering
Stevens Institute of Technology
Castle Point on Hudson,
Hoboken, NJ 07030
USA
E-mail: yan.meng@stevens.edu

Yaochu Jin
Department of Computing
University of Surrey
Guildford, Surrey, GU2 7XH, UK
E-mail: yaochu.jin@honda-ri.de

ISBN 978-3-642-20759-4 e-ISBN 978-3-642-20760-0

DOI 10.1007/978-3-642-20760-0

Studies in Computational Intelligence ISSN 1860-949X

Library of Congress Control Number: 2011927941

Typeset & Cover Design: Scientific Publishing Services Pvt. Ltd., Chennai, India.

Printed on acid-free paper

9 8 7 6 5 4 3 2 1

springer.com

Preface

Robotic systems are increasingly required to work under various dynamic, unpredictable, and unknown environments to accomplish various complex tasks. To address these challenges, self-organizing swarming robots and self-reconfigurable modular robots have been proposed. For example, emerging collective behaviors of large-scale swarm robotic systems can provide high flexibility and adaptation to deal with environmental changes. Self-reconfigurable modular robots can automatically change the morphology of the systems to adapt to complex terrains. To deal with more complex and demanding environments as well as task requirements, autonomous development of cognitive abilities together with the body plan of robots also becomes extremely important nowadays. Compared to traditional, preprogrammed techniques, self-organizing robotic systems are more promising in dynamic, unknown environments, particularly in terms of robustness, self-repair, and self-adaptation. To exhibit the above-mentioned properties, self-organizing systems must be controlled in a distributed manner, ideally through local interactions among individual simple robots without an external global control. Unfortunately, design of distributed self-organizing robotic systems remains one of the most challenging problems in robotics.

Biological systems, from macroscopic swarm systems of social insects to microscopic cellular systems, can generate robust and complex emerging global behaviors through relatively simple local interactions in the presence of various kinds of uncertainty. Borrowing ideas from biological systems for developing self-organizing robotic systems has become increasingly popular and enjoyed considerable success in recent years. For example, swarm intelligence, a novel paradigm for solving complex problems with massively parallel systems, has been inspired by behaviors observed in social insect colonies and flocks of birds. Another self-organizing process in biology is morphogenesis of multi-cellular organisms. Morphogenetic approaches based on computational models of embryogeny to self-organizing robotic systems, which are now known as morphogenetic robotics, have shown to be very promising.

This edited book presents a collection of the most representative research work on biological inspired self-organizing robotic systems. This book is composed of four parts. Part I discusses bio-inspired self-organizing approaches to swarm robotic systems, such as morphogenetic approaches inspired by biological morphogenesis in multi-cellular organisms, swarm intelligence based approaches simulating the behaviors of social insects (e.g., birds, honeybees), hormone-based approaches to robotic organisms, and genetic stigmergy based communication mechanism for swarm robots. In Chapter 1, a new emerging research field in developmental robotics, morphogenetic robotics, is introduced by Jin and Meng. The main philosophy of morphogenetic robotics is to apply development principles to design the morphology and controller of self-organizing robotic systems. The main topics belong to morphogenetic robotics are summarized and the relationship between morphogenetic robotics to evolutionary robotics and epigenetic robotics are discussed. Evolutionary developmental robotics, which is a natural marriage of evolutionary and developmental robotics, is envisaged. Chapter 2, contributed by Schmickl, first presents different swarm robotic systems using controllers inspired from collective behaviors of honeybees and slime model aggregation. Then, a hormone-based control paradigm for multi-modular robotic organisms is discussed. All these systems are self-organized in a distributed manner. In Chapter 3, La and Sheng propose two flocking control algorithms, namely, Multi-CoM-Shrink and Multi-CoM-Cohesion, for multi-robot target tracking in cluttered and noisy environments inspired from flocking behaviors of birds, bees, and fish observed in nature. The stability and scalability of the algorithms have also been investigated theoretically. Inspired by the pheromone-based stigmergy in ant systems, in Chapter 4, Brandoff and Sayama describe an artificial genetic stigmergy for indirect communications among robots in a swarm system, where swarm robots conduct an unknown environment mapping task. In the last chapter of Part I, Garnier,a biologist, shares his points of view on how swarm roboticsinspired from biological self-organization in animal societies can benefit from and contribute back to the study of collective animal behaviors.

Part II introduces several bio-inspired approaches to self-reconfigurable modular robots. Kernbach et al., in Chapter 6, propose constrained-based self-optimization of self-assembly of heterogeneous modular robots, which is mainly inspired from gene regulatory networks observed in molecular organisms. Mechanical and integration constrains of robot modules are taken into account. Inspired by the embryonic development of multi-cellular organisms, hierarchical morphogenetic approaches are presented for self-reconfiguration of two modular robots (i.e., Cross-Cube and Cross-Ball) by Meng and Jin in Chapter 7. This hierarchical framework consists of three layers, where the virtual-cell based layer 1 controller is responsible for automatically generating appropriate target configurations for robots based on environmental constraints, the gene regulatory network based layer 2 controller provides self-reconfiguration plans for modules, and the skeleton-based layer 3 controller guides the modules to move to the target configuration with the mechanical and connectivity constraints of modules. By using this hierarchical morphogenetic framework, the target patterns of the robots can be automatically generated to adapt to changing environments. In Chapter 8, Miyashita et al. first discuss three basic research issues in self-assembling robots for manufacturing 3D micro products, namely, assembly, dynamic, and interaction issues. Then, a case study with passive modules (actuated by permanent magnet) and active modules (actuated by vibration motors) has shown the segregation behaviors of modules in a distributed manner. Entropy analysis is also provided to govern macroscopic self-assembly systems.

Whereas Part I and Part II concentrate on the self-organizing of swarm and modular robots, autonomous mental development of robotic systems is the main focus of Part III. In Chapter 9, Weng presents a developmental network (DN) based general purpose model of the brain for robotic systems. A cell-centered in-place learning scheme is proposed to handle all levels of brain development and operation based on biological genomic equivalence principles, which is automatically built up from five basic brain puzzles: development, architecture, area, space and time. The focus of this chapter is the analysis on how this model deals with temporal contexts.

Two specific applications of self-organizing robotic systems are presented in Part IV. Inspired by the slime mouldPhysarumpolycephalum in biological organisms, Jones et al. propose physarum robots in Chapter 10, where physarum can be considered as a smart computing material. A particle-based computational model is proposed for physarum robots, which spontaneously generate complex oscillatory patterns from simple local interactions in a distributed manner. The authors expect that physarum robots may be used as physical instances of smart materials for the future robotic devices. In the final chapter of the book, Chapter 11, a layered architecture is presented by Hoffmann et al. to build up self-organizing robotic cells for industrial robots. In the proposed system, an organic computing based model is employed to combine the system emergence and self-organization properties.

We believe that this book will provide readers an up-to-date and comprehensive view of bio-inspired self-organizing robotic systems. We hope this book will bridge multi-disciplinary research areas such as robotics, artificial life, cognitive sciences, systems biology, developmental biology and evolutionary computation, thereby inspiring researchers and engineers to generate more creative ideas to further promote this emerging and exciting research field.

We would like to thank all contributors who prepared excellent chapters for this book. We would also like to thank Prof. JanuszKacprzyk, Editor-in-Chief of this book series and Dr. Thomas Ditzinger from Springer for offering us the opportunity to edit the book.

Yan Meng
Department of Electrical and Computer Engineering
Stevens Institute of Technology
Hoboken, NJ 07030, USA

Yaochu Jin
Department of Computing
University of Surrey
Guildford, GU2 7XH, UK

Contents

Part I: Self-Organizing Swarm Robotic Systems

Morphogenetic Robotics - An Evolutionary Developmental Approach to Morphological and Neural Self-Organization of Robotic Systems

Yaochu Jin and Yan Meng

Abstract. Morphogenesis can be considered as a self-organizing process shaped by natural evolution, in which two major adaptation mechanisms found in nature are involved, namely evolution and development. The main philosophy of morphogenetic robotics is to apply evolutionary developmental principles to robotics for designing self-organizing, self-reconfigurable, and self-repairable single- or multi-robot systems. We categorize these methodologies into three areas, namely, morphogenetic swarm robotic systems, morphogenetic modular robots, and co-development of body and brain of robotic systems. In this chapter, we give a brief introduction to morphogenetic robotics. A few examples are also presented to illustrate how evolutionary developmental principles can be applied to swarm robots in changing environment. We also describe computational models for genetically driven neural and morphological development and activity-dependent neural development. As developmental mechanisms are often shaped by evolution both in nature and simulated systems, we suggest that evolutionary developmental robotics is a natural next step to follow.

1 Introduction to Morphogenetic Robotics

The physical development of animals includes the processes that cause the creation of both the body plan and nervous system, including cleavage, gastrulation, neurulation, organogenesis [55]. Some living organisms, such as amphibians, also undergo a biological process known as metamorphosis, during which both the shape and

Yaochu Jin
Department of Computing, University of Surrey, Guildford, Surrey, GU2 7XH, UK
e-mail: yaochu.jin@surrey.ac.uk

Yan Meng
Department of Electrical and Computer Engineering, Stevens Institute of Technology, Hoboken, NJ 07030, USA
e-mail: yan.meng@stevens.edu

Y. Meng and Y. Jin (Eds.): Bio-Inspired Self-Organizing Robotic Systems, SCI 355, pp. 3–23.
springerlink.com

size of the organisms change [5]. The past decade has witnessed rapid technical and theoretical advances in evolutionary developmental biology [10] (often known as *evo-devo*) and systems biology in understanding molecular and cellular mechanisms that control the biological morphogenesis. These advances have not only helped us in understanding biological processes such as human deceases, but also provided us new powerful tools for designing engineered systems. For example, increasing evidence has been revealed that biological morphogenesis can be regarded as a self-organizing and self-assembling process through cellular and molecular interactions under the genetic and environmental control [2, 50]. In addition, biological morphogenesis has also shown a surprising degree of robustness [3]. Due to the attractive properties that biological morphogenesis exhibits, much attention has been paid to employ genetic and cellular mechanisms for designing robotic systems, in particular for self-organizing swarm robotic systems and self-reconfigurable modular robots. In addition, a large body of research has been performed in artificial life and robotics to design the body plan and neural controller of robots using an evolutionary developmental approach [51, 53, 54].

In this chapter, we provide a brief introduction to *morphogenetic robotics*, a terminology that was first coined by the authors in 2009 and reported on a wiki webpage [17], later more formally presented in [20]. In general, morphogenetic robotics denotes the research area dedicated to the application of genetic and cellular mechanisms underlying biological morphogenesis to robotics. Morphogenetic robotics includes the following three main topics:

- Morphogenetic swarm robotic systems that deal with the self-organization of swarm robots using genetic and cellular mechanisms [12, 13, 29, 49].
- Morphogenetic modular robots where modular robots adapt their configurations autonomously based on the current environmental conditions using morphogenetic principles [34, 33, 35].
- Developmental approaches to the design of the body or body parts and its neural controller of robots [15, 26].

Neural development may further be divided into activity-independent [18] and activity-dependent neural development [6]. Activity-independent neural development lays down the initial structure of the nervous system and is mainly regulated by genetic networks, whereas activity-dependent development refines the neural connectivity driven by neural activities. However, a clear boundary between activity-independent and activity-dependent neural development does not exist. Recent findings in neuroscience suggest that early neural development, including neuronal proliferation, migration, differentiation axon growth and dendrite outgrowth are more or less influenced by spontaneous neural activity. On the other hand, activity-dependent neural plasticity involves in changes in gene expression, i.e., activity-dependent neural development is eventually also regulated by genetic networks.

The three areas of morphogenetic robotics are not necessarily fully separated. For example, it can happen that morphogenetic principles are applied to a hybrid of swarm and modular robotic system, where a number of swarm robots may assemble into one single "modular" robot and then dissemble into multiple single robots again. Meanwhile, developmental design of the morphology and controller may be employed for modular robots to accomplish complex tasks.

The motivation to create this new terminology is to fill the gap between developmental robotics and natural development of intelligent animals. In traditional developmental robotics, often known as epigenetic robotics, attention has mainly be paid to cognitive development and the physical development of the body plan and neural system is often neglected.

2 Computational Modeling of Multi-cellular Morphogenesis

2.1 Biological Morphogenesis and Metamorphosis

Morphogenesis of animals can be divided into early embryonic development and later embryonic development [9]. Early embryonic development typically involves cleavage, gastrulation, and axis formation, while later embryonic development is mainly responsible for the development of the nervous systems, starting with the segregation of neural and glial cells from the ectoderm germ layer [47]. An example of morphogenesis of nematostella vectensis is illustrated in Fig. 1.

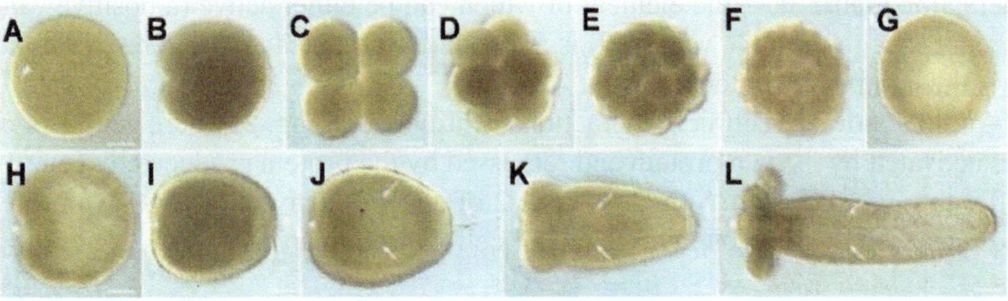

Fig. 1 Morphogenesis of nematostella vectensis. The development stages are: Egg (A), morlula (B-F), blastula (G), gastrula (H), planula (I-J), and polyp (K-L). Taken from [27].

Metamorphosis is another interesting stage of biological development. There are two types of metamorphosis, namely, incomplete and complete metamorphosis. For organisms underlying incomplete metamorphosis, there are three developmental stages, in which nymphs look similar to adults. In contrast, organisms that undergo complete metamorphosis have four developmental stages, in which the shape of the organisms changes drastically.

2.2 Modeling of Developmental Gene Regulatory Networks

Biological morphogenesis is governed by gene regulatory networks (GRNs). To understand the genetic and cellular principles underlying morphogenesis in silico, it is necessary to build up a mathematical model of developmental gene networks. In recent years, much research work has been reported on computational modeling of signal transduction and developmental genetic networks. These network models have been employed either to reconstruct a gene regulatory sub-network in biology based on experimental data, or to analyze the basic properties of biological genetic networks, such as robustness and evolvability.

For describing the morphogenesis of multi-cellular organisms, the interaction between the cells and its influence on gene expression dynamics must be taken into account. Mjolsness et al. [36] has suggested a generalized GRN model that considered diffusion of transcription factors among the cells:

$$\frac{dg_{ij}}{dt} = -\gamma_j g_{ij} + \phi\left[\sum_{l=1}^{n_g} W^{jl} g_{il} + \theta_j\right] + D_j \nabla^2 g_{ij}, \quad (1)$$

where g_{ij} denotes the concentration of j-th gene product (protein) in the i-th cell. The first term on the right-hand side of Equation (1) represents the degradation of the protein at a rate of γ_j, the second term specifies the production of protein g_{ij}, and the last term describes protein diffusion at a rate of D_j. ϕ is an activation function for the protein production, which is usually defined as a sigmoid function $\phi(z) = 1/(1+\exp(-\mu z))$. The interaction between the genes is described with an interaction matrix W^{jl}, the element of which can be either active (a positive value) or repressive (a negative value). θ_j is a threshold for activation of gene expression. n_g is the number of proteins.

An illustration of cell-cell interactions is provided in Fig. 2, where gene 1 of cell 1 is activated by its own protein and repressed by the protein produced by gene 1 of cell 2 through diffusion. Similarly, gene 2 of cell 1 is activated by its own protein, and repressed by the protein of gene 2 of cell 2 through diffusion.

3 Morphogenetic Self-Organization of Swarm Robots

3.1 Swarm Robotic Systems

A swarm robotic system consists of a number of robots. Usually, each robot has only low computational power and limited communication capability, which therefore, can only accomplish simple tasks. However, with a proper control mechanism, the robots can work together to perform complex tasks. Swarm robotic systems with a decenralized control strategy are often believed to be more flexible and robust. Typical applications of swarm robotic systems include group transport, foraging, shape formation, boundary coverage, urban search and rescue, and unknown environment exploration. However, designing a decentralized control algorithm for swarm robotic systems has been a challenging task [30].

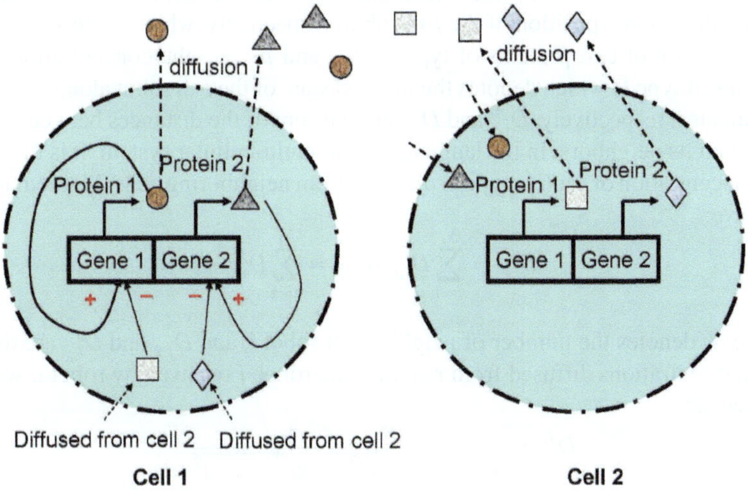

Fig. 2 Illustration of cell signaling in a multi-cellular system.

3.2 *A Metaphor between Swarm Robotic Systems and Multi-cellular Systems*

3.2.1 The Cell-Robot Mapping

To apply genetic and cellular mechanisms in biological morphogenesis to self-organization of swarm robots, it is necessary to establish a metaphor between a multi-cellular system and a multi-robot system. In the metaphor, the most important functions of a robot is mapped to concentrations of a cell. In [13, 18, 31], the location and velocity of the robots are described by the protein concentration of a few genes whose expression is influenced by each other. Typically, for a robot in a three-dimensional space, three proteins are used for denoting the robot's position, and three for the velocity. Note, however, that the mathematical definition of the protein concentrations standing for position and velocity of the robots do not satisfy the exact physical relationship between position and velocity. So the speed in this context can be seen as an internal state of the robots.

Keeping the metaphor between the cells and the robots in mind, the movement dynamics of each robot can be described by a GRN model, where the concentration of two proteins of type G represents the x and y position of a robot, respectively, and that of the proteins of type P representing an internal state of the robot.

$$\begin{aligned} \frac{dg_{i,x}}{dt} &= -az_{i,x} + mp_{i,x} \\ \frac{dg_{i,y}}{dt} &= -az_{i,y} + mp_{i,y} \end{aligned} \tag{2}$$

$$\begin{aligned} \frac{dp_{i,x}}{dt} &= -cp_{i,x} + kf(z_{i,x}) + bD_{i,x} \\ \frac{dp_{i,y}}{dt} &= -cp_{i,y} + kf(z_{i,y}) + bD_{i,y} \end{aligned} \tag{3}$$

where $i = 1, 2, ..., n$. and n is the total number of robots (cells) in the system. $g_{i,x}$ and $g_{i,y}$ are the x and y position of the i-th robot, respectively, which corresponds to the concentration of two proteins of type G. $p_{i,x}$ and $p_{i,y}$ are the concentration of two proteins of type P, which denotes the internal state of the i-th robot along the x and y coordinates, respectively. $D_{i,x}$ and $D_{i,y}$ are the sum of the distances between the i-th robot and its neighbors. In the language of the multi-cellular system, it is the sum of the concentration of protein type G diffused from neighboring cells. Mathematically, we have:

$$D_{i,x} = \sum_{j=1}^{N_i} D^j_{i,x}, \; D_{i,y} = \sum_{j=1}^{N_i} D^j_{i,y} \quad , \tag{4}$$

where N_i denotes the number of neighbors of robot i, and $D^j_{i,x}$ and $D^j_{i,y}$ are the protein concentrations diffused from neighboring robot j received by robot i, which is defined as:

$$D^j_{i,x} = \frac{(g_{i,x} - g_{j,x})}{\sqrt{(g_{i,x} - g_{j,x})^2 + (g_{i,y} - g_{j,y})^2}} \quad , \tag{5}$$

$$D^j_{i,y} = \frac{(g_{i,y} - g_{j,y})}{\sqrt{(g_{i,x} - g_{j,x})^2 + (g_{i,y} - g_{j,y})^2}} \quad . \tag{6}$$

The diffusion term in the regulatory model simulates the cell-cell signaling in multi-cellular systems. For a swarm robotic system, this entails that each robot is able to detect the distance to its neighboring robots, which is practical and easy to realize.

3.2.2 Target Shape Representation Using Morphogen Gradients

In biological morphogenesis, morphogen concentration gradients control cell fate specification and play a key role in understanding pattern formation [1]. In the present gene regulatory model for shape formation of swarm robots, the target shape information is also provided in terms of morphogen gradients, which is defined by f_z in Equation 3. For a two-dimensional target shape, $f(z_i)$ can be defined as follows:

$$\begin{aligned} f(z_{i,x}) &= \frac{1-e^{-z_{i,x}}}{1+e^{-z_{i,x}}} \\ f(z_{i,y}) &= \frac{1-e^{-z_{i,y}}}{1+e^{-z_{i,y}}} \end{aligned} \tag{7}$$

where $z_{i,x}$ and $z_{i,y}$ are the gradients along x-axis and y-axis, respectively, of an analytic function h, which is described as:

$$z_{i,x} = \frac{\partial h}{\partial g_{i,x}}, \; z_{i,y} = \frac{\partial h}{\partial g_{i,y}} \tag{8}$$

where h defines the shape the robots should form.

The map between a multi-cellular system and a multi-robot system used in this work is provided in Fig. 3.

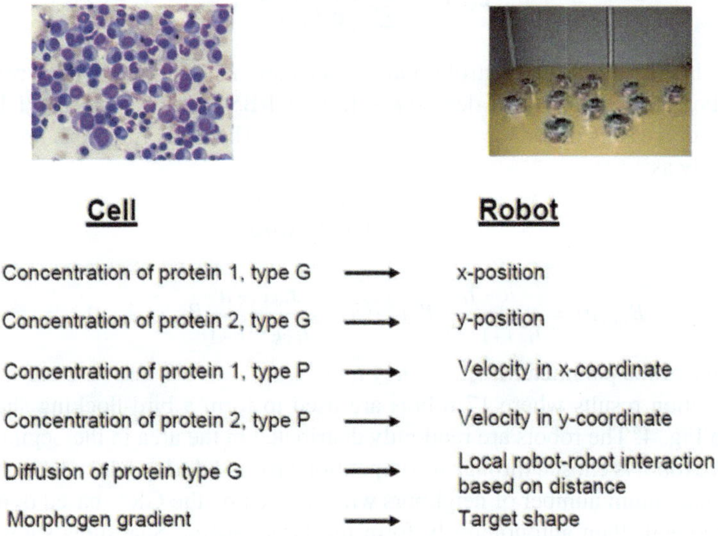

Fig. 3 A metaphor between a multi-cellular system and a multi-robot system.

3.3 *From Analytic to Freeform Shape Representation*

Representation of the target shape of the swarm robotic system in terms of morphogen gradients is an important step in designing morphogenetic self-organizing systems. The most straightforward way to represent a shape is to use an analytic function. For example, if the robots are required to form a unit circle, the following function can be used:

$$h = (g_{i,x}^2 + g_{i,y}^2 - 1)^2. \tag{9}$$

There are potentially three problems with this way of shape representation. First, the complexity of the shapes is limited. In general, analytic functions can describe closed two-dimensional shapes only. Second, the algorithm needs a global coordinate system for describing the shapes, which poses a big problem for decentralized systems. Third, the shape can be formed only on a predefined location. To address these issues, parametrized shape representation models, such as Bézier, B-Spline and non-uniform rational B-Spline (NURBS) can be used [31].

A NURBS curve is defined by its order, a set of weighted control points, and a knot vector. The control points define the shape of the curve, and the knot vector is a set of parameters that determines where and how the control points affect the NURBS curve. A NURBS model can represent both curve and surface in a two- or three-dimensional Cartesian space. Let $B_{i,k}(u)$ be the B-spline basis functions of the NURBS model, where i corresponds to i-th control point, and k denotes the degree of the basis function. In the NURBS model, a curve can be defined as a combination of a set of piecewise rational basis functions with n+1 control points $\mathbf{p}_i$ and the associated weights as follows:

$$\mathbf{c}(u) = \frac{\sum_{i=1}^{n} \mathbf{p}_i w_i B_{i,k}(u)}{\sum_{i=1}^{n} \mathbf{p}_i B_{i,k}(u)}, \tag{10}$$

where n is the number of control points, u is a parameter in the NURBS representation. For basis functions of degree $k-1$, a NURBS curve has $n+k+1$ knots t_i in a non-decreasing sequence: $t_0 \leq t_1 \leq ... \leq t_{n+k}$. The basis functions are defined recursively as:

$$B_{i,k}(u) = \begin{cases} 1, \; t_i \leq u \leq t_{i+1} \\ 0, \text{ otherwise} \end{cases} \tag{11}$$

where

$$B_{i,k}(u) = \frac{u - t_i}{t_{i+k+1} - t_i} B_{i,k-1}(u) + \frac{t_{i+k} - u}{t_{i+k} - t_{i+1}} B_{i+1,k-1}(u). \tag{12}$$

The range of the parameter is $t_{k-1} \leq u \leq t_{k+1}$.

Simulation results where 17 robots are used to form a bird-flocking shape are given in Fig. 4. The robots are randomly distributed in the area in the beginning. A reference robot is chosen through a competition process, during which the robot that has the maximum number of neighbors wins. Driven by the GRN-based dynamics, the robots will then autonomously form the target shape. Snapshots showing 17 robots forming a bird-flocking shape are provided in Fig. 4. More details about this part of the work can be found in [12].

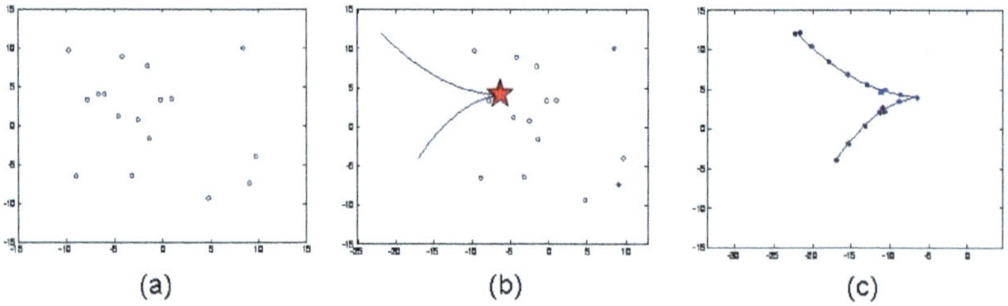

Fig. 4 Snapshots showing the emergence of a pattern from 17 robots similar to bird flocking [12]. (a) Random initialization; (b) Determination of a reference robot (denoted by a star) through competition; (c) Emergence of the target shape.

3.4 From Predefined Target Shape to Adaptive Shape Generation

In the previous models, it is assumed that the target shape to be constructed by the robots is known beforehand and therefore can be pre-defined. Unfortunately, this assumption does not hold if the target shape must change as the environment changes. To address this issue, a hierarchical GRN (H-GRN) network consisting of two layers has been suggested [19], as illustrated in Fig. 5. Layer 1 generates patterns in terms of protein concentration depending on the given environment, for instance, the location of the targets to be followed and entrapped. Layer 2 is responsible for controlling the robot's movement dynamics, where protein types G and P represent the position and internal state vectors of the robots, respectively. If the patterns are

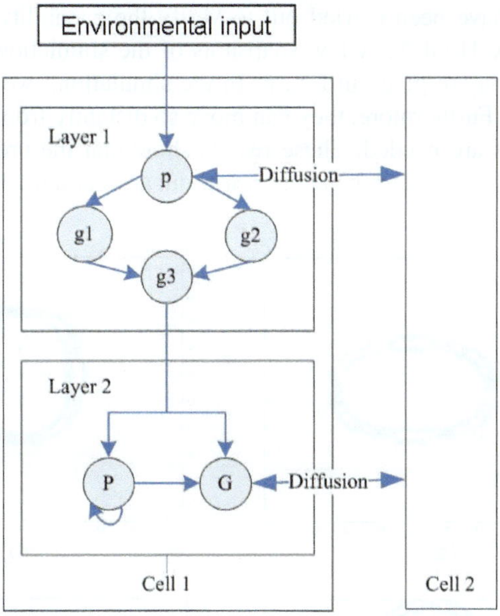

Fig. 5 A diagram of the H-GRN. Layer 1 is a GRN having four proteins, of which protein p can be regulated by environmental inputs, and protein g_3 is the morphogen gradient describing the pattern to be formed. Protein g_3 can influence the production of both proteins G and P, which represents the position and velocity vectors of the robots, respectively.

generated in a two-dimensional (2D) space, the vector length of G and P is two. In a 3D space, the dimension of both position and internal state vectors is three. Note that only protein type G can diffuse into other cells and influence the motion dynamics of other robots. The functionality of GRN layer 2 is similar to the single layer GRN described in the previous section.

From the above discussions, we note that the pattern generated by layer 1 plays a similar role of morphogen gradients in biological morphogenesis. The main advantage of having an additional GRN layer for pattern generation is that it enables the system to generate a desired pattern adaptively in a changing environment, which is impossible to achieve if the target pattern is predefined.

Note that the GRN of layer 1 is activated only when the robot detects a target or multiple targets. Based on the position of the detected targets, a target pattern will be generated. These robots are termed as organizing robots. Once a target pattern is generated in organizing robots, the dynamics of the GRN of layer 2 in these robots are then activated to guide the robots to the target pattern. The robots that do not detect any target will follow the movement of the organizing robots until they detect any targets by themselves. By then, they also become organizing robots: the GRN of layer 1 will be activated, the target pattern will be generated and finally the robots will move to the target pattern guided by the dynamics of the GRN of layer 2.

Simulations have been carried out to verify the capability of adaptive pattern generation of the H-GRN. A few snapshots of the simulation results on adaptive pattern generation are given in Fig. 6. In the simulation, two targets move in the considered area. Furthermore, they can move so distantly from each other that two separate patterns are needed. These results show that the proposed model works properly for adaptive pattern generation without any centralized control.

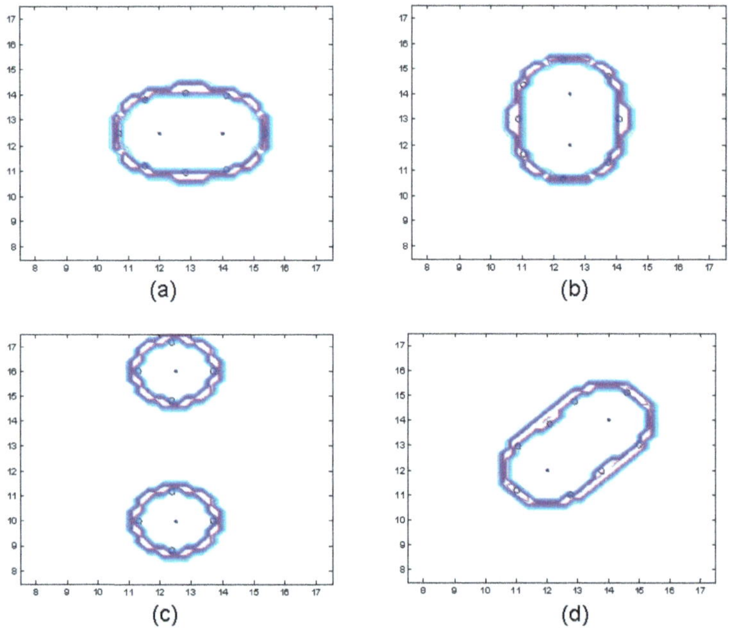

Fig. 6 Snapshots of simulations showing the adaptive pattern generation when the targets to be entrapped move.

Proof-of-concept experiments with physical robots have also been conducted. In the experiments, two targets (e-puck mobile robots covered with a piece of yellow paper) should be entrapped by other eight robots. This function has been achieved, as shown in Fig. 7. After that, one of the target robots moves out the circular shape that the eight robots have formed. This environmental change should be detected automatically and a new pattern must be constructed. Once the new pattern is generated, the eight robots should formulate a new shape that entrap the two target robots again. This adaptation process is shown in Fig. 8.

3.5 *Intermediate Summary*

Compared to existing approaches [16], the morphogenetic approach to self-organizing swarm robotic systems has the following advantages. First, the global behavior, i.e., the target shape in the context of pattern formation, can be embedded

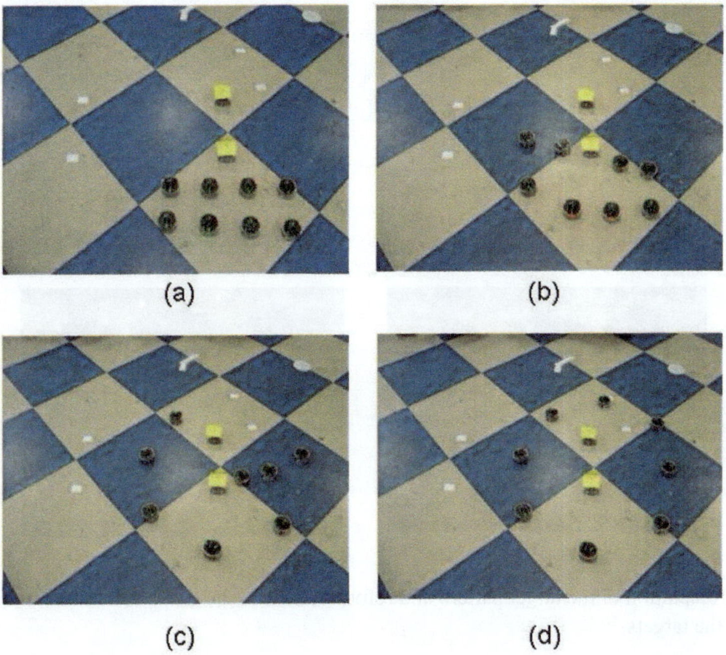

Fig. 7 Snapshots from experimental results where two robots (covered with a piece of paper) are trapped with another eight robots organized by the H-GRN model.

in the robot dynamics in the form of morphogen gradients. In pattern formation, the global shape can be described using parametrized models such as a NURBS model that can represent both analytical and free-form shapes. The GRN model can then generate implicit local interaction rules automatically to generate the global behavior, which can be guaranteed through a rigorous mathematical proof. Second, the morphogenetic approach is robust to perturbations in the system and in the environment. Third, it has also shown that the morphogenetic approach can provide a unified framework for multi-robot shape formation and boundary coverage [12], since the representation of the target shape is independent of a specific global coordination system. Last but not the least, the morphogenetic framework can generate patterns automatically in a changing enviornment, which, to the best of our knowledge, has not been reported in the literature.

4 Morphogenetic Modular Robots for Self-Organized Reconfiguration

Self-reconfigurable modular robots consist of a number of modules and are able to adapt their shape (configuration) by re-arranging their modules to changing environments [39]. Each module is a physical or simulated 'body' containing a controller.

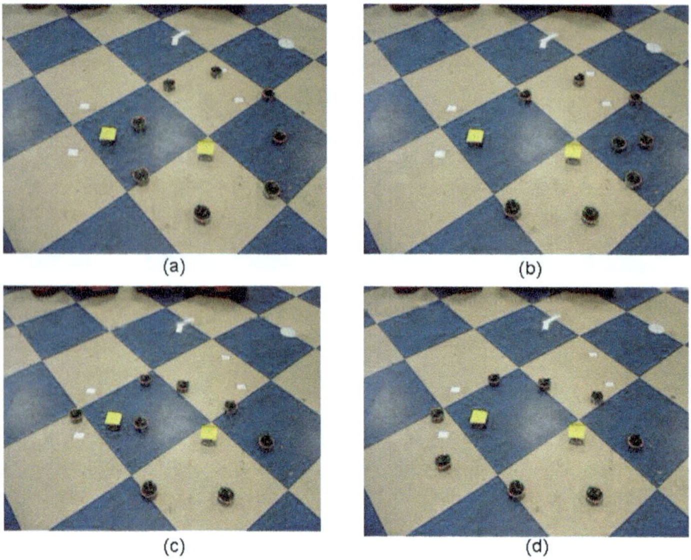

Fig. 8 Adaptation of the target pattern and reformulation of the shape to maintain the entrapping of the targets.

Both physical modular robots, such as M-TRAN [38] and Molecube [40], and simulated 'animats', such as Karl Sims' virtual creature [51] and Framsticks [25] have been constructed for reconfigurable robotic systems.

The connection between reconfigurable modular robots and multi-cellular organisms appears more straightforward. Each unit in modular robots can be seen as a cell, and there are similarities in control, communication and physical interactions between cells in multi-cellular organisms and modules in modular robots. For example, control in both modular robots and multi-cellular organisms is decentralized. In addition, global behaviors of both modular robots and multi-cellular organisms emerge through local interactions of the units, which include mechanic, magnetic and electronic mechanisms in modular robots, and chemical diffusion and cellular physical interactions such as adhesion in multi-cellular organisms. Therefore, it is a natural idea to develop control algorithms for self-reconfigurable modular robots using biological morphogenetic mechanisms [56, 34, 33, 35]. More details can be found in [34, 33, 35].

As we discussed before, there is also a link between swarm robotic systems and modular robots. This happens when individual robots in a swarm robotic system assemble into a modular robots. Vice versa, a modular robot consisting of individual robots can again disassembled into swarm robots.

5 Morphogenetic Brain-Body Co-development

5.1 A GRN Model for Neural and Morphological Development

The growth of the animat morphology is under the control of GRNs and cellular physical interactions. Extended from the cellular growth model for structural design, GRN models for the development of a nervous system [21] and body plan [48] of primitive animals have been proposed. In the genome of the GRN models, each gene consists of a number of structural units (SUs) proceeded by a number of regulatory units (RUs). RUs can be activating (RU^+) or repressive (RU^-). When SUs are activated, they will produce proteins either responsible for cellular behaviors such as cell division, cell death, cell migration, and axon growth, or proteins regulating the activation of the structural units, which are also known as transcription factors (TFs). If a TF can only regulate the genes inside the cell, it is then called an internal TF. If a TF can also diffuse out of the cell and regulate the genes of other cells, it is termed as an external TF. A TF can be both intracellular and intercellular. An example of a chromosome in the cellular model for neural development is provided in Fig. 9. From the figure, we note that single or multiple RUs may regulate the expression of a single or multiple SUs.

Whether a TF can influence an RU is dependent on the degree of match between the affinity value of a TF and that of an RU. If the difference between the affinity values of a TF and a RU is smaller than a predefined threshold ε, the TF can bind to the RU to regulate. The affinity match ($\gamma_{i,j}$) between the i-th TF and j-th RU is defined by:

$$\gamma_{i,j} = \max\left(\varepsilon - \left|\text{aff}_i^{TF} - \text{aff}_j^{RU}\right|, 0\right). \tag{13}$$

If $\gamma_{i,j}$ is greater than zero and the concentration c_i of the i-th TF is above a threshold (ϑ_j) defined in the j-th RU, then the i-th TF influences the j-th RU.

Thus, the activation level contributed by this RU (denoted by $a_j, j = 1,...,N$) amounts to $a_j = \sum_{i=1}^{M} |c_i, -\vartheta_j|$, where M is the number of existing TFs. The expression level of the k-th gene, that is regulated by N RUs, can be defined by

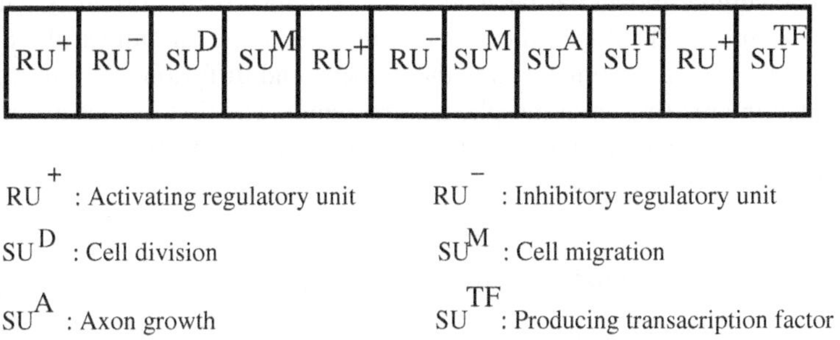

Fig. 9 An example of chromosome for neural development.

$$\alpha_k = 100 \sum_{j=1}^{N} h_j a_j (2s_j - 1), \tag{14}$$

where $s_j \in (0,1)$ denotes the sign (positive for activating and negative for repressive) of the j-th RU and h_j is a parameter representing the strength of the j-th RU. If $\alpha_k > 0$, then the k-th gene is activated and its corresponding behaviors encoded in the SUs are performed.

A SU that produces a TF encodes all parameters related to the TF, such as the affinity value, a decay rate D_i^c, a diffusion rate D_i^f, as well as the amount of the TF to be produced:

$$A = \beta \frac{2}{1+e^{-20 \cdot f \cdot \alpha}} - 1, \tag{15}$$

where f and β are both encoded in the $\mathrm{SU}^{\mathrm{TF}}$.

A TF produced by a SU can be partly internal and partly external. To determine how much of a produced TF is external, a percentage ($p^{\mathrm{ex}} \in (0,1)$) is also encoded in the corresponding gene. Thus, $p^{\mathrm{ex}}A$ is the amount of external TF and $(1 - p^{\mathrm{ex}})A$ is that of the internal TF.

To make it easier for simulating the diffusion of TFs, cells are put in an environment that is divided into a number of grids. External TFs are put on four grid points around the center of the cell, which undergoes first a diffusion (Eqn. 16) and then decay process (Eqn. 17):

$$\mathbf{u}_i(t) = \mathbf{u}_i(t-1) + 0.1 \cdot D_i^f \cdot (\mathbf{G} \cdot \mathbf{u}_i(t-1)), \tag{16}$$

$$\mathbf{u}_i(t) = \min\left((1 - 0.1 \cdot D_i^c)\,\mathbf{u}_i(t), 1\right), \tag{17}$$

where $\mathbf{u}_i$ is a vector of the concentrations of the i-th TF at all grid points and the matrix $\mathbf{G}$ defines which grid points are adjoining.

The SUs encode cellular behaviors and the related parameters. The SU for cell division encodes the angle of division, indicating where the daughter cell is placed. A cell with an activated SU for cell death will die at the end of the developmental time-step.

The above cellular model has been applied to simulate both morphological and neural development [21, 48]. In the experiment to generate an animat like *C. elegans*, two prediffused, external TFs without decay and diffusion are deployed in the computation area (maternal morphogen gradients). The first TF has a constant gradient in the x-direction and the second in the y-direction. A few snapshot of the self-stabilized cellular growth [48] is provided in Fig. 10.

5.2 Activity-Dependent Neural Development

Biological findings indicate that both the structure and connecting weights of the neurons in the brain can change over time depending on the neuronal activities [23], which is resulted from changes in the expression of relevant genes [7]. Based on findings in neuroscience and systems biology, a gene regulatory network model

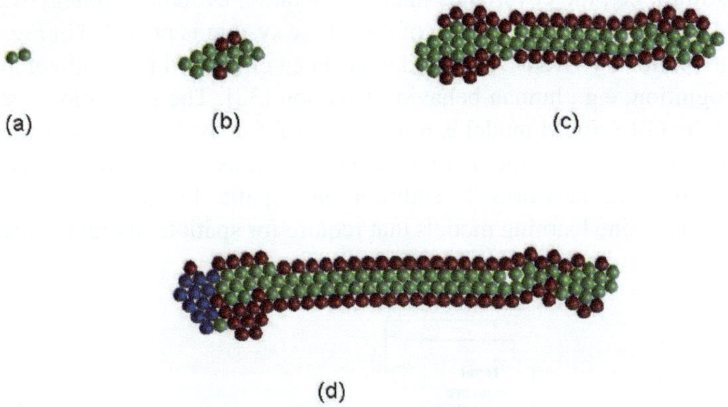

Fig. 10 Self-stabilized cellular growth under the control of a GRN model presented in [48].

is combined with the the Bienenstock, Cooper, and Munro (BCM) spiking neural network [4] to model the synaptic and neural plasticity [32]. BCM-based spiking neural network (SNN) is a graph with weighted, directed edges replacing synapses, as shown in Fig.11 The weight, weight plasticity, and meta-plasticity of the spiking neural network will all be regulated by the GRN, and the GRN will also be influenced by the activity of the neurons it resides in, in a closed loop.

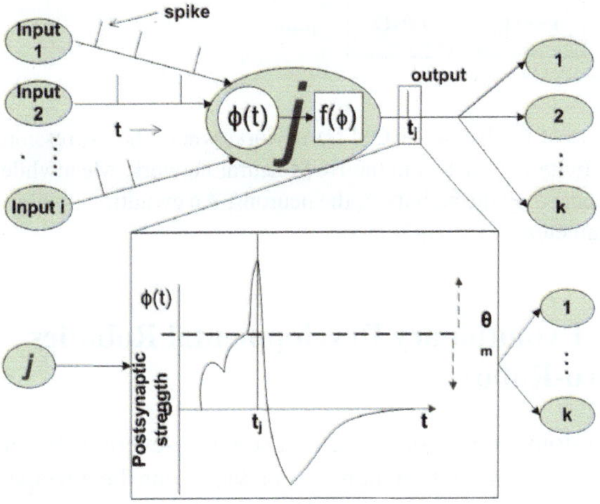

Fig. 11 A diagram of a BCM spiking neural network model.

To optimize the parameters of the gene regulatory network, an efficient evolutionary algorithm, i.e., the covariance matrix adaptation evolution strategy (CMA-ES) [23], [24], is employed. A diagram of the whole system is provided in Fig. 12.

The evolutionary GRN-BCM model has been employed for spatiotemporal pattern recognition, e.g., human behavior detection [32]. The simulation results indicate that the GRN-BCM model is more powerful for spatiotemporal pattern recognition than popular machine learning models such as support vector machines or feedforward neural networks. In addition, only spatial features are used, different from most machine learning models that require for spatiotemporal features.

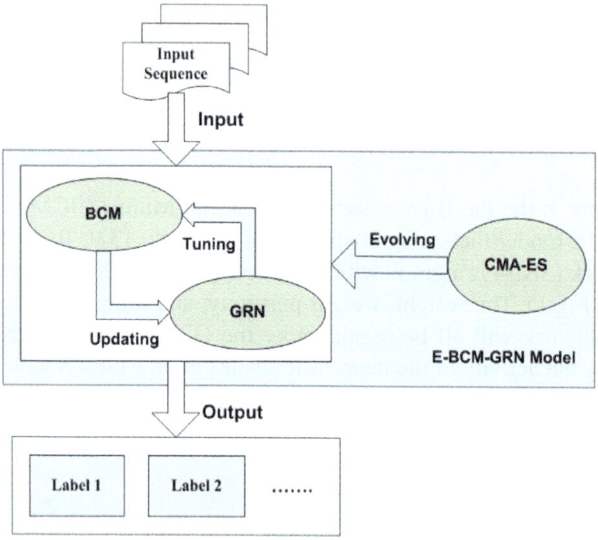

Fig. 12 Illustration of the E-GRN-BCM framework, where the expression level of the GRN regulates the plasticity parameters in the BCM neural network. Meanwhile, the gene expression level is influenced by the activity of the neurons. An evolutionary algorithm is employed to evolve the parameters in the GRN model.

6 Towards Evolutionary Developmental Robotics (Evo-Devo-Robo)

From the discussions in Section 1, we can see that several different but related research lines exist in robotics, which in our view, can be grouped into two categories, namely, evolutionary robotics [46], including coevolutionary robotics [45] and competitive co-evolutionary robotics [8], and developmental robotics [28], including morphogenetic approaches to robotics discussed in this chapter. A natural question is, what is the relationship between evolutionary robotics and developmental robotics?

As pointed out in [24], living systems have three main adaptation mechanisms, i.e., learning, development and evolution. In the context of bio-inspired hardware systems, Sipper et al [52] has provided a nice view on how to combine the three dimensions of natural adaptation, that is, epigenesis, ontogeny, and phylogeny in a unified framework, which is termed as the *POE model*. We will discuss these mechanisms from the robotics perspective.

- *Epigenesis.* Epigenesis can be defined as autonomous, incremental and open-end learning through sensori-motor adaptation, self-exploration, imitation, prediction, and social interactions, which are the main topics of epigenetic robotics. Thus, epigenetic robotics emphasizes on modeling of *mental development*.
- *Ontogenesis / morphogenesis.* Ontogenesis (ontogeny) includes cell growth, cellular differentiation and morphogenesis. Considering the fact that ontogenetic robotics has been used interchangeably with epigenetic robotics and that most computational models of morphogenesis also include cellular differentiation, we suggest that morphogenetic robotics be used to refer to the physical development of the body, including the nervous system. In contrast to epigenetic robotics, morphogenetic robotics covers the *physical development* of living systems.
- *Phylogeny.* In biology, phylogeny refers to the evolutionary relatedness of different species or populations. In robotics, evolution has been a powerful tool for robotics to be adaptable to large environmental changes through genetic variations such as mutation, crossover, and gene duplication.

Obviously, research in epigenetic robotics, morphogenetic robotics and evolutionary robotics cannot be performed separately. First, autonomous mental development would not have been possible without an intrinsic motivation system [42] and genetically wired neural structures for prediction, anticipation and memory. So far in epigenetic robotics, intrinsic motivation systems have often been predefined [41]. We hypothesize that the most basic components of such intrinsic motivation systems have been endowed by evolution. Second, the body plan of the robots are a result of morphogenetic development, on which mental development is based through interaction with the environment. Particularly, activity-dependent and activity-independent development of neural networks are closely coupled, which suggests that a synergy between epigenetic robotics and morphogenetic robotics is indispensable. Finally, development can not only bias the direction of evolution, but also enhance evolvability [22], while learning can influence evolution [14]. For example, it has been shown in [43], learning can attribute to genetic diversity in changing environments, and evolution is able to find an optimal balance in allocating adaptation resources for evolution and learning [44].

Evolutionary developmental biology has revolutionized our understanding of the morphological and neural development of living organisms [37]. In addition, the *evo-devo* approach has also helped us gain deeper insights into human cognitive development, resulting a new discipline known as *evolutionary developmental psychology* [11].

To summarize, we believe that evolutionary robotics and developmental robotics, two distinct yet complementary disciplines in robotics, should also integrate and form a new discipline: *evolutionary developmental robotics* (evo-devo-robo).

7 Conclusions

This chapter introduces a morphogenetic approach to self-organizing robotic systems, which focuses on employing genetic and cellular mechanisms in biological morphogenesis for developing self-organizing, self-reconfigurable and self-adaptive robotic systems, covering a wide range of robotic systems such as swarm robotic systems, modular robots and intelligent robots. While epigenetic robotics concentrates on the cognitive development of robotic systems, morphogenetic robotics focuses on the growth process of the body plan and nervous system. Therefore, we believe that morphogenetic robotics is complementary to epigenetic robotics and fills the gap between epigenetic robotics and developmental robotics in that developmental robotics should include both neural, morphological and cognitive development. We also expect that we will benefit from the synergies between morphogenetic and epigenetic robotics, as neural and morphological development lay the neuro-physiological foundation for cognitive development. Finally, we advocate to go from developmental robotics to *evolutionary developmental robotics*, thus systematically embedding the three main adaptation mechanisms of natural intelligence, i.e., evolution, development and learning, in robotic systems.

Acknowledgments

The authors would like to thank Hongliang Guo, Yuyang Zhang, Jun Yin, Matthew Conforth, and Lisa Schramm for the illustrative examples used in this book chapter.

References

1. Ashe, H.L., Briscoe, J.: The interpretation of morphogen gradients. Development 133, 385–394 (2007)
2. Beloussov, L.V.: Integrating self-organization theory into an advanced course on morphogenesis at Moscow State University. Int. J. Dev. Biol. 47, 177–181 (2003)
3. Ben-Amor, H., Cadau, S., Elena, A., Dhouailly, D., Demongeot, J.: Regulatory networks analysis: Robustness in morphogenesis. In: 2009 Int. Conf. on Advanced Information Networking and Application Workshops, Bradford, UK, pp. 924–928 (2009)
4. Bienenstock, E.L., Cooper, L.N., Munro, P.W.: Theory for the development of neuron selectivity: orientation specificity and binocular interaction in visual cortex. The Journal of Neuroscience 2(1), 32–48 (1982)
5. Bishop, C.D., Erezyilmaz, D.F., Flatt, T., Georgiou, C.D., Hadfield, M.G., Heyland, A., Hodin, J., Jacobs, M.W., Maslakova, S.A., Pires, A., Reitzel, A.M., Santagata, S., Tanaka, K., Youson, J.H.: What is metamophosis? Integrative and Comparative Biology 46, 655–661 (2006)
6. Cohen, D.J., Cicchetti, D.: Developmental Neuroscience. Wiley, Chichester (2006)

7. Flavell, S., Greenberg, M.E.: Signaling mechanisms linking neuronal activity to gene expression and plasticity of the nervous system. Annual Review of Neuroscience 31, 563–590 (2008)
8. Floreano, D., Nolfi, S., Mondana, F.: Competitive co-evolutionary robotics: from theory to practice. From Animal to Animats 5, 515–525 (1998)
9. Gilbert, S.F.: Developmental Biology. Sinauer Associates (2003)
10. Gilbert, S.F.: The morphogenesis of evolutionary developmental biology. Int. Journal of Developmental Biology 47, 467–477 (2003)
11. Griffiths, P.E.: Evo-devo meets the mind: Towards a developmental evolutionary psychology. In: Sanson, R., Brandon, R.N. (eds.) Integrating Development and Evolution. Cambridge University Press, Cambridge (2007)
12. Guo, H., Jin, Y., Meng, Y.: A framework for self-organized multi-robot pattern formation and boundary coverage inspired from morphogenesis. ACM Transactions on Autonomous and Adaptive Systems (2010) (accepted)
13. Guo, H., Meng, Y., Jin, Y.: A cellular mechanism for multi-robot construction via evolutionary multi-objective optimization of a gene regulatory network. BioSystems 98(3), 193–203 (2009)
14. Hinton, G.E., Nowlan, S.J.: How learning can guide evolution. Complex Systems 1, 495–502 (1987)
15. Hornby, G.S., Pollack, J.B.: Creating high-level components with a generative representation for body-brain evolution. Artificial Life 8, 223–246 (2002)
16. Hsieh, M.A., Kumar, V.: Pattern generation with multiple robots. In: International Conference on Robotics and Automation. IEEE Press, Los Alamitos (2006)
17. Jin, Y.: Morphogenetic robotics (2010), `http://en.wikipedia.org/wiki/~Morphogenetic_robotics`
18. Jin, Y., Guo, H., Meng, Y.: Robustness analysis and failure recovery of a bio-inspired self-organizing multi-robot system. In: Third IEEE International Conference on Self-Adaptive and Self-organizing Systems, pp. 154–164. IEEE Press, Los Alamitos (2009)
19. Jin, Y., Guo, H., Meng, Y.: A hierarchical gene regulatory network model for adaptive pattern formation in changing environment. IEEE Transactions on Robotics (submitted 2011)
20. Jin, Y., Meng, Y.: Morphogenetic robotics: An emerging new field in developmental robotics. IEEE Transactions on Systems, Man, and Cybernetics, Part C: Reviews and Applications (2010) (accepted)
21. Jin, Y., Schramm, L., Sendhoff, B.: A gene regulatory model for the development of primitive nervous systems. In: Köppen, M., Kasabov, N., Coghill, G. (eds.) ICONIP 2008. LNCS, vol. 5506, pp. 48–55. Springer, Heidelberg (2009)
22. Jin, Y., Trommler, J.: A fitness-independent evolvability measure for evolutionary developmental systems. In: IEEE Symposium on Computational Intelligence in Bioinformatics and Computational Biology (2010) (accepted)
23. Kalat, J.W.: Biological Psychology. Wadsworth Publishing, Belmont (2008)
24. Kodjabachian, J., Meyer, J.-A.: Development, learning and evolution in animats. In: From Perception to Action. IEEE Press, New York (1994)
25. Komosinski, M., Ulatowski, S.: Framsticks: Towards a simulation of a nature-like world, creatures and evolution. In: European Conference on Artificial Life, pp. 261–265 (1999)
26. Lee, J.A., Sitte, J.: Morphogenetic evolvable hardware controllers for robot walking. In: 2nd International Symposium on Autonomous Minirobots for Research and Edutainment (February 2003)

27. Lee, P.N., Kumburegama, S., Marlow, H.Q., Martindale, M.Q., Wikramanayake, A.H.: Asymmetric developmental potential along the animal-vegetal axis in the anthozoan cnidarian, nematostella vectensis, is mediated by dishevelled. Developmental Biology 310(1), 169–186 (2007)
28. Lungarella, M., Metta, G., Pfeifer, R., Sandini, G.: Developmental robotics: A survey. Connection Sciences 15, 151–190 (2003)
29. Mamei, M., Vasirani, M., Zambonelli, M.: Experiments in morphogenesis in swarms of simple mobile robot. Applied Artificial Life 18, 903–919 (2004)
30. Meiner, D.: Swarm robotics algorithm: A survey. Technical report, University of Maryland (May 2007)
31. Meng, Y., Guo, H., Jin, Y.: A morphogenetic approach to flexible and robust shape formation for swarm robotic systems. Robotics and Autonomous Systems (2010) (submitted)
32. Meng, Y., Jin, Y., Yin, J., Conforth, M.: Human activity detection using spiking neural networks regulated by a gene regulatory network. In: International Joint Conference on Neural Networks, pp. 2232–2237 (2010)
33. Meng, Y., Zhang, Y., Jin, Y.: A morphogenetic approach to self-reconfigurable modular robots using a hybrid hierarchical gene regulatory network. In: 12th International Conference on the Synthesis and Simulation of Living Systems (ALIFE XII), pp. 765–772 (2010)
34. Meng, Y., Zhang, Y., Jin, Y.: Autonomous self-reconfiguration of modular robots using a hierarchical mechanochemical mode. IEEE Computational Intelligence Magazine 6(1), 43–54 (2011)
35. Meng, Y., Zhang, Y., Sampath, A., Jin, Y., Sendhoff, B.: Cross-ball: a new morphogenetic self-reconfigurable modular robot. In: IEEE International Conference on Robotics and Automation (ICRA) (accepted 2011)
36. Mjolsness, E., Sharp, D.H., Reinitz, J.: A connectionist model of development. Journal of Theoretical Biology 52, 429–453 (1991)
37. Müller, G.B.: Evo-devo: extending the evolutionary synthesis. Nature Review Genetics, 943–949 (2007)
38. Murata, S., Kakomura, K., Kurokawa, H.: Toward a scalable modular robotic system - navigation, docking, and integration of M-TRAN. IEEE Robotics & Automation Magazine 14, 56–63 (2008)
39. Murata, S., Kurokawa, H.: Self-reconfigurable robots. IEEE Robotics &Automation Magazine, 71–78 (March 2007)
40. Mytilinaios, E., Marcus, D., Desnoyer, M., Lipson, H.: Designed and evolved blueprints for physical artificial life. In: Ninth Int. Conf. Artificial Life (ALIFE IX), pp. 15–20 (2004)
41. Oudeyer, P.-Y., Kaplan, F.: How can we define intrinsic motivation? In: Proceedings of the 8th International Conference on Epigenetic Robotics. LUCS, Brighton (2008)
42. Oudeyer, P.-Y., Kaplan, F., Hafner, V.V.: Intrinsic motivation systems for autonomous mental development. IEEE Transactions on Evolutionary Computation 11(1), 265–286 (2007)
43. Paenke, I., Branke, J., Jin, Y.: On the influence of phenotype plasticity on genotype diversity. In: 2007 IEEE Symposium on Foundations of Computational Intelligence, pp. 33–40. IEEE Press, Los Alamitos (2007)
44. Paenke, I., Jin, Y., Branke, J.: Balancing population and individual level of adaptation in changing environments. Adaptive Behavior 17(2), 153–174 (2009)
45. Pollack, J., Lipson, H., Funes, P., Ficici, S., Hornby, G.: Coevolutionary robotics. In: Evolvable Hardware, pp. 208–216 (1999)
46. Nolfi, S., Floreano, D.: Evolutionary Robotics. The MIT Press, Cambridge (2004)

47. Sanes, D.H., Reh, T.A., Harris, W.A.: Development of Nervous Systems, 2nd edn. Academic Press, London (2006)
48. Schramm, L., Jin, Y., Sendhoff, B.: Emerged coupling of motor control and morphological development in evolution of multi-cellular animates. In: 10th European Conference on Artificial Life (2009)
49. Shen, W., Will, P., Galstyan, A.: Hormone-inspired self-organization and distributed control of robotic swarms. Autonomous Robots 17, 93–105 (2004)
50. Simons, K., Karsenti, E., St Johnston, D., Wijer, C., Swarminathan, S. (eds.): Self-Organization and Morphogenesis in Biological Systems, Schloss Ringberg, Tegersee, Germany (December 2006)
51. Sims, K.: Evolving 3D morphology and behavior by competition. Artificial Life 1 (1994)
52. Sipper, M., Sanchez, E., mange, D., Tomassini, M., Perez-Uribe, A., Stuffer, A.: A phylogenetic, ontogenetic, and epigenetic view of bio-inspired hardware systems. IEEE Transactions on Evolutionary Computation 1(1), 83–97 (1997)
53. Stanley, K.O., Miikkulainen, R.: A taxonomy for artificial embryogeny. Artificial Life, 93–130 (2003)
54. Taylor, T., Massey, C.: Recent developments in the evolution of morphologies and controllers for physically simulated creatures. Artificial Life 7(1), 77–87 (2001)
55. Wolpert, L.: Principles of Development. Oxford University Press, Oxford (2002)
56. Yu, C.-H., Haller, K., Ingber, D., Nagpal, R.: Morpho: A self-deformable modular robot inspired by cellular structure. In: Proceedings of the 2008 IEEE/RSJ International Conference on Intelligent Robots and Systems, pp. 3571–3578. IEEE Press, Los Alamitos (2008)

How to Engineer Robotic Organisms and Swarms?*

Bio-Inspiration, Bio-Mimicry, and Artificial Evolution in Embodied Self-Organized Systems

Thomas Schmickl

Abstract. In large-scale systems composed of autonomous embodied agents (e.g., robots), unpredictability of events, sensor noise and actuator imperfection pose significant challanges to the designers of control software. If such systems tend to self-organize, emergent phenomena prevent classical engineering approaches per se. In recent years, the Artificial Life Lab at the University of Graz has investigated a variety of methods to synthesize such control algorithms used in multi-modular robotics and in swarm robotics. These methods either translate mechanisms directly from biology to the engineering domain (bio-mimicry, bio-inspiration) or generates such controllers through artificial evolution from scratch. In this article I first discuss distributed control algorithms, which determine the collective behavior of autonomous robotic swarms. These algorithms are derived from collective behavior of honeybees and from slime mold aggregation. One of these algorithms is inspired by inter-adult food exchange in honeybees ('trophallaxis') another one from chemical signaling in slime molds. In addition to the control of robot swarms, control paradigms for multi-modular robotic organisms are presented, which are again based on simulated fluid exchange (hormones) among compartments of robotic organisms. In both domains –swarms and organisms– the control system is self-organized and consists of many homeostatic sub-systems which adapt to each other on the individual (module) and on the collective level (organism, swarm). Additionally, I discuss the importance of distributed feedback networks, as well as the benefits and drawbacks of bio-inspiration and bio-mimicry in collective robotics.

Thomas Schmickl
Artificial Life Lab of the Department for Zoology, Karl-Franzens University Graz,
Universitä"tsplatz 2, 8010 Graz, Austria
e-mail: `thomas.schmickl@uni-graz.at`

* This work is supported by the following grants: EU-IST-FET 'SYMBRION', no. 216342; EU-ICT 'REPLICATOR', no. 216240; EU-IST FET 'I-SWARM', no. 507006; FWF (Austrian Science Fund), no. P19478-B16.

Y. Meng and Y. Jin (Eds.): Bio-Inspired Self-Organizing Robotic Systems, SCI 355, pp. 25–52.
springerlink.com

1 Introduction

Even though much progress was made in the field of robotic engineering in the past, it is still a challenging task to develop control software for robots that are aimed to work in heterogeneous, dynamic, and sometimes even unpredictable environments. This challenge is even harder whenever the aim is to control not only one singular robotic unit but several – sometimes even masses – of robot modules.

The field of collective robotics can be roughly split into two domains: (1) Swarm robotics: Robotic modules are fully autonomous in control and in their physical connectedness. (2) Multi-modular robotics and reconfigurable robotics: Robotic modules are physically coupled. I think that systems of both domains can be seen as being quite similar. It is just the degree of (physical) connectedness between modules that differs between systems of these two domains. Thus I interpret a swarm as being a 'very loosely coupled' variant of an organism, or, in other words, I call an organism to be a very tightly connected swarm of modules. In the work of the Artificial Life Lab in Graz (Austria), we draw bio-inspiration from natural swarm systems and from simple multi-cellular organisms. The basic principles that govern the self-organization of the natural systems are then translated into algorithms that produce comparable self-organization in robotic systems. However, as the 'substrate' changes significantly, this translational step has to carefully consider the capabilities and constraints of the technical target systems. In figure 1 the potential 'flows of inspiration' between biological research domains and engineering research domains, which are relevant for my work group's research paradigm, are depicted.

Within the community of biology-interested engineers, the term 'bio-mimicry' and 'bio-inspiration' are used very often synonymously. I think there is significant difference between these two research fields: In bio-inspiration, a mechanism is imported from the biological source of inspiration to the technical target system. Very often, the biological mechanism is either a chemical or a behavioural mechanism, which is converted into a sort of algorithm that reflects the key aspects of the inspiration. Prominent examples are genetic algorithms & evolution strategies [22, 12], particle swarm optimization [14] or ant colony optimization [7]. Although these algorithmic methodologies clearly reflect key features of the biological counterpart, they are very abstract, disembodied and do not incorporate any biological constraints into the technical world. Thus they act similar to the biological counterpart but they do not look similar to it. In contrast to that, 'bio-mimicry' is a field of research where the engineer tries to copy the look of biological systems as closely as possible. For example, the walking of a modern humanoid robot might look similar to a human's walk, but it is achieved by servo motors, hydraulics and similar mechanisms, which do not at all resemble the biological mechanisms associated with walking. Thus, in 'bio-mimicry', products look similar to biological examples but act in a different way. However, there are many examples that aim for approach an engineering problem from both sides, thus which exploit 'bio-inspiration' and and 'bio-mimicry' in parallel. I consider these approaches to be typical Artificial Life approaches, as the estimative product of such an approach will be in-discriminable from a living organism: It will act and look like a natural organism. The re-creation of life forms

Fig. 1 Biological sources of inspiration for swarm robotic and for multi-modular robotics organisms. The arrows indicate how knowledge in one domain could potentially influence research on another domain. (b) 'Volvox', by Dr. Ralf Wagner, reprinted from the 'wikimedia commons' library.

is a fundamental goal of Artificial Life, thus 100% successful 'bio-inspiration' and 'bio-mimicry' might lead to a technical singularity: the first artificially created life form.

In the article at hand, I give a short overview how such systems, robotic swarms and robotic organisms, can be controlled in a bio-inspired way, that means that control software is written that functionally resembles a known control mechanism in a comparable natural system. Of course, abstraction is needed to achieve this conversion from the natural domain to the artificial. This is a critical issue, because the more the control mechanism gets abstracted the lower is the linkage between the natural system and the artificial one. This means that high abstraction prevents inspiration of further biological research, converting the potential symbiosis between biology and engineering into a rather parasitic relationship. To prevent our research to be such a one-way street of knowledge-flow, we keep on our biological research even after the moment of bio-inspiration. By performing in parallel 'bio-mimicry' kind of research, our work group tries to transfer engineering success back into biologically relevant research topics.

Another approach to generate control software for robotic swarms and robot organisms is open-ended evolution. In this approach, which is frequently used in my lab for the projects SYMBRION and REPLICATOR [16, 38, 23], we use a bio-inspired pluri-potent (maybe also Turing-complete) control system which is subject to artificial evolution algorithms. This represents a three-fold combination of bio-inspiration in parallel: Morphology, control (physiology) and selection are well known domains in biology and are important aspects in our way to automatically generate controllers, body shapes and interaction patterns of robots. In the article at hand, both of my major scientific approaches mentioned above are described exemplary in two case studies.

In a final discussion, I review the relative positions of the methods and projects described in this article in the 'bio-inspiration/bio-mimicry' feature space, thus I discuss how close these research tracks have approached the fundamental technological singularity of artificial life.

2 Bio-Inspiration and Bio-Mimicry in Swarm Robotics

2.1 Bio-Inspiration

The aim of bio-inspiration is to find solutions to a problem by looking at comparable natural systems where natural selection has already favored genetic, morphologic or physiologic variants of organisms that are able to solve the particular problem. As described above, in my interpretation the aim of bio-inspiration is not to produce a technological copy of these biological systems. Instead, the aim of bio-inspiration is to understand *why* and *how* the focal biological system works in an efficient manner. After these questions are answered, a mechanism is developed that is efficient in the technical entity in a comparable way, following similar sets of governing rules. In my work group, we perform laboratory experiments with those animals that we take as source of inspiration. In those experiments we find out the key components of the natural solution to identify all relevant mechanisms. Afterwards, these key mechanisms are abstracted and analyzed in mathematical models and simulation studies to identify a suitable simplified model of the actual task and of the nature-inspired solution. Based on these models, a robot controller and a suitable robot arena setup are constructed, which is the major 'translational process' to convert the algorithmic core of the observed natural mechanism into a robot algorithm. For additional studies, a model and a simulation tool that depicts the robotic setup is constructed, which allows parameter optimization and evolutionary computation on the robotic system. At this point in time, the investigated task is scientifically studied on various levels and with various tool sets: real animal experiments and robotic experiments, simulation of biological entities and of the robotic system.

Initially, the bio-inspired robotic system is usually able to produce qualitatively similar behaviors as the natural system. However, we usually observe lowered efficiency or unwanted side-effects in the robots' collective behaviors, as a straightforward translation of behaviors and mechanisms from the biological to the technical domain will always be sub-optimal. To compensate for this, we perform additional experiments, modeling and computational parameter optimization. In some cases, the biological system is not studied in our lab. This is the case when the interesting behavioral mechanisms in the natural system are already well researched and well described in existing literature.

Based on biological inspiration, we have developed a set of bio-inspired control algorithms for swarm robotics: vector-based algorithm [40], slime-mould-inspired algorithm [26], trophallaxis-inspired algorithm [30, 27], and the BEECLUST algorithm [31]. These algorithms are described in the following to demonstrate the different levels of detail at which nature can inspire robotic algorithms.

2.1.1 A Benchmark Scenario for Swarm Algorithms

To compare algorithms and to tune the performance of swarm algorithms, it is important to develop realistic benchmark scenarios. During the EU IST-FET FP6 project *I-Swarm*, two types of robots were designed: The small three-legged *I-Swarm* robot, communicating with four LEDs pointing into 4 directions (inter-beam angle 90°) and the *Jasmine* robot (2 wheels, 6 LEDs for communication at an angle of 60°). The I-swarm robot is able to pick up small dust particles with an electrostatic lever and also the *Jasmine* robot was initially planned to be equipped with an magnetic gripper. Having these two hardware platforms in mind, we developed a scenario which can be best described by a collective approach to foraging: foraging for dirt, which means collective cleaning. The swarm of autonomous robots is distributed randomly in the arena, all robots start unloaded. In the arena, there are designated 'dump' regions, where individually collected particles should be dropped by robots. These particles are deposited initially in designated 'dirt' regions, which have to be discovered by the robots. This is a hard task, as the robots have no sensors that can report dirt or dump from a distance, they just can observe the arena floor located directly below themselves. Moreover, communication to other robots as well as obstacle detection, is also restricted to a short distance around the robots (1-2 robot diameters).

The collective tasks are: The robots have to explore the arena collectively in an efficient manner. After some robots have found the dirt areas, they have to recruit other robots to these places and –simultaneously– pick up dirt particles and carry them to the dump region in an efficient manner. This involves a sort of 'find-the-shortest-path' task which should be solved collectively. This scenario has the advantage that efficiency measures are obvious, e.g.: How many particles are delivered by x robots within a period of y time steps? The 'difficulty' of this collective task can be adjusted by increasing the distance between dirt and dump areas or by placing obstacles in the way. Using this benchmark, we were able to compare different swarm algorithms. In addition, it allowed us to investigate important swarm properties, for example, the critical minimum swarm density.

In the following, I describe 4 different swarm algorithms, beginning with the communication-extensive and rather technical 'vector-based' strategy. This strategy has little bio-inspiration and is more or less a classical engineering approach to the posed set of problems. The second algorithm is the 'trophallaxis-inspired' algorithm, which still requires a lot of communication (several float numbers), which is inspired by social-insect inter-adult feedings. The third algorithm is the 'slime-mold' algorithm, which has lower requirements concerning communication, as only one-bit signals have to be exchanged. Finally, the 'BEECLUST' algorithms is described. This algorithm is almost communication-less. However, as no more trail formation of robots can be observed in swarms using this algorithm, it was investigated in an aggregation scenario which is just a subset of the above-described cleaning scenario.

2.1.2 Vector-Based Algorithm

The first algorithm we investigated is called 'vector-based algorithm', because it is based on inter-robot communication of vector information and on collective vector summarization. It was analyzed in several publications [40, 6], and will be shortly summarized in the following. Using this algorithm, a robot that found a target (dirt or dump) by chance in its random exploration mode turns on a binary signal that indicates the location of the target to nearby other robots. These robots, which now can calculate the angle between their current heading and the signaling robot, turn on another binary signal and send this angular information to other robots nearby. The vector calculation of other robots is made possible as the sending robots, which send the message through their LEDs, encode also the direction of the sending LED into the message. This way, the receiving robot can assume the relative heading of the other robot and the communicated angle between the heading of the sending robot and the foraging target. By vector addition, this interpreted information is passed throughout the swarm, ideally telling all robots their bearing towards the target.

As can be seen in figure 2a, simulations predict trail formation of swarm robots that transport particles from dirt areas to dump areas. Using this benchmarks, we analyzed also the impact of swarm size (density) onto important efficiency measures of the swarm (see figure 2b). However, we realized many shortcomings of this algorithm: As soon as we implemented noise into robot-to-robot communication

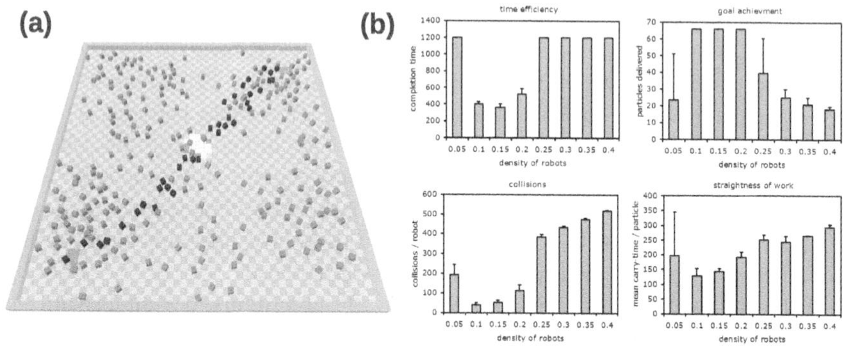

Fig. 2 (a) Emerging transportation trails in the cleaning scenario. Loaded robots (black boxes) carry their particles on the shortest path from the dirt areas (dark grey floor) to the central dump area (light grey floor). Unloaded robots (grey boxes) help in navigation through vector communication. (b) Analysis of the dependency of several efficiency measures on the swarm density. We clearly found an optimal density (size) of the robot swarm. Reprinted from [6].

and angular measurements, the errors due to this noise were summed up and finally impaired the swarms from performing well. Another shortcoming of this algorithm was that we observed that the communicated vectors were always pointing directly towards the foraging targets, even when obstacles blocked the way. This impaired the robots from circumventing such obstacles. A third shortcoming was that old, thus outdated, information never leaves the system, which prevents the swarm from reacting to environmental changes. First we tried to fix these issues by additionally using hop-counts for determining the 'age' of communicated vectors and allowed the robots to accept only these information that were newer than the information they already carried. But even with these improvements, the performance and robustness of the swarm behavior was not of the desired quality. Thus, we started to look for swarm algorithms which are more robust to noise and which require less communication bandwidth. As natural selection shaped natural systems into efficient and robust configurations, we aimed for increasing the level of bio-inspiration, as can be seen by the algorithms described in the following.

2.1.3 Trophallaxis-Inspired Algorithm

In contrast to the vector-based strategy, the next algorithm does not require any 'vector'-calculations which can lead to an aggregation of calculation errors (due to noise) within the swarm. It requires less computational power of the robots and requires the communication of two floating point numbers as 'messages' between neighboring robots. The trophallaxis-based algorithm uses the mechanisms found in of bee-to-bee nectar feedings to regulate the behavior of a robot swarm. It is used to generate a distributed map and this way it is also some kind of collective perception of the swarm. In this distributed algorithm the agents can generate a shared gradient map by adding and consuming virtual nectar volumes to their internal memory. This shared gradient map is then locally used by agents for target oriented navigation. In the cleaning scenario, which resembles social insects' foraging task, one source of virtual nectar is a place with dirt particles which should be picked up by the agents and dropped off at a dump. The dump is another source for a different sort of virtual nectar. Like honeybees, the agents also consume a small part of their virtual nectar loads when moving, which results in decay of old information. Nectar is also shared with neighbors, like social insects do in 'trophallaxis'. From these locally executed rules, two gradients emerge within the swarm, one pointing towards the dirt particles, the other one pointing towards the dump. Fig. 3 shows a visualization of the shared collective maps. In the depicted scenario, we placed walls in the environment, which blocked robot movement and also robot communication. This is clearly reflected in the shared map.

The trophallaxis-inspired algorithm is based on three important features: (1) Positive feedback (gradient & uphill movement of agents) recruits the robots to their targets. (2) Negative feedback (consumption of nectar) prevents overcrowding and removes outdated information from the system. (3) 'Trophallaxis'-like communication leads to diffusion of information within the swarm, allowing collective adaptation to environmental fluctuations by modulating the emergent gradient map.

Basically, the shared gradient map dynamically encodes those local steady-states that are established in a homeostatic way by the individual agents' behaviors and by their interactions with local neighbors and with the environment. Thus, multi-level distributed and behavior-based homeostasis is a key component of this swarm algorithm.

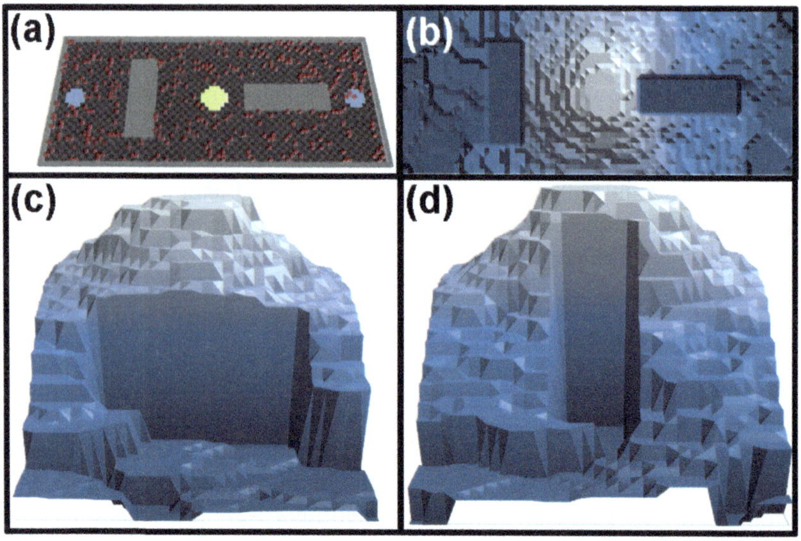

Fig. 3 Simulation of a robot swarm using the throphallaxis-inspired algorithm. a) Screenshot of the cleaning scenario arena setup: Robots (red) try to find the shortest path from the two places with dirt particles (blue) to the central dump (yellow). The path is blocked by 2 walls (gray). b) Top view of the emerged gradient pointing towards the dump. c) Side view of the same gradient from the left side. d) Side view of the same gradient from the right side. Reprinted from [30].

2.1.4 Slime Mold Algorithm

The slime-mold algorithm [26] is an algorithm that reduces the amount of required communication but still produces collective behaviors comparable to the two algorithms described above. It is inspired by the collective aggregation of amoebas of the slime mold species *dictyostelium discoideum*. In this algorithm, robots emit signals whenever they locate themselves either on the dirt area or on the dump area by chance due to the basic random motion that all uninformed robots perform in all of our swarm algorithms. This binary signal can be perceived by local neighbors which in turn emit a similar signal and then switch to a pause mode (= refractory period) for some time in which they do not respond to any signals. The emerging chain reaction is well known from many examples in nature, usually referred to as 'excitable media'. In slime mold, amoebas that want to aggregate emit a chemical stimulus frequently, fireflies emit light signals and even in soccer stadiums, 'Laola'-waves are produced by similar mechanisms. In our robot swarm, blinking wave propagation throughout the swarm can be observed, leading to pulsating waves emitted from dirt areas and dump areas. Depending on whether individual robots are currently loaded or unloaded, they turn their heading against the dirt-originating wave or against the dump-originating wave, thus they are guided towards their foraging targets. It is known from other domains that such waves yield interesting properties: They automatically circumvent barriers/obstacles and they annihilate each other as soon as such waves collide, which allows interesting systems: If there are multiple ways from the origin to the target, a consistent wave propagation occurs only on the shortest path, as all longer path will encounter colliding waves and thus wave annihilation. This allowed our robotic swarm to always find the shortest path from dirt to dump, as it is shown in figure 4.

Fig. 4 Simulation of a robot swarm using the slime-mould-inspired algorithm showing the cumulative paths of loaded robots from the dirt area in lower left corner of the arena to the dump area in the upper right corner of the arena. (a) First the swarm selects the shortest path through the central door in the barrier, only a small fraction of the swarm chooses the second door which allows a slightly longer way from dirt to dump. (b) After the central door was closed, the swarm automatically selects the lower door which offers now the shortest path. (c) After this door was closed too, the swarm chooses the only remaining longest path. Reprinted from [26].

2.1.5 BEECLUST Algorithm

However, all algorithms described above rely heavily on communication, which is sometimes hard to establish in bigger swarms. Therefore, we developed another swarm algorithm which works almost without explicit communication and which was also inspired by honeybee behavior.

The swarm algorithm 'BEECLUST' is inspired by the aggregation behavior of young honeybees and resulted in an very simple, yet robust and flexible aggregation algorithm for robot swarms. The idea for this algorithm originates from the observation of young honeybees in the beehive where the freshly emerged honeybees have a preferred temperature of approx. $36\,^{\circ}C$ [11]. These young bees tend to locate themselves in a collective way in the warmest central areas of the hive. Experiments with single young honeybees in a temperature gradient (approx. $30\,^{\circ}C$ - $36\,^{\circ}C$) showed that most bees cannot locate themselves in the warmest zone permanently. Instead, most of them wander around aimlessly and frequently leave warm areas soon after they have encountered them (see Fig. 5a). Thus, a 'swarm effect' seems to be responsible for the bees' well-functioning collective temperature-finding behavior. Further experiments with a specialized arena offered insight to this behavior [15]: Single honeybees usually wandered around in the arena randomly but stopped when colliding with another bee and then waited at this place for a duration that correlated with the temperature at this place. Low temperatures resulted in a short waiting-time of the bee, whereas warmer temperatures resulted in longer waiting-times. Thus, clusters of bees formed all over the arena, but in the warmer zone these clusters lasted longer than in colder zones. Finally all clusters merged into one big cluster near the global temperature optimum (Fig. 5b-d).

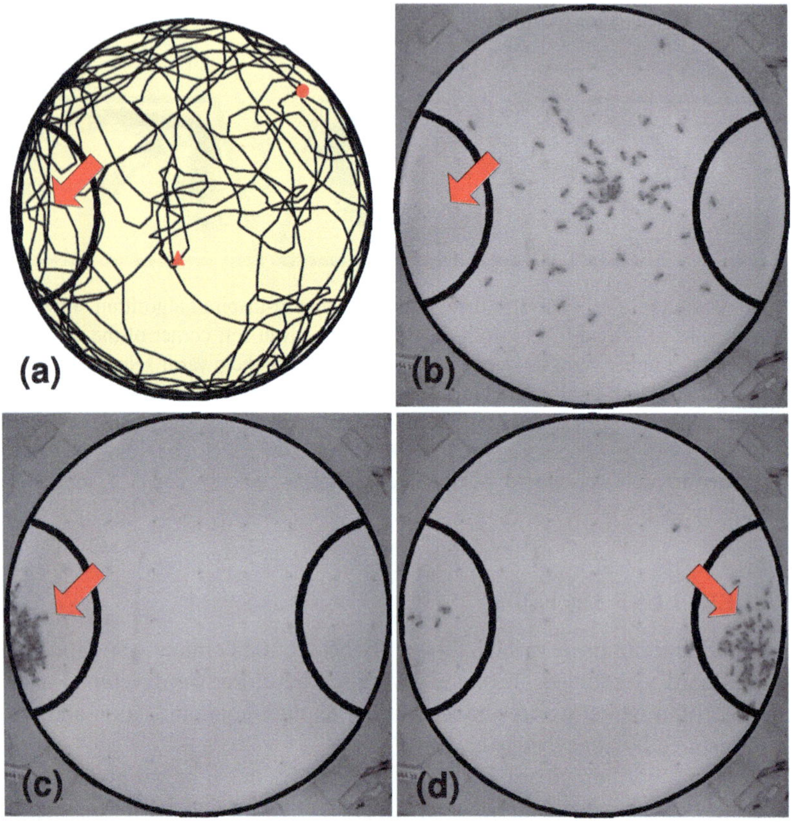

Fig. 5 Experiment with bees in a specialized arena. a) A single bee does not find the 36 $^\circ C$ temperature optimum to the left, indicated by the red arrow. b) Initial state of an experiment with 64 bees. The 36 $^\circ C$ (global) optimum is to the left, indicated by the red arrow. The 32 $^\circ C$ sub-optimum is to the right. c) Bees collectively clustered at the optimum. d) After the 36 $^\circ C$ optimum to the left was switched off, the bees were able to re-decide and cluster at the new 32 $^\circ C$ (global) optimum to the right, indicated by the red arrow. Ambient temperature: approx. 30 $^\circ C$.

This behavior of the young honeybees was then abstracted into an algorithm (called 'BEECLUST algorithm') and analyzed in a multi-agent simulation. Simulations showed that this algorithm is not only able to aggregate robots at a zone of interest, but is also able to enable the swarm to differentiate between zones of different qualities. Furthermore, it also allows the swarm to adapt to quality changes of the target zones.

The BEECLUST algorithm works as described by the following rules that are executed one after another in the microprocessors of the autonomous robots or in the executing loops of other (non-embodied) types of agents:

1. All agents move in straight lines and constantly check for collisions.
2. If they sense a collision, they stop.
3. If the collision happened with an obstacle which is not another agent, the agent turns around and continues with step 1.
4. If the collision happened with another agent, the focal agent measures the local quality of the environment. The higher the quality is, the longer it waits at this spot.
5. After the waiting period is over, the agent continues with step 1.

After the simulation studies, the BEECLUST algorithm was ported to swarm robots and adapted for light spot finding behavior with Jasmine robots [13]. In these experiments, the temperature gradient that we used in bee experiments was replaced by light gradients, as the Jasmine robot has a luminance sensor but no temperature sensor. These experiments posed very low requirements for the robots' hardware, as they only need to be able to avoid collisions, to discern other robots from obstacles or walls and to measure the local luminance.

In our experiments robot swarms executing BEECLUST show a collective behavior that is very similar to the behaviour of swarms of young honeybees [31]: The robots are able to optimally distribute themselves in the arena, resulting in more robots clustering at a brighter light spot and fewer robots clustering at a dimmed light spot. Moreover, the robots were able to quickly redistribute themselves after these different light spots changed places (see Fig. 6).

2.2 Evolutionary Adaptation of Swarm Algorithms

Swarms of robots that run a bio-inspired algorithm usually show a qualitatively similar, but quantitatively different behavior compared to the natural swarm systems that acted as a source of inspiration. We use evolutionary computation techniques to significantly optimize our swarm algorithms [27]. In these studies, the key parameters of the swarm (swarm density, robot speed, collision-avoidance distances, ...) were modulated by an evolution strategy [22], whereby the whole swarm of robots was the unit of selection. Thus, we tested populations of swarms in a competitive scenario, one swarm against each other.

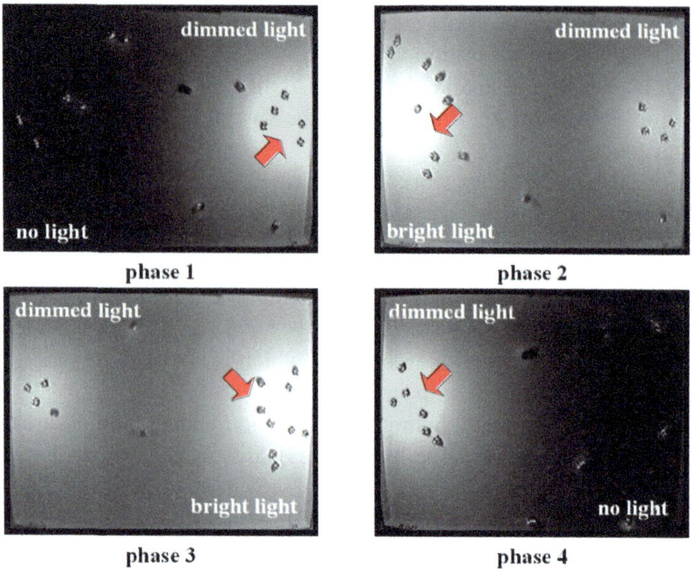

Fig. 6 Photographs of an experiment with real swarm robots using the BEECLUST algorithm. The red arrows indicate the global optimum. Reprinted from [31].

2.3 Bio-Mimicry

In contrast to bio-inspiration, the process of bio-mimicry shapes a technical entity (algorithm, robot) in a way that it resembles the biological counterpart. Although it looks like the biological organism (from the outside) it does not necessarily have to use the same internal mechanisms. However, in most cases this means that the interface to the outer world (for example sensors and actuators) closely resemble the biological source of inspiration. This can be important for understanding real organisms, because bio-mimicry forces the engineer or scientist to 'see the world through the eyes of the biological organism'. Even though inside mechanisms can differ from biology, they are fed with data similar to the data a biological organism might perceive and, in case of bio-mimetic actuation, motion-principles have to be finally transformed into patterns that are close to those observed in biology.

2.3.1 Bio-Mimicry of the BEECLUST Algorithm

In addition to abstract models and light-finding Jasmine swarms, we want to investigate the BEECLUST-algorithm also in real temperature fields, which is the kind of stimulus that mainly drives honeybee aggregation in nature. Therefore we designed an add-on for the Hemisson robot (Fig. 7) which allows such a robot to navigate in such a temperature gradient. This add-on consists of a set of temperature sensors that are mounted on two artificial antennae, similar to the configuration of temperature sensors in honeybees. We called these robots 'Thermobots'. In contrast to light, heat has different physical characteristics concerning the diffusion of heat in the air, the stability of the gradient and the time delay of the measurement, thus we expect this task to be more challenging than aggregation in a light gradient.

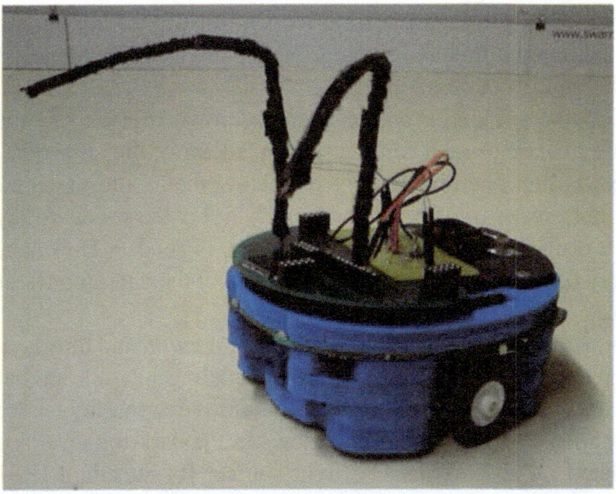

Fig. 7 Thermobot: The robot 'Hemisson' with additional antennae which hold the sensors that measure local temperature.

Experiments with three robots in a temperature gradient field showed that robots executing the swarm intelligent BEECLUST algorithm have a higher success rate compared to a standard gradient ascent algorithm. We expect that working with an even bigger swarm will further increase the success rate for the BEECLUST-algorithm, due to the fact that more robots lead to more collisions and so more measurements are taken.

2.3.2 Bio-Mimicking Ant Pheromone Trails

Ants use pheromone trails in order to navigate efficiently between potential food sources and the nest. We investigate this behavior in the ANTBOTS project [18]. There have been several approaches to model the pheromone trail laying and following of ants using robots:

By means of chemical sensors and alcohol-depositing robots [25]. This is a very realistic imitation of the pheromone-based trails of ants. However, the chemical sensors used in this setup and the combination of robotics and substances such as alcohol have been shown to be very unreliable and not very practical.

Drawing lines onto the floor using pen and paper [37]. In this scenario each robot is equipped with a pen, with which it is able to draw solid thin lines onto the ground. Although a decay of these trails is archived by using a special kind of disappearing ink, the trails laid by these robots remain thin in comparison to the robots. This does not provide a close analogy to the biologically inspired behavior of ants.

Laying trails of heat [24]. This method promises an extremely flexible way to model the foraging behavior of ants by laying trails of residual heat onto normal surfaces such as carpets or tiles. One problem is that the electrical generation of heat is not possible even on bigger mobile robots because of constraints in battery power. The researchers stored heat in the form of paraffin wax to lay trails instead. This presents an additional difficulty for experimental use and dynamically adjusting the strength of the trail is not possible.

Using robot-tracking and a projector setup, in which each robot is able to lay trails by being tracked using a camera suspended above the arena [9]. A computer superimposes 'virtual pheromones' by projecting them onto the arena floor. This system does not present a fully autonomous way for the robots to lay and follow trails, and a a central unit, an external computer, is needed. However, this system provides a very flexible way to modify parameters of the pheromones, such as decay and diffusion.

Emitting ultraviolet light onto a phosphorescent paint, and thus laying luminous green trails on the arena floor. This method of modeling ant trails has been published for use in an artistical context [1]. In this setup, the arena floor is coated with a special phosphorescent glow-paint that glows in the dark for several minutes after being stimulated by an external UV light source. By attaching UV-LEDs to the mobile robots, they can leave glowing trails on the ground. The idea is that because of the constant decay in brightness, the green glow that emanates from the floor can be seen as an analogy to the evaporating pheromones ants utilize in their trail following.

For our 'ANTBOTS' we have extended the approach of using phosphorescent paint for the emulation of pheromone streets. In our experimental setup we use the e-puck robot developed at the EPFL Lausanne [2] with two add-on boards to allow for trail laying and navigation to a light source outside of the arena. By emitting ultraviolet light onto the arena floor, robots are able to lay trails on the arena floor. Trail following is achieved using the on-board camera of the e-puck robot. In order to navigate back to the nest we use six photodiodes to measure the light intensity from a light source in a corner of the arena (sun compass and sun).

We have conducted several experiments with a single robot in order to test the feasibility of our setup using repeated trail laying and following cycles with a single robot. In these experiments we placed transparent plastic enclosures with red and blue LEDs in the arena, representing nest and food source respectively. The robots task was to navigate back and forth between the two target zones on its own trail repeatedly. The results show that our newly developed sensors are reliable enough for the robot to navigate to the two spots for longer periods of time with a distance of about 1.3m (Fig. 8).

In addition to these single robot trials we have developed a multi-agent simulation closely resembling our experimental setup in order to test if the efficiency is enhanced when the robots have the ability to lay trails on the arena floor versus a

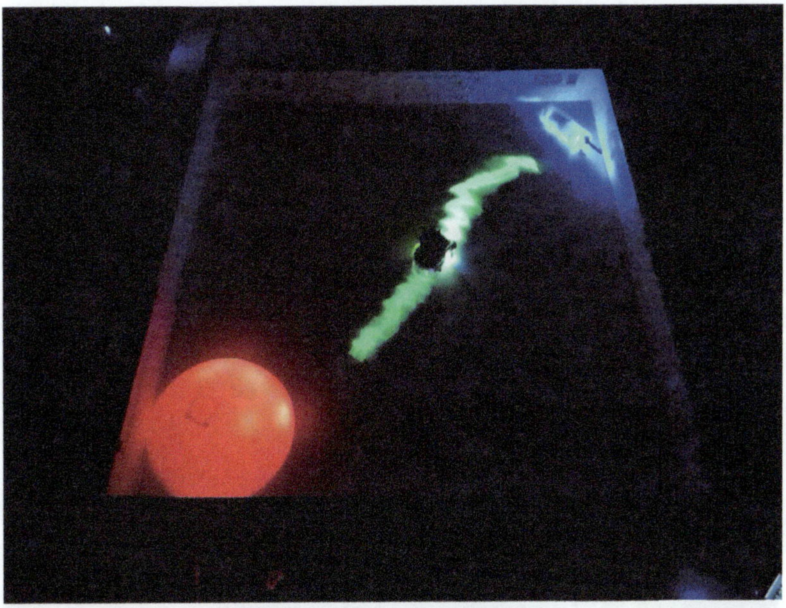

Fig. 8 Photograph showing the trail laid by the robot in a 1m×1m arena and a distance of 1.3m between food (top-right) and nest (bottom-left). The robot is on its way back from the food source pointing towards the nest and is following its own line.

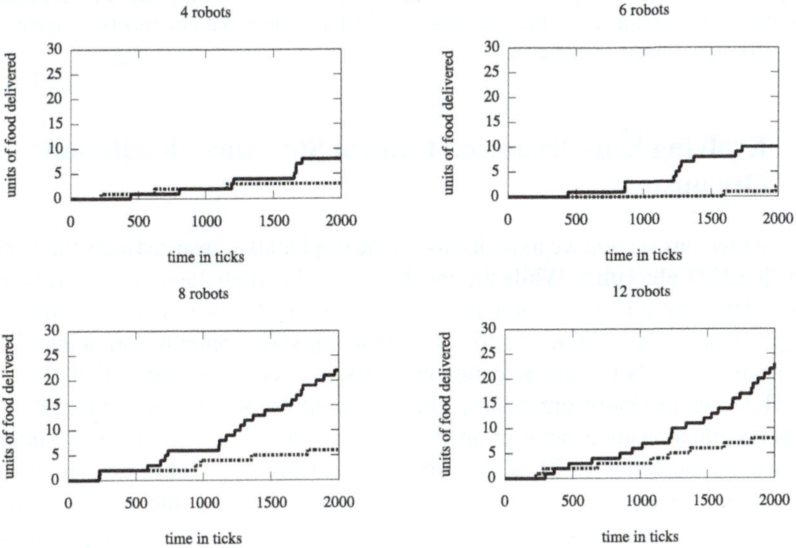

Fig. 9 Efficiency measurements for 4 different simulation runs. Top-left shows 4 robots, top-right 6, bottom-left 8, bottom-right 12. Solid lines are with pheromones turned on, dashed lines represent runs without pheromones. The plots show the sum of units of food delivered to the nest. The efficiency of the agents that could utilize the pheromone trails is greatly increased.

'normal' arena without phosphorescent paint. Our results show that in this simulation the efficiency is greatly enhanced with the use of trails (Fig. 9).

In the future we plan to do experiments with multiple robots in order to fully emulate the ant foraging behavior with autonomous robots in real life. Figure 10 shows a contrived photograph of how this could look like.

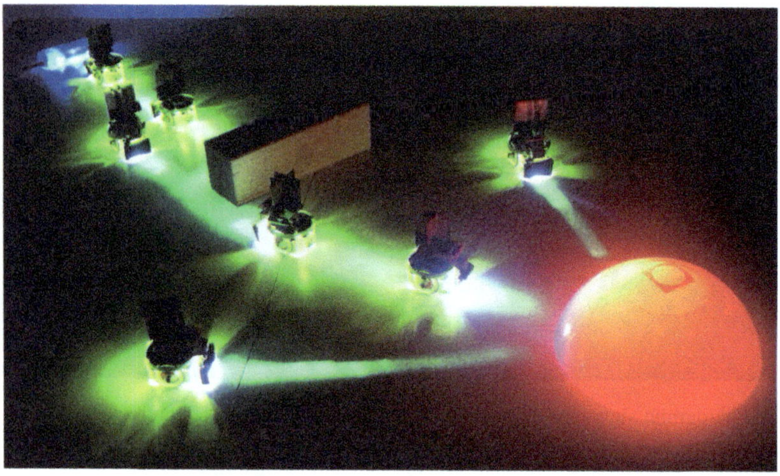

Fig. 10 Contrived photograph of how the glowing floor and our sensors should be used in the future. Two robots leave the nest to search for food, the remaining robots navigate to and from the nest around an obstacle.

3 Evolving Self-Organized Control Structures for Robotic Organisms

In the precious section we have discussed the trophallaxis-inspired algorithm and the BEECLUST algorithm. While the capabilities of the trophallaxis-inspired algorithm were impressive in simulation studies (see Fig. 3) it was found to be difficult to implement in a real robotic swarm due to the immense communication interference of nearest-neighbor-communication in such systems (e.g., sound or IR-light).

However, in robotic organisms, where the modules are closely connected to each other, communication between modules is less defective and also the bandwidth is usually much higher. Thus, for robotic organisms, we want to explore again the potential of a network of homeostatic sub units which all interact autonomously and which generate collective information (maps, waves, ...) on the organism level. In addition, the system should not be hand-coded per se, it should represent an (almost) open-ended dynamical system which can encode many processes in parallel. Again looking into comparable counterparts in nature, we found that signal processing in unicellular organisms and the homeostatic hormone control in multicellular organisms provide a good model for such an open-ended system.

To allow evolutionary computation operators to act on the configuration of such robotic organisms, the parametrization of the underlying dynamical system was encoded into a data-structure called 'genome' of the organism, which is subject to selection, mutation and inheritance. We named our control system Artificial Homeostatic Hormone Systems (AHHS), basically defined and investigated in [10, 29, 28, 36].

3.1 AHHS for Robot Control

In AHHS, a physiological model depicts the inner space of the robot and this model is controlling the robot's behaviors: Sensors trigger hormone excretions, which increase hormone concentrations in the robot's virtual inner body. These hormones diffuse, integrate, decay, interact and finally, affect actuators. The virtual inner body is partitioned into several compartments, whereas each compartment is associated with a specific part of the real robot's body to facilitate the emergence of complex behaviors. Each sensor and actuator is associated with one of these compartments. Thus, we achieve a kind of embodiment [21] that enforces an appropriate spatio-temporal context.

The principle functionality of AHHS is based on homeostatic processes that are interrupted by dynamic sensory stimuli. The hormone concentrations of an AHHS are initially set to zero. Once the system is started, an initial intrinsic dynamics is instantiated. Hormone concentrations increase and hormones begin to interact. This happens even without any initial sensor input. From a system theoretic point of view, the system undergoes a transient until an equilibrium, or preciser, an attractor (fixed point, oscillation, or even a chaotic attractor) is reached. This can be interpreted as homeostasis. Once the sensors report non-zero input the current equilibrium is disturbed. A new equilibrium associated to the current sensor input is pursued and finally reached, if the sensor input was constant for long enough. Typically each robot module will execute a copy of a common controller. These controllers communicate implicitly via hormones that diffuse from module to module.

The AHHS is implemented by a system of ordinary differential equations (which are discretized in the microprocessor). Hence, the execution of an AHHS controller is mathematically interpreted as the numerical forward integration in time of an initial value problem with time-variant disturbances (sensor inputs).

An important issue in evolutionary robotics and multi-modular robotics is to achieve systems with high evolvability, that is, fast synthesis of controllers by artificial evolution. A good model for understanding the underlying processes of this evolution is the concept of fitness landscapes. The fitness landscape of a controller synthesis scenario is defined by the mapping from the high-dimensional space of features, that describe the controller, to the actual fitness value that is obtained by

(a) 3 modules

(b) 5 modules

Fig. 11 Example benchmark scenario of 'gait learning' in modular robotics. In both configurations, an AHHS evolved that moved the robot efficiently within several tens generations.

executing the controller and observing the resulting behavior. The shape of such landscapes is defined by a variety of influences, such as the robot's task, robot's hardware, its environment, the used mutation operators, and the controller design itself. The leading design concept of AHHS is to generate smooth fitness landscapes, that is, there is a high causality of the mutation operator (small changes in the controller result in small changes of the behavior). Several effects, for example the trade-off between evolvability and an increase of the search space, are investigated currently.

First studies of the AHHS in the context of multi-modular robotics and comparisons to other controller approaches have been made [10, 29]. One of the benchmarks was the so-called 'gait learning' in modular robotics (see Fig. 11). One of the results, that was reported in [10], is shown in Fig. 12. It shows a comparison of the best fitness obtained by artificial evolution for $N = 12$ independent runs per controller approach. The superiority of AHHS over a simple artificial neural network approach is significant.

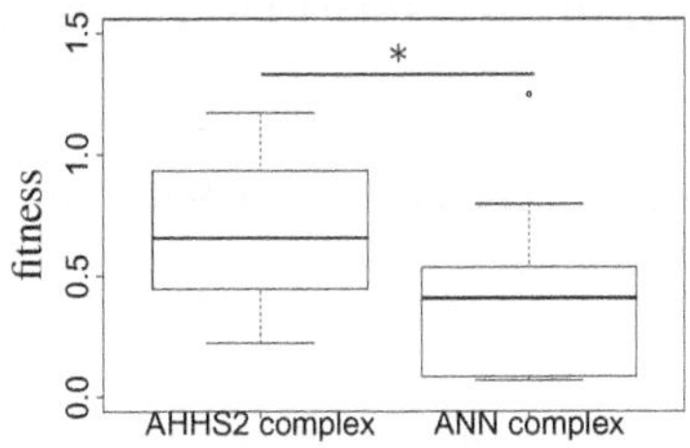

Fig. 12 Comparison between the AHHS controller and a simple artificial neural network for a gait learning task with three modules ($N = 12$) [10].

3.2 *Comparison of AHHS to Other Controller Types*

The controller described above was analyzed concerning its evolvability and adaptability. As a first benchmark test a scenario was chosen in which a maze had to be explored: A robot controlled by the AHHS was put in a simulated 2D-arena and evolutionary runs were performed. Behavior which resulted in moving around in the arena was rewarded with good fitness values. The population in the evolutionary runs consisted of 100 individuals and evolution took 500 generations. For comparison these evolutionary runs were repeated for the same task with standard types of artificial neural network (ANN) controllers. The results revealed that maximum and average fitness gained by the two the controller families did not differ after having evolved for 500 generations. Additional analyses on the time needed to evolve satisfying behavior (75% of the overall maximum fitness) were performed. Those studies revealed that after 20 to 30 generations this value was reached by both controller families and the observed behavior was a kind of wall following behavior (Fig.13). Interestingly, the behavior between the two controller families differed significantly qualitatively. In contrast to the ANN controller the AHHS steered the robot in straight lines and very smooth curves (see Fig.13(a)).

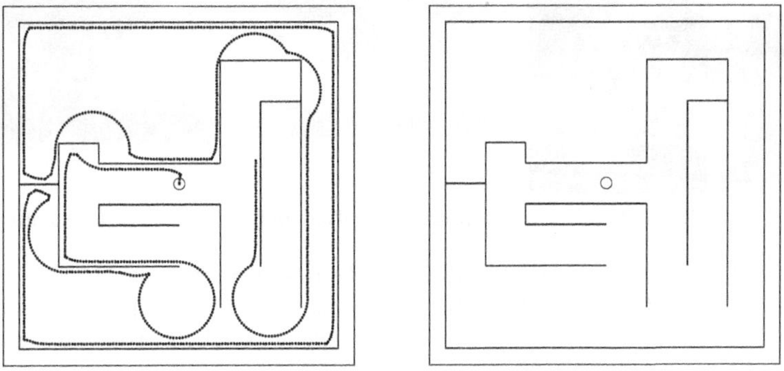

(a) AHHS (b) ANN

Fig. 13 Trajectory of the best evolved individual of (a) AHHS and (b) ANN controller in an "exploring the maze" scenario.

4 Evolutionary Shaping of Network Topology of Controllers to Body Shapes

Neural networks are a state-of-the-art technology in evolutionary robotics. ([8]). One advantage is that they are easy to mutate and quite powerful and efficient in a computational sense. In contrast to out-of-the-box ANNs, living organisms show highly structured neural networks, often split up into brains and ganglia. Often, regular patterns of such ganglia are found, for example in a ventral nerve cord. In such structured networks, specific ganglia (densely clustered aggregations of neural cells) can specialize on specific local tasks, for example coordinating the motion of a nearby limb. An engineering approach to generate such patterns of artificial neural networks is Hyperneat [34], an extension of Neat [35], which was used successfully in several applications [5, 4]. To enhance our evolutionary robotic systems, we aim for a similar pattern-generating system which is inspired by the growth process of multi-cellular organisms (see Fig. 14). To achieve this, we developed of virtual embryogenesis (VE) to generate various topologies autonomously and dynamically [39, 16]. The main idea of VE is the simulation of EvoDevo-like processes [3]. These processes are observable in nature during the developmental phase of multicellular lifeforms e.g. *Drosophila m.* [19]. EvoDevo-inspired processes will allow us to evolve network patterns and multi-modular robot shapes in parallel (see Fig. 15(b)), thus expecting a joint evaluation of controller and corresponding body shape.

The embryo developed by VE (see Fig. 14(b)) consists of many individual cells. The virtual embryogenetical growth process is controlled again by a genome (see Fig. 15). There is high similarity between the rule-set of AHHS and VE, which is also reflected by high similarities in the genome syntax and grammar. In VE the genome encodes the reactions of a cell to different concentrations of

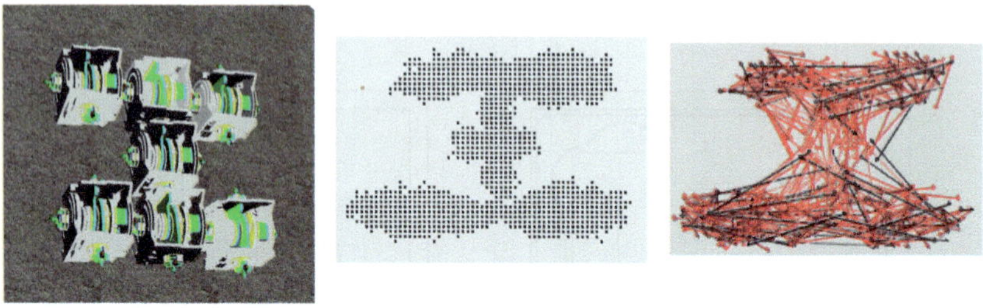

(a) Robot shape (b) Shape of evolved embryo (c) Shape of resulting ANN

Fig. 14 Evolution of structured ANNs. (a): The robot's given shape, which is part of the fitness-function of the artificial evolutionary process. (b): Evolved shape of the virtual embryo which constructs the ANN network topology. (c): Example of one ANN topology, grown in this embryo.

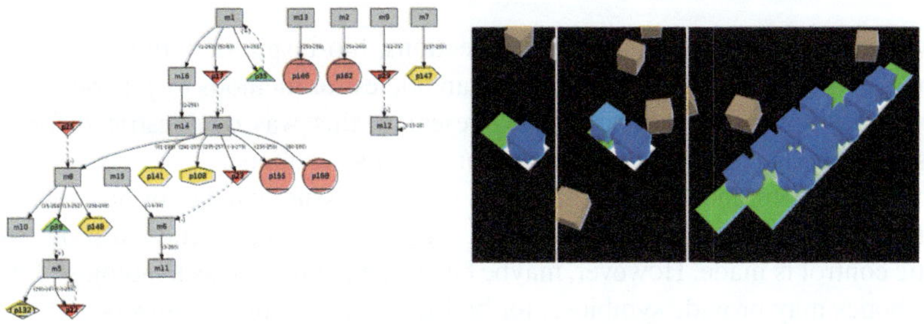

(a) Genome (b) Multi-modular robot

Fig. 15 a): Evolved genome structure that controls the growth process of the embryo shown in Fig. 14b. Each gene of this genome is able to produce proteins, which in turn can activate other genes, produce morphogens, change the receptivity of the cell for morphogens, or build neural links to other cells. Genes and proteins are indicated by geometrical shapes. Interactions of proteins and genes are indicated by arrows. b): Formation of a multi-modular robot using VE in a simulation environment. In contrast to the VE mentioned above, the cellular duplication process is realized by docking robotic modules to the robotic organism. Blue boxes indicate docked robotic modules that are part of a multicellular robotic organism. Brown boxes indicate robotic modules, that are available for the docking process. Green patches indicate positions where a "free" robotic module can dock the robotic organism. Left subfigure: start of the process. The robotic organism consists of a single module, which waits for an other module to dock. Middle subfigure: A robotic module docks, the robotic organism now consists of 2 modules. Right subfigure: Body-formation in progress. Several modules have already docked together.

morphogens (virtual chemical substances diffusing throughout the embryo, emitted by the cells), which could be compared to the flow of hormones in AHHS. Each cell of an embryo has the same genome during the whole embryogenetical process. The genome is not changed during the life-time of the cell. Offspring-cells have the same properties as their ancestor-cells, so that tissue specialization can emerge. Fig. 14 exemplary shows one target robotic organism, the evolved embryo and the ANN topology grown by this virtual embryo. Evolution operates on the genome only, which is depicted in Fig. 15, which also indicates the feedbacks that exist between gene products. It is important to understand this feedback network to understand the evolutionary path that can be observed during evolution. This understanding links back to biological research were similar ways of thought are current topics of state-of-the-art research. AHHS and VE are both derived from biology in a classical bio-inspired way to solve engineering problems. However, both models can additionally fertilize new understanding and novel research also in the biological domain, as the course of evolution in the development of our robotic organisms might well resemble the 'how?' and the 'why?' of their natural counterparts' evolution.

5 Discussion

As was shown throughout the previous sections, bio-inspiration, bio-mimicry and artificial evolution are clearly promising candidates for methods to generate control software for collective robotics. In my research, this was demonstrated for both types of collective robotics (swarms and organisms of robots).

From a pure engineering perspective it is not an issue whether or not biological research draws benefit from bio-inspired robotics, as long as enough progress in robotic control is made. However, maybe because I am trained as a biologist, I think bio-robotics may provide symbiosis for biologists and robotic engineers.

I try to achieve this symbiosis by a two-fold approach: On the one hand my lab performs comparable research scenarios with robots and with animals in parallel, thus allowing two groups of researchers to influence each other. This happens, for example, whenever one research group inspires the other for performing a new experimental setup. On the other hand, we generate 'bio-mimicking' robots that are driven by nature-derived 'bio-inspired' control software that further supports such a potential scientific symbiosis. In addition, we follow another method of joint research which is abstract (macroscopic) modeling. By generating such models we can excavate the 'algorithmic core' of both the natural system and the freshly emerged bio-inspired robotic system. This way, a process of abstraction and generalization allows us to draw new scientific insights of the similarities and intrinsic causations within and between both systems.

In most of our approaches we use a network of positive and negative feedbacks and delays to govern the behavior of the collective system (swarm or organism). Especially in the trophallaxis-inspired algorithm, but also in AHHS and VE, a major component is behavior-based, distributed, multi-level homeostasis in which every single module and the collective system are both designed as autonomous homeostatic systems. Similar concepts for collective robotics were suggested in [32] for robot organisms and in [33] for swarm systems. Although significant differences exist, our approach of AHHS and VE has similarities to [33], which are subject to evolutionary adaptation as was shown in [20]. AHHS and VE also have aspects that can be interpreted as gene regulatory networks (GRN), as they were investigated in [17]. In AHHS and VE, these GRN-like functionality is acting locally and is closely bound to the shape of the robot (or of the simulated physics of the virtual embryo), thus our research paradigm contains significant aspects of embodiment, as it is discussed in [21].

As pointed out in the introduction section, I see important differences between the approach of 'bio-inspiration' and 'bio-mimicry'. Bio-inspiration means deduction of mechanisms and/or functionality from biological sources of inspiration which can be converted significantly to work in a technical application in a similar way. For example wing shapes of airplanes are in fact inspired by wings of birds, but concerning their embodiment significant differences are obvious: Wings of birds are bendable, are made of a stiff skeleton with several joints and are covered by very light and soft feathers. In contrast, wings of airplanes are solid without feather-like structures, they have no joints and their basic skeleton is of median stiffness. Thus they act (function) similar to some birds wings, but they do not look and feel similar.

The approach of bio-mimicry is more about producing technical entities that look and feel like living organisms but are internally exploiting totally different mechanisms. Prominent examples are humanoid robots, toy robots (dogs, dinosaurs, ...), and scarecrows.

In figure 16 I depict a feature space that spans along two axes: The top axis indicates the level of bio-inspiration while the right axis indicates the level of bio-mimicry. In the lower left corner (B), we see classical technical entities like cars, cell phones, computers and satellites which incorporate mainly non-bioinspired mechanisms and which also do not resemble biological organisms very much. In the upper left corner (A) we see typical forms of bio-mimicry, which look similar to living organisms but are based on classical technical mechanisms. In the lower right corner, we find bio-inspired algorithms. They are strongly inspired by mechanisms found

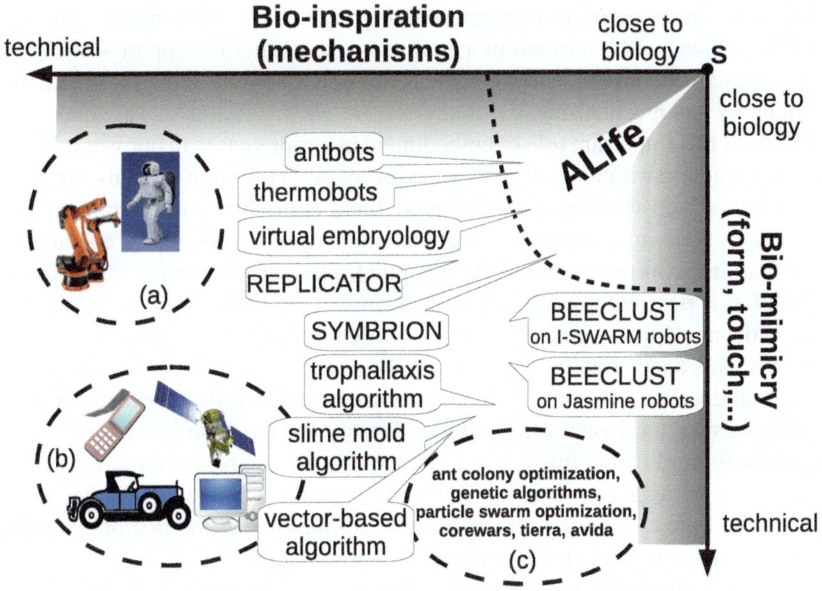

Fig. 16 Bio-inspiration/bio-mimicry feature space. Top axis: level of bio-inspiration. Technical applications that use nature-like mechanisms are locate towards the right side of this axis. Right axis: Level of bio-mimicry. The closer a technical entity resembles a biological counterpart, the higher it is located on this axis. S: The ALife singularity, indicating the highest simultaneous level of bio-inspiration and bio-mimicry. Technological products located at this point are indiscriminable form natural organisms from inside (mechanisms) and from outside (bio-mimicry). For discussion of the examples located in this graph see text. Picture source: wikimedia commons.

in nature but, as they are computer algorithms, they have a totally different physical nature (bits). Also wings of airplanes could probably be grouped here.

In the upper right corner, we find the field of ALife research, which tries to achieve the creation of life-like structures and mechanisms, thus it tries to maximize both aspects in this feature-space. In the very-most upper right corner, we find the before-mentioned ALife singularity which is reached as soon as one succeeded in producing a robot that achieves the highest level along both axes: Such an entity would be indiscriminable from a natural organism.

As is shown in this figure, the work of my lab is approaching the Alife corner step-by-step. While earlier robotic algorithms (trophallaxis-algorithm, slime mold algorithms, BEECLUST in multi agent simulation or in non-bio-mimicking robots) are quite distant from the ALife area, newer research (e.g., SYMBRION, thermobots, antbots) are getting much closer to the upper right corner. However, our projects are still not located quite deep inside of the ALife field, telling us that we are still quite far away from recreating living organisms. The reason for this is also shown in figure 16: It shows two shaded fields along both axis. These fields indicate a sort of 'no-access' area. I think that a certain degree of bio-inspiration is unachievable without a high degree of bio-mimicry and vice versa. This means that for a technical entity to resemble a living organism's mechanisms, also form and material have to be adapted in a well adjusted manner to support – or even to allow – these mechanisms. Only specific forms and materials allow for nature-near functionality and mechanisms.

I realized this important role of embodiment in various cases in our work: Whenever we found a (swarm-)intelligent behavior by observation of animals, we translated it into a well-working computer model and algorithm. But as soon as these mechanisms were implemented on real embodied agents (robots), we encountered significant drops in performance. At this point, there are two ways to go: On the one hand, it is possible to adapt the mechanisms (algorithms) in a way that they fit better to the new form of embodiment. This is done in my lab by exploiting artificial evolution. This adaptation increases the performance of the technical system significantly, but does not bring the system closer to the upper right corner (S) in figure 16, as it does not increase the level of bio-mimicry. On the other hand, it is possible to alter the form of embodiment, what is to increase the degree of bio-mimicry. This step involves hardware engineering and the testing of new materials. Although this way is more time consuming and more resource-intensive as the first one, it brings the system closer to or deeper into the field of artificial life.

For swarm intelligent systems, embodiment is an important challenge and opportunity [21]. Many algorithms that work perfectly in non-embodied systems, like optimization algorithms [22, 12, 7] show significantly lowered efficiency as soon as they are executed by embodied agents. Also, many macroscopic models (ODE, PDE) which basically treat entities as volumeless points in space are not sufficient to derive microscopic behavioral rules (algorithms) necessary for swarms or organisms of embodied autonomous agents (robots). Frequently agent collisions, forces applied by one agent to another, and other forms of physical interferences tend to inhibit algorithms from functioning well in real-world embodied systems. For example, our antbots still have problems with collision handling on their ant trails which do not appear in well known ODE models of ant pheromone trails. The more the antbots algorithm is extended with collision handling procedures the further the overall antbots' algorithm deviates from mechanisms described in these ODE models.

However, the existance of physics does not only pose problems to the field of bio-inspired/bio-mimicking robotics. Exploitation of physical phenomena provides also an important opportunity for engineering such collective systems: Emergent phenomena of self-organized systems can be exploited by optimization procedures like artificial evolution, exhibiting collective behaviors and solutions that are unreachable in abstract models, because one of the major components of such mechanisms is missing there in most cases: *physics*. Physical properties and mechanisms based on physical interactions can provide functionality that replaces algorithmic mechanisms that are executed by software. In my opinion, each embodied agent executes a sort of 'master-algorithm' which is the summarization of the sensor system, the software algorithm (incl. BIOS, operating system, hardware-abstraction layers and middleware), the actuators, the physics of the agents and all physical interaction with the environment (including all other agents). If one of these components is altered, the 'master-algorithm' is altered. Also here, emergent phenomena arise from the interplay of sub-components, again suggesting a favouring of evolutionary approaches over engineering approaches, as emergence is still not engineerable.

Finally in this discussion, I want to point out a major shortcoming of our current mechatronic systems in the field of bio-inpired/bio-mimicking robotics: *Reproduction*. In figure 16, the path towards the ALife singularity becomes more and more narrow due to the converging two 'non-access' areas. One essential functionality of living systems is the ability to reproduce. While this is not a problem in non-embodied systems (e.g. genetic algorithms [12] or evolution strategies [22]), real-world replication of embodied autonomous agents is currently not achievable in mechatronics. To achieve embodied replication, material science has to bring new techniques and materials to the field of robotics and ALife research has to solve many replicator-related problems that would allow for autonomous replication on such a level. Currently, many researchers are discussing this issue (personal communication), pushing the ALife field more towards research with bacteria or even bio-molecules. In these systems, the issue of replication is already solved by nature, thus scientific progress in such research will not improve our understanding in how larger embodied compounds can be made replicating. Thus, as the robots' basic goal is to serve and assist humans in dangerous, unpleasant or boring tasks, progress in larger scale robotics (at least cm-range) is still a desirable challenge to tackle. Such demands will not be satisfied by altering bacteria or molecules, as they cannot execute most of the real-world jobs robots are thought to perform. In conclusion, I think that ALife research with mechatronic devices should be continued with high intensity, of course in parallel with further and novel research in self-organizing, evolving and complexity-generating life forms.

References

1. Blow, M.: 'stigmergy': Biologically-inspired robotic art. In: Proceedings of the Symposium on Robotics, Mechatronics and Animatronics in the Creative and Entertainment Industries and Arts, pp. 1–8 (2005)
2. Bonani, M., Raemy, X., Pugh, J., Mondana, F., Cianci, C., Klaptocz, A., Magnenat, S., Zufferey, J.-C., Floreano, D., Martinoli, A.: The e-puck, a robot designed for education in engineering. In: Proc. of the 9th Conference on Autnomous Robot Systems and Competitions, vol. 1, pp. 59–65 (2009)
3. Carrol, S.B.: Endless Forms Most Beautiful: The New Science of Evo Devo. W. W. Norton, New York (2006)
4. Clune, J., Beckmann, B., Ofria, C., Pennock, R.: Evolving coordinated quadruped gaits with the hyperneat generative encoding. In: Proceedings of the IEEE Congress on Evolutionary Computing Special Section on Evolutionary Robotics, Trondheim, Norway (2009)
5. Clune, J., Beckmann, B.E., Ofria, C., Pennock, R.T.: Evolving coordinated quadruped gaits with the hyperneat generative encoding. In: Proceedings of the 2009 IEEE Congress on Evolutionary Computation (CEC), IEEE, New York (2009)
6. Corradi, P., Schmickl, T., Scholz, O., Menciassi, A., Dario, P.: Optical networking in a swarm of microrobots. Lecture Notes of the Institute for Computer Sciences, Social Informatics and Telecommunications Engineering 3(2), 107–119 (2009)
7. Dorigo, M., Stützle, T.: Ant Colony Optimization. The MIT Press, Cambridge (2004)
8. Floreano, D., Dürr, P., Mattiussi, C.: Neuroevolution: From architectures to learning. Evolutionary Intelligence 1, 47–62 (2008)
9. Garnier, S., Tache, F., Combe, M., Grimal, A., Theraulaz, G.: Alice in pheromone land: An experimental setup for the study of ant-like robots. In: Swarm Intelligence Symposium, SIS 2007, pp. 37–44. IEEE, New York (2007)
10. Hamann, H., Stradner, J., Schmickl, T., Crailsheim, K.: A hormone-based controller for evolutionary multi-modular robotics: From single modules to gait learning. In: Proceedings of the IEEE Congress on Evolutionary Computation (CEC 2010), pp. 244–251 (2010)
11. Heran, H.: Untersuchungen über den Termperatursinn der Honigbiene (Apis mellifica) unter besonderer Berücksichtigung der Wahrnehmung strahlender Wärme. Zeitschrift für vergleichende Physiologie 34, 179–206 (1952)
12. Holland, J.H.: Adaptation in Natural and Artificial Systems. Univ. Michigan Press, Ann Arbor (1975)
13. Jasmine. Swarm robot - project website (2010), `http://www.swarmrobot.org/`
14. Kennedy, J., Eberhart, R.C.: Particle swarm optimization. In: Proceedings IEEE International Conference on Neural Networks, vol. 4 (1995)
15. Kernbach, S., Thenius, R., Kornienko, O., Schmickl, T.: Re-embodiment of honeybee aggregation behavior in an artificial micro-robotic swarm. Adaptive Behavior 17, 237–259 (2009)
16. Levi, P., Kernbach, S. (eds.): Symbiotic Multi-Robot Organisms: Reliability, Adaptability, Evolution. Springer, Heidelberg (2010)
17. Mattiussi, C., Floreano, D.: Analog genetic encoding for the evolution of circuits and networks. IEEE Transactions on evolutionary computation 11, 596–607 (2007)

18. Mayet, R., Roberz, J., Schmickl, T., Crailsheim, K.: Antbots: A feasible visual emulation of pheromone trails for swarm robots. In: Dorigo, M., Birattari, M., Di Caro, G.A., Doursat, R., Engelbrecht, A.P., Floreano, D., Gambardella, L.M., Groß, R., Şahin, E., Sayama, H., Stützle, T. (eds.) ANTS 2010. LNCS, vol. 6234, pp. 84–94. Springer, Heidelberg (2010)
19. Meinhardt, H., Gierer, A.: Pattern formation by local self-activation and lateral inhibition. Bioessays 22, 753–760 (2000)
20. Nolfi, S., Floreano, D.: Evolutionary Robotics: The Biology, Intelligence, and Technology of Self-Organizing Machines. MIT Press, Cambridge (2004)
21. Pfeiffer, R., Bongard, J.C.: How the Body Shapes the Way We Think. MIT Press, Cambridge (2006)
22. Rechenberg, I.: Evolutionsstrategie 1994. Frommann Holzboog (1994)
23. REPLICATOR. Project website (2010), `http://www.replicators.eu`
24. Russell, R.A.: Heat trails as short-lived navigational markers for mobile robots. In: Proceedings of International Conference on Robotics and Automation, 1997, vol. 4, pp. 3534–3539 (1997)
25. Russell, R.A.: Ant trails – an example for robots to follow? In: Proceedings of IEEE International Conference on Robotics and Automation, 1999, vol. 4, pp. 2698–2703 (1999)
26. Schmickl, T., Crailsheim, K.: A Navigation Algorithm for Swarm Robotics Inspired by Slime Mold Aggregation. In: Şahin, E., Spears, W.M., Winfield, A.F.T. (eds.) SAB 2006 Ws 2007. LNCS, vol. 4433, pp. 1–13. Springer, Heidelberg (2007)
27. Schmickl, T., Crailsheim, K.: Trophallaxis within a robotic swarm: bio-inspired communication among robots in a swarm. Autonomous Robots 25(1-2), 171–188 (2008)
28. Schmickl, T., Crailsheim, K.: Modelling a hormone-based robot controller. In: 6th Vienna International Conference on Mathematical Modelling, MATHMOD 2009 (2009)
29. Schmickl, T., Hamann, H., Stradner, J., Crailsheim, K.: Hormone-based control for multi-modular robotics. In: Levi, P., Kernbach, S. (eds.) Symbiotic Multi-Robot Organisms: Reliability, Adaptability, Evolution. Springer, Heidelberg (2010)
30. Schmickl, T., Möslinger, C., Thenius, R., Crailsheim, K.: Individual adaptation allows collective path-finding in a robotic swarm. International Journal of Factory Automation, Robotics and Soft Computing 4, 102–108 (2007)
31. Schmickl, T., Thenius, R., Möslinger, C., Radspieler, G., Kernbach, S., Crailsheim, K.: Get in touch: Cooperative decision making based on robot-to-robot collisions. Autonomous Agents and Multi-Agent Systems 18(1), 133–155 (2008)
32. Shen, W.-M., Salemi, B., Will, P.: Hormone-inspired adaptive communication and distributed control for CONRO self-reconfigurable robots. IEEE Trans. on Robotics and Automation 18(5), 700–712 (2002)
33. Shen, W.-M., Will, P., Galstyan, A., Chuong, C.-M.: Hormone-inspired self-organization and distributed control of robotic swarms. Autonomous Robots 17, 93–105 (2004)
34. Stanley, K.O., D'Ambrosio, D.B., Gauci, J.: A hypercube-based encoding for evolving large-scale neural networks. Artificial Life 15(2), 185–212 (2009)
35. Stanley, K.O., Miikkulainen, R.: Competitive coevolution through evolutionary complexification. Journal of Artificial Intelligence Research 21(1), 63–100 (2004)
36. Stradner, J., Hamann, H., Schmickl, T., Thenius, R., Crailsheim, K.: Evolving a novel bio-inspired controller in reconfigurable robots. In: 10th European Conference on Artificial Life (ECAL 2009). LNCS, Springer, Heidelberg (2010)

37. Svennebring, J., Koenig, S.: Building terrain-covering ant robots: A feasibility study. Autonomous Robots 16(3), 313–332 (2004)
38. SYMBRION. Project website (2010), `http://www.symbrion.eu`
39. Thenius, R., Schmickl, T., Crailsheim, K.: Novel concept of modelling embryology for structuring an artificial neural network. In: Troch, I., Breitenecker, F. (eds.) Proceedings of the MATHMOD (2009)
40. Valdastri, P., Corradi, P., Menciassi, A., Schmickl, T., Crailsheim, K., Seyfried, J., Dario, P.: Micromanipulation, communication and swarm intelligence issues in a swarm microrobotic platform. Robotics and Autonomous Systems 54, 789–804 (2006)

Flocking Control Algorithms for Multiple Agents in Cluttered and Noisy Environments

Hung Manh La and Weihua Sheng

Abstract. Birds, bees, and fish often flock together in groups based on local information. Inspired by this natural phenomenon, flocking control algorithms are designed to coordinate the activities of multiple agents in cluttered and noisy environments, respectively. First, to allow agents to track and observe a target better in cluttered environments, an adaptive flocking control algorithm is proposed. With this algorithm, all agents can track the target better and maintain a similar formation and connectivity. Second, to deal with noisy measurements we proposed two flocking control algorithms, *Multi-CoM-Shrink* and *Multi-CoM-Cohesion*. Based on these algorithms, all agents can form a network and maintain connectivity, even with noisy measurements. We also investigate the stability and scalability of our algorithms. Simulations and real experiments are conducted to demonstrate the effectiveness of the proposed approach.

Keywords: Flocking control, multi-agent systems, mobile sensor networks.

1 Introduction

Flocking is a natural phenomenon in which a number of agents move together and interact with each other. In nature, schools of fish, birds, ants, and bees, etc. demonstrate the phenomena of flocking. Flocking control for multiple mobile agents has been studied in recent years [1, 2], and it is designed based on three basic flocking

Hung Manh La
The School of Electrical and Computer Engineering, Oklahoma State University,
Stillwater, OK 74078, USA
e-mail: hung.la@okstate.edu

Weihua Sheng
The School of Electrical and Computer Engineering, Oklahoma State University,
Stillwater, OK 74078, USA
e-mail: weihua.sheng@okstate.edu

Y. Meng and Y. Jin (Eds.): Bio-Inspired Self-Organizing Robotic Systems, SCI 355, pp. 53–79.
springerlink.com

rules proposed by Reynolds [3]: flock centering , collision avoidance, and velocity matching. The problems of flocking have also attracted many researchers in physics [4], mathematics [5], biology [6] and especially in control science in recent years [1, 2, 7].

Flocking control has wide applications in mobile robots and mobile sensor networks. Early work on flocking control includes [1, 2]. Tanner *et al.* [1] studied the stability properties of a system of multiple mobile agents with double integrator dynamics in the case of fixed and dynamic topologies. However, in their work the target tracking problem and sensing errors are not considered. In the context of target tracking Olfati-Saber [2] proposed the theoretical framework for the design and analysis of distributed flocking algorithms, which solve the flocking problem in free space and in the presence of obstacles. Based on his flocking control algorithms, all agents can flock together and track the target quite well in free space. However, the target tracking performance is not satisfactory in the obstacle space. Moreover, every agent is assumed to know the position and velocity of the target precisely. To relax this assumption, he developed a distributed Kalman filter in [8] for each agent to estimate the target's position. Due to the measurement errors, the target tracking performance is not very good. In addition, there is another assumption in his flocking algorithm [2, 8] that all agents need the information of the target to maintain the cohesion or avoid the fragmentation. To deal with this situation, an extension of the flocking control algorithm in [2] with a virtual leader in the case of a minority of informed agents and varying velocity of the virtual leader was presented in [7]. However, their work does not consider the tracking problems in cluttered and noisy environments.

In this paper we propose three new flocking control algorithms to deal with more realistic environments. The main differences between our algorithms and those of the above related work are listed below.

1. In cluttered environments, the agents usually get stuck behind the obstacles and sometimes can not follow the target [2]. To handle this problem we present a new algorithm to flocking control of multi-agent systems to track a moving target while avoiding obstacles. The main motivation of this algorithm is to make the agents flock together in an adaptive and distributed fashion. In this way the agents can track the moving target better and maintain connectivity in cluttered environments where the normal flocking control algorithms [2, 7, 8] have poor tracking performance and connectivity loss.

2. In real flocking control environments, noise handling is always an important issue since the noise usually causes broken network or connectivity loss. This problem exists in most of the previous work on flocking control [1, 2, 7, 8]. To make the flocking control more applicable in real applications we consider the effect of position and velocity measurement errors of the agent itself, the agent's neighbors and the target. None of the flocking control algorithms in the above related work considers this noise issue. We propose two flocking control algorithms, *Multi-CoM-Shrink* and *Multi-CoM-Cohesion*, which are based on the extensions of the *Multi-CoM* flocking control algorithm in our previous work [9]. Our algorithms allow the flocks to preserve connectivity, avoid collision, and follow the target in such noisy

environments. We demonstrate that by applying our algorithms the agents can flock together in the presence of noise with better performances such as connectivity and tracking performance.

The rest of this paper is organized as follows. In the next section we present the background of flocking control and the problem formulation. Section 3 describes the adaptive flocking control algorithm for tracking and observing a moving target in noise-free environments. Section 4 presents flocking control algorithms, *Multi-CoM-Shrink* and *Multi-CoM-Cohesion*, for tracking a moving target in noisy environments. Section 5 shows the main results on stability analysis of flocking control in both noise-free and noisy environments. Section 6 demonstrates the experimental results. Finally, Section 7 concludes this paper.

2 Flocking Backgrounds and Problem Formulation

In this section we present the flocking control background and the problems in existing flocking control algorithms.

We consider n agents moving in an m ($m = 2,3$) dimensional Euclidean space. The dynamic equations of each agent are described as:

$$\begin{cases} \dot{q}_i = p_i \\ \dot{p}_i = u_i, \; i = 1,2,...,n. \end{cases} \tag{1}$$

here $q_i, p_i \in R^m$ are the position and velocity of node i, respectively, and u_i is the control input of agent i.

To describe the topology of flocks we consider a dynamic graph G consisting of a vertex set $\vartheta = \{1,2...,n\}$ and an edge set $E \subseteq \{(i,j) : i,j \in \vartheta, j \neq i\}$. In this topology each vertex denotes one member of the flock, and each edge denotes the communication link between two members.

We know that during the movement of agents, the relative distance between them may change, hence the neighbors of each agent also change. Therefore, we can define a neighborhood set of agent i as follows:

$$N_i^{\alpha} = \left\{ j \in \vartheta : \|q_j - q_i\| \leq r, \; \vartheta = \{1,2,...,n\}, j \neq i \right\}, \tag{2}$$

here r is an active range (radius of neighborhood circle in the case of two dimensions, $m = 2$, or radius of neighborhood sphere in the case of three dimensions, $m = 3$), and $\|.\|$ is the Euclidean distance. The superscript α indicates the actual neighbors (α neighborhood agents) of agent i that is used to distinguish with virtual neighbors (β neighborhood agents) in the case of obstacle avoidance discussed later.

The geometry of flocks is modeled by an α-lattice [2] that meets the following condition:

$$\|q_j - q_i\| = d, j \in N_i^{\alpha}, \tag{3}$$

here d is a positive constant indicating the distance between agent i and its neighbor j. However, at singular configuration ($q_i = q_j$) the collective potential used to construct the geometry of flocks is not differentiable. Therefore, the set of algebraic constrains in (3) is rewritten in term of σ - norm [2] as follows:

$$\|q_j - q_i\|_\sigma = d^\alpha, j \in N_i^\alpha, \tag{4}$$

here the constraint $d^\alpha = \|d\|_\sigma$ with $d = r/k_c$, where k_c is the scaling factor. The σ - norm, $\|.\|_\sigma$, of a vector is a map $R^m \Longrightarrow R_+$ defined as

$$\|z\|_\sigma = \frac{1}{\varepsilon}[\sqrt{1+\varepsilon\|z\|^2} - 1], \tag{5}$$

here $\varepsilon > 0$. Unlike the Euclidean norm $\|z\|$, which is not differentiable at $z = 0$, the σ - norm $\|z\|_\sigma$, is differentiable every where. This property allows to construct a smooth collective potential function for agents.

The flocking control law in [2] controls all agents to form an α-lattice configuration. This algorithm consists of three components as follows:

$$u_i = f_i^\alpha + f_i^\beta + f_i^\gamma. \tag{6}$$

The first component of (6) f_i^α, which consists of a gradient-based component and a consensus component (more details about these components see [10], [11], [12]), is used to regulate the potentials (repulsive or attractive forces) and the velocity among agents.

$$f_i^\alpha = c_1^\alpha \sum_{j \in N_i^\alpha} \phi_\alpha(\|q_j - q_i\|_\sigma) n_{ij} + c_2^\alpha \sum_{j \in N_i^\alpha} a_{ij}(q)(p_j - p_i), \tag{7}$$

where c_1^α and c_2^α are positive constants, and each term in (7) is computed as follows [2]:

1. The action function $\phi_\alpha(z)$ that vanishes for all $z \geq r^\alpha$ with $r^\alpha = \|r\|_\sigma$ is defined as follows:

$$\phi_\alpha(z) = \rho_h(z/r_\alpha)\phi(z - d^\alpha) \tag{8}$$

with the uneven sigmoidal function $\phi(z)$ defined as $\phi(z) = 0.5[(a+b)\sigma_1(z+c) + (a-b)]$, here $\sigma_1(z) = z/\sqrt{1+z^2}$ (z is an arbitrary variable), and parameters $0 < a \leq b$, $c = |a-b|/\sqrt{4ab}$ to guarantee $\phi(0) = 0$. The bump function $\rho_h(z)$ with $h \in (0,1)$ is

$$\rho_h(z) = \begin{cases} 1, & z \in [0, h) \\ 0.5[1 + cos(\pi(\frac{z-h}{1-h}))], & z \in [h, 1) \\ 0, & otherwise. \end{cases} \tag{9}$$

2. The vector along the line connecting q_i to q_j is

$$n_{ij} = (q_j - q_i)/\sqrt{1+\varepsilon\|q_j - q_i\|^2}. \tag{10}$$

3. The elements $a_{ij}(q)$ of the adjacency matrix $[a_{ij}(q)]$ are defined as

$$a_{ij}(q) = \begin{cases} \rho_h(\|q_j - q_i\|_\sigma / r_\alpha), & if\ j \neq i \\ 0, & if\ j = i. \end{cases} \tag{11}$$

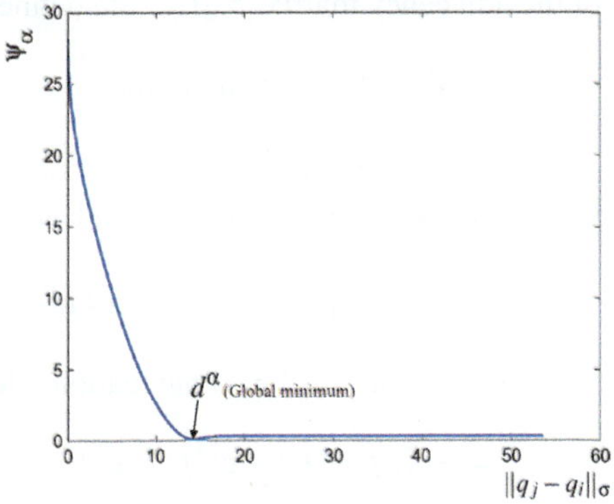

Fig. 1 Smooth pairwise potential function $\Psi_\alpha(\|q_j - q_i\|_\sigma)$.

The pairwise attractive/repulsive potential $\Psi_\alpha(z)$ is defined as:

$$\Psi_\alpha(z)) = \int_{d_\alpha}^{z} \phi_\alpha(s)ds, \tag{12}$$

and this function is illustrated in Figure 1.

Then we have the smooth collective potential function in the following form

$$V_\alpha(q) = \frac{1}{2}\sum_i \sum_{j \neq i} \Psi_\alpha(\|q_j - q_i\|_\sigma). \tag{13}$$

The second component of Equation (6) f_i^β is used to control the mobile agents to avoid obstacles,

$$f_i^\beta = c_1^\beta \sum_{k \in N_i^\beta} \phi_\beta(\|\hat{q}_{i,k} - q_i\|_\sigma)\hat{n}_{i,k} + c_2^\beta \sum_{k \in N_i^\beta} b_{i,k}(q)(\hat{p}_{i,k} - p_i) \tag{14}$$

where c_1^β and c_2^β are positive constants, and the set of β neighbors (virtual neighbors) of agent i at time t with k obstacles is

$$N_i^\beta(t) = \left\{k \in \vartheta_\beta : \|\hat{q}_{i,k} - q_i\| \leq r', \vartheta_\beta = \{1, 2, ..., k\}\right\} \tag{15}$$

here r' is selected to be less than r, in our simulations $r' = 0.6r$. ϑ_β is a set of obstacles. $\hat{q}_{i,k}, \hat{p}_{i,k}$ are the position and velocity of agent i projecting on the obstacle k, respectively. The virtual neighbors are used to generate the repulsive force to push the agents away from the obstacles.

Similar to vector n_{ij} defined in Equation (10), vector $\hat{n}_{i,k}$ is defined as

$$\hat{n}_{i,k} = (\hat{q}_{i,k} - q_i)/\sqrt{1 + \varepsilon\|\hat{q}_{i,k} - q_i\|^2}. \tag{16}$$

The elements $b_{i,k}(q)$ of the adjacency matrix $[b_{i,k}(q)]$ are defined as

$$b_{i,k}(q) = \rho_h(\|\hat{q}_{i,k} - q_i\|_\sigma / d_\beta) \tag{17}$$

where $d_\beta = \|r'\|_\sigma$.

The repulsive action function of β neighbors is defined as

$$\phi_\beta(z) = \rho_h(z/d_\beta)(\sigma_1(z - d_\beta) - 1). \tag{18}$$

The third component of (6) f_i^γ is a distributed navigational feedback.

$$f_i^\gamma = -c_1^\gamma(q_i - q_\gamma) - c_2^\gamma(p_i - p_\gamma) \tag{19}$$

where c_1^γ and c_2^γ are positive constants, and the γ - agent (q_γ, p_γ) is the virtual leader (more information of virtual leader, see [13]) defined as follows

$$\begin{cases} \dot{q}_\gamma = p_\gamma \\ \dot{p}_\gamma = f_\gamma(q_\gamma, p_\gamma) \end{cases} \tag{20}$$

Finally, the Olfati-Saber flocking control algorithm [2] is proposed as:

$$\begin{aligned} u_i = {} & c_1^\alpha \sum_{j \in N_i^\alpha} \phi_\alpha(\|q_j - q_i\|_\sigma) n_{ij} + c_2^\alpha \sum_{j \in N_i^\alpha} a_{ij}(q)(p_j - p_i) \\ & + c_1^\beta \sum_{k \in N_i^\beta} \phi_\beta(\|\hat{q}_{i,k} - q_i\|_\sigma) \hat{n}_{i,k} + c_2^\beta \sum_{k \in N_i^\beta} b_{i,k}(q)(\hat{p}_{i,k} - p_i) \\ & - c_1^\gamma(q_i - q_\gamma) - c_2^\gamma(p_i - p_\gamma). \end{aligned} \tag{21}$$

Observing the algorithm (21), we see that in cluttered environments or obstacles spaces:

- It is hard for the agents to follow the target because of repulsive forces generated from the obstacles.
- The tracking performance is not good.
- The agents sometimes get stuck around the obstacles.
- The network sometimes gets broken.

In addition, this algorithm works under the following assumptions:

- Each agent can sense its own position and velocity precisely (without noises).
- Each agent can obtain its neighbor's position and velocity via sensing or communication precisely.
- Each agent can sense the target position and velocity precisely.

However, in reality these assumptions are not valid because sensing errors always exist. Motivated by these observations we will study how to design distributed flocking control algorithms which can still perform well when the agents are in cluttered environments, and the measurements are affected by noises.

3 Adaptive Flocking Control for Tracking a Moving Target

We consider the γ agent as the moving target. Hence, we slightly modify the flocking control algorithm (21) as

$$\begin{aligned} u_i = & c_1^\alpha \sum_{j\in N_i^\alpha} \phi_\alpha(\|q_j - q_i\|_\sigma)n_{ij} + c_2^\alpha \sum_{j\in N_i^\alpha} a_{ij}(q)(p_j - p_i) \\ & + c_1^\beta \sum_{k\in N_i^\beta} \phi_\beta(\|\hat{q}_{i,k} - q_i\|_\sigma)\hat{n}_{i,k} + c_2^\beta \sum_{k\in N_i^\beta} b_{i,k}(q)(\hat{p}_{i,k} - p_i) \\ & - c_1^t(q_i - q_t) - c_2^t(p_i - p_t), \end{aligned} \tag{22}$$

here q_t and p_t are the position and velocity of the moving target, respectively, and c_1^t, c_2^t are positive constants. In this control algorithm, we assume that each agent has the ability to sense the position and velocity of the moving target.

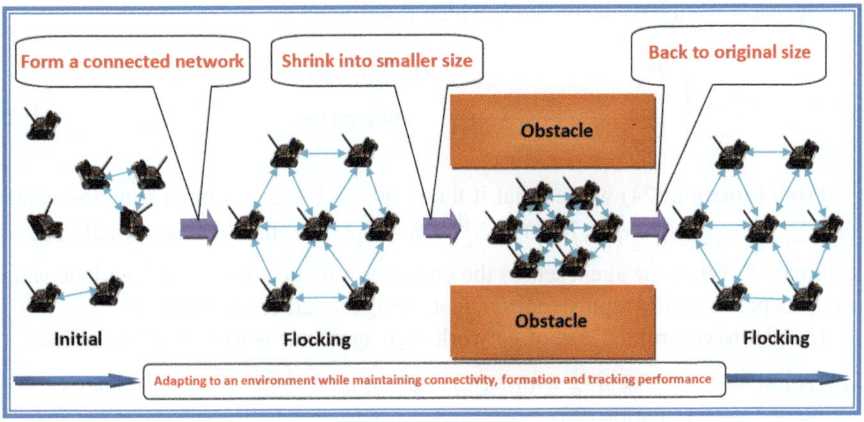

Fig. 2 Illustration of the adaptive flocking control.

The problem here is how to cooperatively control the size of the network in an adaptive and decentralized fashion in order to maintain the network's connectivity, similar formation and tracking performance in the presence of obstacles. One example of such flocking control is illustrated in Figure 2.

To control the size of the network, we need to control the set of algebraic constraints in Equation (4), which means that if we want the size of the network to be smaller to pass the narrow space then d^α should be smaller. This raises the question of how small the size of network should be reduced and how to control the size in a decentralized and dynamic fashion.

To control the constraint d^α one possible method is to use the knowledge of obstacle obtained by any agent in the network, which will broadcast a new d^α to all other agents. However, it is difficult for a single agent to learn the size of the passage due to its limited sensing range. To overcome this problem we propose a method based on the repulsive force, $\sum_{k\in N_i^{\beta}} \phi_{\beta}(\|\hat{q}_{i,k} - q_i\|_{\sigma})$, which is generated by the β-agent (virtual agent) projected on the obstacles. If any agent in the network gets this repulsive force it will shrink its own d_i^{α}. If this repulsive force is big (agent is close to obstacle(s)) d_i^{α} will be further reduced. Then, in order to maintain the neighborhood (topology) the active range of each agent is re-designed. To achieve an agreement on the relative distance and active range among agents, a consensus or a local average update algorithm is proposed. Furthermore, to keep the connectivity each agent maintains with an adaptive weight of attractive force from the target and an adaptive weight of interaction force from its neighbors so that the network reduces or recovers the size gradually. That is, if an agent has weak connection to the network it should have a big weight of attraction force to the target and a small weight of interaction force from its neighbors.

Firstly, we control the set of algebraic constraints as

$$\|q_j - q_i\|_{\sigma} = d_i^{\alpha}, j \in N_i^{\alpha}, \tag{23}$$

and let each agent have its own d_i^{α}, which is designed as

$$d_i^{\alpha} = \begin{cases} d^{\alpha}, & if\ \sum_{k\in N_i^{\beta}} \phi_{\beta}(\|\hat{q}_{i,k} - q_i\|_{\sigma}) = 0 \\ \frac{d^{\alpha}}{\sum_{k\in N_i^{\beta}} |\phi_{\beta}(\|\hat{q}_{i,k} - q_i\|_{\sigma})| + 1}, & otherwise. \end{cases} \tag{24}$$

From Equation (24) we see that if the repulsive force generated from the obstacles $\sum_{k\in N_i^{\beta}} \phi_{\beta}(\|\hat{q}_{i,k} - q_i\|_{\sigma}) = 0$ or $N_i^{\beta} = \emptyset$ (empty set) then the agent will keep its original d^{α}. When the agent senses the obstacles it reduces its own d_i^{α}, and the value of d_i^{α} depends on the repulsive force that the agent gets from obstacles.

In order to control the size of network each agent needs its own r_i^{α} that relates to d_i^{α} as follows: $r_i^{\alpha} = \|k_c d\|_{\sigma}$ with $\|d\|_{\sigma} = d_i^{\alpha}$ or $d = \sqrt{\frac{(\varepsilon d_i^{\alpha}+1)^2-1}{\varepsilon}}$. Explicitly, r_i^{α} is computed as in Equation (25).

$$r_i^{\alpha} = \begin{cases} r^{\alpha}, & if\ \sum_{k\in N_i^{\beta}} \phi_{\beta}(\|\hat{q}_{i,k} - q_i\|_{\sigma}) = 0 \\ \frac{1}{\varepsilon}[\sqrt{k_c^2 \frac{(\varepsilon d_i^{\alpha}+1)^2-1}{\varepsilon} + 1} - 1], & otherwise. \end{cases} \tag{25}$$

Similar to computing r_i^{α}, r_i is computed as

$$r_i = \begin{cases} r, & if\ \sum_{k\in N_i^{\beta}} \phi_{\beta}(\|\hat{q}_{i,k} - q_i\|_{\sigma}) = 0 \\ \sqrt{\frac{1}{\varepsilon}[(\varepsilon r_i^{\alpha} + 1)^2 - 1]}, & otherwise. \end{cases} \tag{26}$$

It should be pointed out that the active range r_i is different from the physical communication range. The active range is the range that each agent decides its neighbors to talk with, but the physical communication range is the range defined by the RF module. This implies that even a robot can communicate with many other robots in the network, it will only talk (interact) with robots in its active range. That is why we want to control the active range of each robot in order to reduce the communication and maintain the similar formation when the network shrinks.

To achieve an agreement on d_i^α, r_i^α and r_i among agents in the connected network we use the following update algorithm based on the local averages for d_i^α, r_i^α and r_i, respectively:

$$\begin{cases} d_i^\alpha = \frac{1}{|N_i^\alpha|+1} \sum_{j=1}^{|N_i^\alpha \cup i|} d_j^\alpha \\ r_i^\alpha = \frac{1}{|N_i^\alpha|+1} \sum_{j=1}^{|N_i^\alpha| \cup i} r_j^\alpha \\ r_i = \frac{1}{|N_i^\alpha|+1} \sum_{j=1}^{|N_i^\alpha \cup i|} r_j, \end{cases} \tag{27}$$

here $|N_i^\alpha|$ is the number of neighbors of agent i, and $|N_i^\alpha \cup i|$ is the inclusive set of agent i and its neighbors.

The coefficients of the interaction forces $(c_1^\alpha,\ c_2^\alpha)$, $(c_1^\beta,\ c_2^\beta)$ and attractive force $(c_1^t,\ c_2^t)$ which deliver desired swarm-like behaviour are used to adjust the weight of interaction forces and attractive force. The pair $(c_1^\alpha,\ c_2^\alpha)$ is used to adjust the attractive/repulsive forces among agent i and its actual neighbors (α-agent), and the pair $(c_1^\beta,\ c_2^\beta)$ is used to adjust the repulsive forces among agent i and its virtual neighbors (β-agent) that is generated from agent i when it sees the obstacles, and the pair $(c_1^t,\ c_2^t)$ is used to adjust the attractive forces between agent i and its target. The bigger c_1^t and c_2^t the faster convergence to the target. However if c_1^t and c_2^t are too big the center of mass (CoM) as defined in Equation (28)

$$\begin{cases} \overline{q} = \frac{1}{n} \sum_{i=1}^n q_i \\ \overline{p} = \frac{1}{n} \sum_{i=1}^n p_i \end{cases} \tag{28}$$

oscillates around the target, and the formation of network is not guaranteed.

From the above analysis we see that these adaptive weights allow the network to reduce and recover the size gradually. They also allow the network to maintain the connectivity in obstacle space. Therefore, we let each agent have its own weight of the interaction forces as in Equation (29)

$$c_1^\alpha(i) = \begin{cases} c_1^\alpha, \ if \ |N_i^\alpha| \geq 3 \\ c_1^{\alpha'}, \ if \ |N_i^\alpha| < 3, \end{cases} \tag{29}$$

here $c_1^{\alpha'} < c_1^\alpha$, $c_2^\alpha(i) = 2\sqrt{c_1^\alpha(i)}$, and $i = 1, 2, ..., n$.

In addition, we let each agent have its own weight of the attractive force to the target as in Equation (30)

$$c_1^t(i) = \begin{cases} c_1^t, \ if \ |N_i^\alpha| \geq 3 \\ c_1^{t'}, \ if \ |N_i^\alpha| < 3, \end{cases} \tag{30}$$

here $c_1^{t'} > c_1^t$, $c_2^t(i) = 2\sqrt{c_1^t(i)}$, and $i = 1, 2, ..., n$.

In the α-lattice configuration if the agent has less than 3 neighbors it is considered as having a weak connection to the network. This means that this agent is on the border of network, or far from the target hence it should have bigger weight of attractive force from its target and smaller weight of interaction forces from its neighbors to get closer to the target. This design also has the benefit of making the whole network track the target faster.

Now, the neighborhood of agent i ($N_i^{'\alpha}$), the new adjacency matrix $a_{ij}^{'}(q)$ and the new action function $\phi_{\alpha}^{'}(z)$ are redefined as follows:

$$N_i^{'\alpha} = \{j \in \vartheta : \|q_j - q_i\| \le r_i,\ \vartheta = \{1,2,...,n\},\ j \ne i\}; \tag{31}$$

$$a_{ij}^{'}(q) = \begin{cases} \rho_h(\|q_j - q_i\|_\sigma / r_i^\alpha), & if\ j \ne i \\ 0, & if\ j - i, \end{cases} \tag{32}$$

$$\phi_{\alpha}^{'}(\|q_j - q_i\|_\sigma) = \rho_h(\|q_j - q_i\|_\sigma / r_i^\alpha)\phi(\|q_j - q_i\|_\sigma - d_i^\alpha). \tag{33}$$

Finally, the adaptive flocking control law for dynamic target tracking is

$$\begin{aligned} u_i = {} & c_1^\alpha(i) \sum_{j \in N_i^{'\alpha}} \phi_{\alpha}^{'}(\|q_j - q_i\|_\sigma) n_{ij} + c_2^\alpha(i) \sum_{j \in N_i^{'\alpha}} a_{ij}^{'}(q)(p_j - p_i) \\ & + c_1^\beta \sum_{k \in N_i^\beta} \phi_\beta(\|\hat{q}_{i,k} - q_i\|_\sigma)\hat{n}_{i,k} + c_2^\beta \sum_{k \in N_i^\beta} b_{i,k}(q)(\hat{p}_{i,k} - p_i) \\ & - c_1^t(i)(q_i - q_t) - c_2^t(i)(p_i - p_t). \end{aligned} \tag{34}$$

4 Flocking Control for Multiple Agents in Noisy Environments

The above flocking control algorithms are designed under the following assumptions: each agent can sense the position and velocity of itself, the neighbors and the target precisely. However, in reality these assumptions are not valid because sensing always has noise. Motivated by this observation we study how to design distributed flocking control algorithms which can still perform well when the measurements are corrupted by noises.

In this section we are going to design two algorithms in noisy environments. The first one is the *Multi-CoM-Shrink* flocking control algorithm. The main idea of this algorithm is to shrink the size of the network in oder to keep the connectivity. The second one is the *Multi-CoM-Cohesion* flocking control algorithm, and its main idea is based on the position and velocity cohesion feedbacks to create the strong cohesion between the agent and the network. Both algorithms are based on the *Multi-CoM* flocking control algorithm presented in our previous work [9]. The *Multi-CoM* flocking control algorithm is shown below

$$\begin{aligned} u_i = {} & c_1^\alpha \sum_{j \in N_i^\alpha} \phi_\alpha(\|q_j - q_i\|_\sigma) n_{ij} + c_2^\alpha \sum_{j \in N_i^\alpha} a_{ij}(q)(p_j - p_i) \\ & - c_1^t(q_i - q_t) - c_2^t(p_i - p_t) - c_1^l(\overline{q}_i - q_t) - c_2^l(\overline{p}_i - p_t), \end{aligned} \tag{35}$$

here c_1^l and c_2^l are positive constants. $\overline{q}_i$ and $\overline{p}_i$ are the local average of position and velocity, respectively for each agent i defined as:

$$\begin{cases} \overline{q}_i = \frac{1}{|N_i^\alpha|+1}\sum_{j=1}^{|N_i^\alpha \cup i|} q_j \\ \overline{p}_i = \frac{1}{|N_i^\alpha|+1}\sum_{j=1}^{|N_i^\alpha \cup i|} p_j. \end{cases} \tag{36}$$

In this control algorithm, the first two terms are used to control the formation (α-lattice configuration) and to allow agents to avoid collision [2]. The terms $-c_1^t(q_i - q_t) - c_2^t(p_i - p_t)$ and $-c_1^l(\overline{q}_i - q_t) - c_2^l(\overline{p}_i - p_t)$ allow each agent and its neighbors to closely follow the target [9].

4.1 Multi-CoM-Shrink Algorithm

Assume that the estimates of the position and velocity of agent i are: $\hat{q}_i = q_i + \varepsilon_q^i$ and $\hat{p}_i = p_i + \varepsilon_p^i$, where ε_q^i and ε_p^i are the position and velocity measurement errors, respectively. Then we have:
$\hat{q}_i - \hat{q}_j = q_i - q_j + \varepsilon_q^{ij}$; $\hat{p}_i - \hat{p}_j = p_i - p_j + \varepsilon_p^{ij}$, here $\varepsilon_q^{ij} = \varepsilon_q^i - \varepsilon_q^j$ and $\varepsilon_p^{ij} = \varepsilon_p^i - \varepsilon_p^j$.

Similarly, the estimates of the position and velocity of the target are: $\hat{q}_t = q_t + \varepsilon_q^t$ and $\hat{p}_t = p_t + \varepsilon_p^t$, where ε_q^t and ε_p^t are the position and velocity measurement errors, respectively. Then we have:
$\hat{q}_i - \hat{q}_t = q_i - q_t + \varepsilon_q^{it}$; $\hat{p}_i - \hat{p}_t = p_i - p_t + \varepsilon_p^{it}$, here $\varepsilon_q^{it} = \varepsilon_q^i - \varepsilon_q^t$ and $\varepsilon_p^{it} = \varepsilon_p^i - \varepsilon_p^t$.

If all noises are bounded, one possible method to maintain connectivity in noisy environments is to shrink the size of the network. We assume that the noise ε_q^i satisfies $\|\varepsilon_q^i\| \le r_w$ as shown in Figure 3.

Let us denote $d_a = \|q_i - q_j\|$ to be the actual distance between agent i and agent j. Then to maintain the connectivity and no collision among agents we need

$$0 < d_a \le r. \tag{37}$$

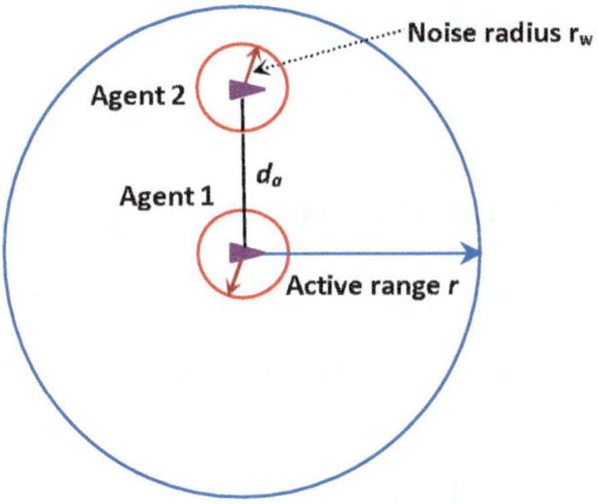

Fig. 3 Agent 2 is considered as a neighbor of agent 1 because the estimated distance $\hat{d}_a$ is less than the active range r.

Denote $\hat{d}_a$ to be the estimate of the actual distance d_a, then we have

$$\hat{d}_a = \|\hat{q}_i - \hat{q}_j\| \leq \|q_i - q_j\| + \|\varepsilon_q^{ij}\|. \tag{38}$$

Since $\|\varepsilon_q^i\| \leq r_w$ we have $\|\varepsilon_q^{ij}\| \leq 2r_w$, and we obtain

$$\|q_i - q_j\| - 2r_w \leq \hat{d}_a \leq \|q_i - q_j\| + 2r_w. \tag{39}$$

With $\|q_i - q_j\| = d_a$ we have

$$d_a - 2r_w \leq \hat{d}_a \leq d_a + 2r_w, \tag{40}$$

or,

$$\hat{d}_a - 2r_w \leq d_a \leq \hat{d}_a + 2r_w. \tag{41}$$

Since the control algorithm (22) guarantees that $\hat{d}_a$ converges to the desired distance d. Then from (41) we obtain

$$d - 2r_w \leq d_a \leq d + 2r_w. \tag{42}$$

From (37) and (42) we should have

$$\begin{cases} d - 2r_w > 0 \\ d + 2r_w \leq r. \end{cases} \tag{43}$$

Hence from (43) we obtain d to be

$$2r_w < d \leq r - 2r_w. \tag{44}$$

Equation (44) shows that we need to design the distance d within the range $(2r_w, r - 2r_w]$ to maintain connectivity and no collision among agents. However if we select d to be smaller than $r - 2r_w$ then each agent will have more neighbors than necessary. Hence, we choose $d = r - 2r_w$.

Now, from (5) we obtain d_{new}^{α} as

$$d_{new}^{\alpha} = \|d\|_\sigma = \frac{1}{\varepsilon}[\sqrt{1 + \varepsilon(r - 2r_w)^2} - 1]. \tag{45}$$

From (8) we obtain a new action function $\phi_\alpha^{new}(\|\hat{q}_j - \hat{q}_i\|_\sigma)$ as follows:

$$\phi_\alpha^{new}(\|\hat{q}_j - \hat{q}_i\|_\sigma) = \rho_h(\|\hat{q}_j - \hat{q}_i\|_\sigma / r_\alpha)\phi(\|\hat{q}_j - \hat{q}_i\|_\sigma - d_{new}^{\alpha}). \tag{46}$$

From (36) we have the local average of position and velocity for each agent i, $\hat{\bar{q}}_i$ and $\hat{\bar{p}}_i$ with noise computed as

$$\begin{cases} \hat{\bar{q}}_i = \frac{1}{|N_i^\alpha|+1} \sum_{j=1}^{|N_i^\alpha \cup i|} \hat{q}_j \\ \hat{\bar{p}}_i = \frac{1}{|N_i^\alpha|+1} \sum_{j=1}^{|N_i^\alpha \cup i|} \hat{p}_j, \end{cases} \tag{47}$$

From (10) and (11) we obtain $\hat{n}_{ij}$ and $\hat{a}_{ij}(q)$ as

$$\hat{n}_{ij} = (\hat{q}_j - \hat{q}_i)/\sqrt{1+\varepsilon\|\hat{q}_j - \hat{q}_i\|^2} \tag{48}$$

$$\hat{a}_{ij}(q) = \begin{cases} \rho_h(\|\hat{q}_j - \hat{q}_i\|_\sigma / r_\alpha), & if\ j \neq i \\ 0, & if\ j = i, \end{cases} \tag{49}$$

Now, we propose a *Multi-CoM-Shrink* algorithm with d^{α}_{new} as

$$\begin{aligned} u_i = {} & c_1^\alpha \sum_{j \in N_i^\alpha} \phi_\alpha^{new}(\|\hat{q}_j - \hat{q}_i\|_\sigma)\hat{n}_{ij} + c_2^\alpha \sum_{j \in N_i^\alpha} \hat{a}_{ij}(q)(\hat{p}_j - \hat{p}_i) \\ & - c_1^t(\hat{q}_i - \hat{q}_t) - c_2^t(\hat{p}_i - \hat{p}_t) - c_1^l(\hat{\overline{q}}_i - \hat{q}_t) - c_2^l(\hat{\overline{p}}_i - \hat{p}_t). \end{aligned} \tag{50}$$

4.2 Multi-CoM-Cohesion Algorithm

In this subsection we describe the *Multi-CoM-Cohesion* algorithm. The main idea of the *Multi-CoM-Cohesion* algorithm is that each agent should have a strong cohesion to the network so that the connectivity is maintained. In order to do that we introduce local position and velocity cohesion feedbacks to each agent.

Before presenting the algorithm, we have the following definitions:

$d_{il} = q_i - \overline{q}_i$ is the relative distance between node i and its local average of position;

$v_{il} = p_i - \overline{p}_i$ is the relative velocity between node i and its local average of velocity;

However, because agent i senses its own position and velocity with noise, hence the estimates $\hat{d}_{il}$ and $\hat{v}_{il}$ are also corrupted by noise $(\varepsilon_d^i, \varepsilon_v^i)$ as:

$$\begin{cases} \hat{d}_{il} = \hat{q}_i - \hat{\overline{q}}_i = q_i + \varepsilon_q^i - (\overline{q}_i + \overline{\varepsilon}_q^i) = d_{il} + \varepsilon_d^i \\ \hat{v}_{il} = \hat{p}_i - \hat{\overline{p}}_i = p_i + \varepsilon_p^i - (\overline{p}_i + \overline{\varepsilon}_p^i) = v_{il} + \varepsilon_v^i, \end{cases} \tag{51}$$

here $\varepsilon_d^i = \varepsilon_q^i - \overline{\varepsilon}_q^i$ with $\overline{\varepsilon}_q^i = \frac{1}{|N_i^\alpha|+1}\sum_{i=1}^{|N_i^\alpha \cup i|} \varepsilon_q^i$,

and $\varepsilon_v^i = \varepsilon_p^i - \overline{\varepsilon}_p^i$ with $\overline{\varepsilon}_p^i = \frac{1}{|N_i^\alpha|+1}\sum_{i=1}^{|N_i^\alpha \cup i|} \varepsilon_p^i$.

Based on the above definitions, we design a distributed flocking control law, *Multi-CoM-Cohesion* , in noisy environments as:

$$\begin{aligned} u_i = {} & c_1^\alpha \sum_{j \in N_i^\alpha} \phi_\alpha(\|\hat{q}_j - \hat{q}_i\|_\sigma)\hat{n}_{ij} + c_2^\alpha \sum_{j \in N_i^\alpha} \hat{a}_{ij}(q)(\hat{p}_j - \hat{p}_i) \\ & - c_{pos}\hat{d}_{il} - c_{ve}\hat{v}_{il} \\ & - c_1^t(\hat{q}_i - \hat{q}_t) - c_2^t(\hat{p}_i - \hat{p}_t) - c_1^l(\hat{\overline{q}}_i - \hat{q}_t) - c_2^l(\hat{\overline{p}}_i - \hat{p}_t), \end{aligned} \tag{52}$$

here $\hat{d}_{il}$, $\hat{v}_{il}$ are the estimates of d_{il} and v_{il}, respectively, and c_{pos} and c_{ve} are positive constants. The terms $-c_{pos}\hat{d}_{il}$ and $-c_{ve}\hat{v}_{il}$ are called local position and velocity cohesion feedbacks, respectively. The role of these negative feedbacks is to maintain position and velocity cohesions. This means that each agent tries to stay close to the local average of position and minimize the velocity mismatch between its velocity and the local average of velocity in noisy environments.

In this algorithm, to make it simpler in the stability analysis provided later we dropped the obstacle avoidance term. However, in real applications, to allow each agent to avoid both static and dynamic obstacles we only need to add the second component (14) to the control algorithm (52). In general, this component does not affect the properties of the global stability of the whole system.

5 Stability Analysis

5.1 Stability Analysis of Adaptive Flocking

By applying the control law (34), the CoM (defined in Equation (28)) of positions and velocities of all mobile agents in the network will exponentially converge to the target in both free space and obstacle space. In addition, the formation (collision free and velocity matching among mobile agents) will maintain in the process of the target tracking.

Let us consider adaptive flocking control in free space and obstacle space, respectively.

Case 1 (Free space): In free space, $\sum_{k \in N_i^{\beta}} \phi_{\beta}(\|\hat{q}_{i,k} - q_i\|_{\sigma}) = 0$, hence we can rewrite the control law (34) by ignoring constants c_{η}^{ν} (for $\forall \eta = 1,2$ and $\nu = \alpha, \beta$) as follows:

$$\begin{aligned} u_i = & - \sum_{j \in N_i^{\alpha}} \nabla_{q_i} \psi_{\alpha}(\|q_j - q_i\|_{\sigma}) + \sum_{j \in N_i^{\alpha}} a_{ij}(q)(p_j - p_i) \\ & - c_1^t(q_i - q_t) - c_2^t(p_i - p_t) \end{aligned} \tag{53}$$

where $\psi_{\alpha}(z) = \int_{d_{\alpha}}^{z} \phi_{\alpha}(s) ds$ is the pairwise attractive/repulsive potential function. From (53), by considering the symmetry of pair $(\psi_{\alpha}, a(q))$ we can obtain the average of control law u as follows:

$$\overline{u} = -c_1^t(\overline{q} - q_t) - c_2^t(\overline{p} - p_t). \tag{54}$$

Equation (54) implies that

$$\begin{cases} \dot{\overline{q}} = \overline{p} \\ \dot{\overline{p}} = -c_1^t(\overline{q} - q_t) - c_2^t(\overline{p} - p_t). \end{cases} \tag{55}$$

The solution of (55) indicates that the CoM of positions and velocities exponentially converge to those of the target.

The formation (collision-free and velocity matching among mobile agents) is maintained in the free space tracking because the gradient-based term and the consensus term are considered in this situation (more details please see [2]).

Case 2 (Obstacle space): In the obstacle space d_i^α is designed to be reduced when each agent senses the obstacles. Therefore, when the agent network has to pass through the narrow space between two obstacles its size will shrink gradually, and when the network already passed this narrow space it grows back to the original size gradually. This reduces the impact of the obstacle on the network hence the speed of agents can be maintained, or the CoM keeps tracking the target. Also, the connectivity and similar formation can be maintained in this scenario.

5.2 Stability Analysis of Flocking in Noisy Environments

Before analyzing the stability of the flocking control algorithm, *Multi-CoM-Cohesion*, we build the error dynamic model of the flocking system in noisy environments in the next subsection.

5.2.1 Error Dynamic Model

To study the stability properties, we have the error dynamics of the system given as follows:

$$\begin{cases} \dot{d}_{ig} = v_{ig} \\ \dot{v}_{ig} = u_i - \frac{1}{n}\sum_{j=1}^{n} u_j = u_i - \overline{u},\ i = 1,2,...,n. \end{cases} \tag{56}$$

here $\overline{u} = \frac{1}{n}\sum_{j=1}^{n} u_j$.

We have following definitions:

$d_{ig} = q_i - \overline{q}$ is the relative distance between node i and its global average of position;

$v_{ig} = p_i - \overline{p}$ is the relative velocity between node i and its global average of velocity;

Then we have the following relations:

$$\begin{aligned} d_{il} &= q_i - \overline{q}_i = d_{ig} + \overline{q} - \frac{1}{|N_i^\alpha|+1}\sum_{j=1}^{|N_i^\alpha \cup i|} q_j \\ &= d_{ig} + \overline{q} - \frac{1}{|N_i^\alpha|+1}\sum_{j=1}^{|N_i^\alpha \cup i|} (d_{jg} + \overline{q}) = d_{ig} - \frac{1}{|N_i^\alpha|+1}\sum_{j=1}^{|N_i^\alpha \cup i|} d_{jg}. \end{aligned} \tag{57}$$

Then similar to d_{il}, v_{il} is obtained as follows:

$$v_{il} = v_{ig} - \frac{1}{|N_i^\alpha|+1}\sum_{j=1}^{|N_i^\alpha \cup i|} v_{jg}. \tag{58}$$

The estimates of the local average of position and velocity, respectively in (47) is rewritten as

$$\hat{\overline{q}}_i = q_i - d_{ig} + \frac{1}{|N_i^\alpha|+1}\sum_{j=1}^{|N_i^\alpha \cup i|} d_{jg} + \overline{\varepsilon}_q^i. \tag{59}$$

$$\hat{\bar{p}}_i = p_i - v_{ig} + \frac{1}{|N_i^\alpha|+1}\sum_{j=1}^{|N_i^\alpha \cup i|} v_{jg} + \bar{\varepsilon}_p^i. \tag{60}$$

Now, we can rewrite the control law (52) with considering (51), (59) and (60):

$$\begin{aligned}
u_i &= c_1^\alpha \sum_{j\in N_i^\alpha} \phi_\alpha(\|\hat{q}_j - \hat{q}_i\|_\sigma)\hat{n}_{ij} + c_2^\alpha \sum_{j\in N_i^\alpha} \hat{a}_{ij}(q)(\hat{p}_j - \hat{p}_i)\\
&+(c_1^l - c_{pos})(d_{ig} - \frac{1}{|N_i^\alpha|+1}\sum_{j=1}^{|N_i^\alpha \cup i|} d_{jg}) + (c_2^l - c_{ve})(v_{ig} - \frac{1}{|N_i^\alpha|+1}\sum_{j=1}^{|N_i^\alpha \cup i|} v_{jg})\\
&-(c_1^t + c_1^l)(q_i - q_t) - (c_2^t + c_2^l)(p_i - p_t) - c_{pos}\varepsilon_d^i - c_{ve}\varepsilon_v^i - c_1^l\bar{\varepsilon}_q^i - c_2^l\bar{\varepsilon}_p^i\\
&-(c_1^t + c_1^l)\varepsilon_q^{it} - (c_2^t + c_2^l)\varepsilon_p^{it}
\end{aligned} \tag{61}$$

The average of control law for composite system is

$$\begin{aligned}
\bar{u} &= \frac{c_1^\alpha}{n}\sum_{i=1}^{n}[\sum_{j\in N_i^\alpha} \phi_\alpha(\|\hat{q}_j - \hat{q}_i\|_\sigma)\hat{n}_{ij}] + \frac{c_2^\alpha}{n}\sum_{i=1}^{n}[\sum_{j\in N_i^\alpha} \hat{a}_{ij}(q)(\hat{p}_j - \hat{p}_i)]\\
&+(\frac{c_1^l - c_{pos}}{n})\sum_{i=1}^{n}(d_{ig} - \frac{1}{|N_i^\alpha|+1}\sum_{j=1}^{|N_i^\alpha \cup i|} d_{jg})\\
&+(\frac{c_2^l - c_{ve}}{n})\sum_{i=1}^{n}(v_{ig} - \frac{1}{|N_i^\alpha|+1}\sum_{j=1}^{|N_i^\alpha \cup i|} v_{jg})\\
&-(\frac{c_1^t + c_1^l}{n})\sum_{i=1}^{n}(q_i - q_t) - (\frac{c_2^t + c_2^l}{n})\sum_{i=1}^{n}(p_i - p_t)\\
&-\frac{1}{n}\sum_{i=1}^{n}[c_{pos}\varepsilon_d^i + c_{ve}\varepsilon_v^i + c_1^l\bar{\varepsilon}_q^i + c_2^l\bar{\varepsilon}_p^i + (c_1^t + c_1^l)\varepsilon_q^{it} + (c_2^t + c_2^l)\varepsilon_p^{it}]
\end{aligned} \tag{62}$$

Substitute u_i in (61) and $\bar{u}$ in (62) into (56) we obtain:

$$\begin{aligned}
\dot{v}_{ig} &= c_1^\alpha \sum_{j\in N_i^\alpha} \phi_\alpha(\|\hat{q}_j - \hat{q}_i\|_\sigma)\hat{n}_{ij} - \frac{c_1^\alpha}{n}\sum_{i=1}^{n}[\sum_{j\in N_i^\alpha} \phi_\alpha(\|\hat{q}_j - \hat{q}_i\|_\sigma)\hat{n}_{ij}]\\
&+c_2^\alpha \sum_{j\in N_i^\alpha} \hat{a}_{ij}(q)(\hat{p}_j - \hat{p}_i) - \frac{c_2^\alpha}{n}\sum_{i=1}^{n}[\sum_{j\in N_i^\alpha} \hat{a}_{ij}(q)(\hat{p}_j - \hat{p}_i)]\\
&-(\frac{c_1^l - c_{pos}}{|N_i^\alpha|+1})\sum_{j=1}^{|N_i^\alpha \cup i|} d_{jg} - (\frac{c_2^l - c_{ve}}{|N_i^\alpha|+1})\sum_{j=1}^{|N_i^\alpha \cup i|} v_{jg}\\
&-(\frac{c_1^l - c_{pos}}{n})\sum_{i=1}^{n}(d_{ig} - \frac{1}{|N_i^\alpha|+1}\sum_{j=1}^{|N_i^\alpha \cup i|} d_{jg})\\
&-(\frac{c_2^l - c_{ve}}{n})\sum_{i=1}^{n}(v_{ig} - \frac{1}{|N_i^\alpha|+1}\sum_{j=1}^{|N_i^\alpha \cup i|} v_{jg})
\end{aligned}$$

$$
\begin{aligned}
&-(c_{pos}-c_1^l)d_{ig}-(c_{ve}-c_2^l)v_{ig}-(c_1^t+c_1^l)d_{ig}-(c_2^t+c_2^l)v_{ig}\\
&-c_{pos}\varepsilon_d^i-c_{ve}\varepsilon_v^i-c_1^l\overline{\varepsilon}_q^i-c_2^l\overline{\varepsilon}_p^i-(c_1^t+c_1^l)\varepsilon_q^{it}-(c_2^t+c_2^l)\varepsilon_p^{it}\\
&+\frac{1}{n}\sum_{i=1}^{n}[c_{pos}\varepsilon_d^i+c_{ve}\varepsilon_v^i+c_1^l\overline{\varepsilon}_q^i+c_2^l\overline{\varepsilon}_p^i+(c_1^t+c_1^l)\varepsilon_q^{it}+(c_2^t+c_2^l)\varepsilon_p^{it}]\\
=&-(c_1^t+c_{pos})d_{ig}-(c_2^t+c_{ve})v_{ig}+\Phi_i+\Omega_i(V)+\zeta_i,
\end{aligned}
\tag{63}
$$

where

$$
\begin{aligned}
\Phi_i=&\,c_1^\alpha\sum_{j\in N_i^\alpha}\phi_\alpha(\|\hat{q}_j-\hat{q}_i\|_\sigma)\hat{n}_{ij}-\frac{c_1^\alpha}{n}\sum_{i=1}^{n}[\sum_{j\in N_i^\alpha}\phi_\alpha(\|\hat{q}_j-\hat{q}_i\|_\sigma)\hat{n}_{ij}]\\
&+c_2^\alpha\sum_{j\in N_i^\alpha}\hat{a}_{ij}(q)(\hat{p}_j-p_i)-\frac{c_2^\alpha}{n}\sum_{i=1}^{n}[\sum_{j\in N_i^\alpha}\hat{a}_{ij}(q)(\hat{p}_j-p_i)];
\end{aligned}
$$

$$
\begin{aligned}
\Omega_i(V)=&-(\frac{c_1^l-c_{pos}}{|N_i^\alpha|+1})\sum_{j=1}^{|N_i^\alpha\cup i|}d_{jg}-(\frac{c_2^l-c_{ve}}{|N_i^\alpha|+1})\sum_{j=1}^{|N_i^\alpha\cup i|}v_{jg}\\
&-(\frac{c_1^l-c_{pos}}{n})\sum_{i=1}^{n}(d_{ig}-\frac{1}{|N_i^\alpha|+1}\sum_{j=1}^{|N_i^\alpha\cup i|}d_{jg})\\
&-(\frac{c_2^l-c_{ve}}{n})\sum_{i=1}^{n}(v_{ig}-\frac{1}{|N_i^\alpha|+1}\sum_{j=1}^{|N_i^\alpha\cup i|}v_{jg});
\end{aligned}
$$

$$
\begin{aligned}
\zeta_i=&\,\frac{1}{n}\sum_{i=1}^{n}[c_{pos}\varepsilon_d^i+c_{ve}\varepsilon_v^i+c_1^l\overline{\varepsilon}_p^i+c_2^l\overline{\varepsilon}_p^i+(c_1^t+c_1^l)\varepsilon_q^{it}+(c_2^t+c_2^l)\varepsilon_p^{it}]\\
&-[c_{pos}\varepsilon_d^i+c_{ve}\varepsilon_v^i+c_1^l\overline{\varepsilon}_q^i+c_2^l\overline{\varepsilon}_p^i+(c_1^t+c_1^l)\varepsilon_q^{it}+(c_2^t+c_2^l)\varepsilon_p^{it}]
\end{aligned}
$$

here, we define $V_i=[d_{ig}\ v_{ig}]^T$ and $V=[V_1,V_2,...,V_n]^T$.

Rewrite (63) in state space representation

$$
\begin{bmatrix}\dot{d}_{ig}\\ \dot{v}_{ig}\end{bmatrix}=\begin{bmatrix}0 & I\\ -k_1I & -k_2I\end{bmatrix}\begin{bmatrix}d_{ig}\\ v_{ig}\end{bmatrix}+\begin{bmatrix}0\\ I\end{bmatrix}(\Phi_i+\Omega_i(V)+\zeta_i),
\tag{64}
$$

here $k_1=(c_1^t+c_{pos})$, $k_2=(c_2^t+c_{ve})$, and I is an m x m identity matrix.

Then we can rewrite (64) as

$$
\dot{V}_i=\begin{bmatrix}0 & I\\ -k_1I & -k_2I\end{bmatrix}V_i+\begin{bmatrix}0\\ I\end{bmatrix}(\Phi_i+\Omega_i(V)+\zeta_i)
\tag{65}
$$

Let the matrix $A_i=\begin{bmatrix}0 & I\\ -k_1I & -k_2I\end{bmatrix}$, then we have the characteristic equation as:

$$det(\lambda I - A_i) = (\lambda^2 + k_2\lambda + k_1)^m = 0. \tag{66}$$

Since $k_1 > 0$, $k_2 > 0$, and if $k_2 < 2\sqrt{k_1}$ then all roots of the characteristic equation (66) have negative real parts ($Re(\lambda_i) < 0$).

5.2.2 Stability Analysis of the *Multi-CoM-Cohesion* algorithm

In this subsection we will analyze the stability of the flocking control algorithm, *Multi-CoM-Cohesion*, in noisy environments based on the Lyapunov approach.

We assume that the errors of sensing position and velocity have linear relationship with the magnitude of the state of the error system. That is because two agents are far away from each other, the sensing errors will usually increase. Hence, we have

$$\begin{cases} \|\varepsilon_d^i(t)\| \le c_{ed_1}^i \|V_i(t)\| + c_{ed_2}^i \\ \|\varepsilon_v^i(t)\| \le c_{ev_1}^i \|V_i(t)\| + c_{ev_2}^i,\ i = 1,2,...,n. \end{cases} \tag{67}$$

We also assume that the noise ε_q^{it} and ε_p^{it} on the target tracking terms (negative feedbacks) are bounded as

$$\begin{cases} \|\varepsilon_q^{it}(t)\| \le c_{eq}^i \\ \|\varepsilon_p^{it}(t)\| \le c_{ep}^i,\ i = 1,2,...,n, \end{cases} \tag{68}$$

and the noise $\overline{\varepsilon}_q^i$ and $\overline{\varepsilon}_p^i$ on the estimates of local average of position and velocity are bounded as

$$\begin{cases} \|\overline{\varepsilon}_q^i(t)\| \le \overline{c}_{eq}^i \\ \|\overline{\varepsilon}_p^i(t)\| \le \overline{c}_{ep}^i,\ \overline{i} = 1,2,...,n. \end{cases} \tag{69}$$

here $\overline{c}_{eq}^i = \frac{1}{|N_i^\alpha|+1}\sum_{i=1}^{|N_i^\alpha \cup i|} c_{eq}^i$, and $\overline{c}_{ep}^i = \frac{1}{|N_i^\alpha|+1}\sum_{i=1}^{|N_i^\alpha \cup i|} c_{ep}^i$.

Theorem 1. Consider a system of n mobile agents with dynamics (1) and controlled by (52), and all noise are bounded by (67), (68) and (69) . Let

$$c_{pv}^1 = \frac{(c_{pos}+1)^2 + c_{ve}^2}{2c_{pos}c_{ve}} + \sqrt{(\frac{c_{pos}+c_{ve}^2-1}{2c_{pos}c_{ve}})^2 + \frac{1}{c_{pos}^2}},$$

and if

$$c_{pos}c_{ed_1}^i + c_{ve}c_{ev_1}^i \le \frac{1}{c_{pv}^1},$$

and the parameters are such that

$$\sum_{j=1}^{m} \frac{2c_{pv}^1[\sqrt{(c_1^l - c_{pos})^2 + (c_2^l - c_{ve})^2} - \frac{1}{n}(c_{pos}c_{ed_1}^i + c_{ve}c_{ev_1}^i)]}{(1-\varepsilon_i)[1 - c_{pv}^1(c_{pos}c_{ed_1}^i + c_{ve}c_{ev_1}^i)]} < 1,$$

here $0 < \varepsilon_i < 1$ for $\forall i$, then the trajectories of (65) are bounded.

Proof. To study the stability of the error dynamics (65), one possible choice is to choose the Lyapunov function for each agent as

$$L_i(V_i) = V_i^T P V_i, \tag{70}$$

here $P = P^T$ is a $2m$ x $2m$ positive-definite matrix ($P > 0$). Then, the Lyapunov function for the composite system is

$$L(V) = \sum_{i=1}^{n} V_i^T P V_i.$$

From (70) we have

$$\dot{L}_i(V_i) = V_i^T P \dot{V}_i + \dot{V}_i^T P V_i. \tag{71}$$

Then, substitute $\dot{V}_i$ in (65) into (71) we obtain

$$\begin{aligned}\dot{L}_i(V_i) &= V_i^T (PA_i + A_i^T P) V_i + 2V_i^T PB(\Phi_i + \Omega_i(V) + \zeta_i) \\ &= -V_i^T C V_i + 2V_i^T PB(\Phi_i + \Omega_i(V) + \zeta_i),\end{aligned}$$

here $B = \begin{bmatrix} 0 \\ I \end{bmatrix}$, and $C = -(PA_i + A_i^T P)$.

The remaining part of this proof is to show $\dot{L}_i(V_i) < 0$. The detailed proof of $\dot{L}_i(V_i) < 0$ is similar to that in the reference [14].

6 Experimental Results

In this section we are going to test our proposed algorithms, adaptive flocking control (34), *Multi-CoM-Shrink* (50), and *Multi-CoM-Cohesion* (52). Then we compare our algorithms with the existing one (22) in terms of network connectivity, formation and tracking performance. First we discuss how to evaluate the connectivity of the network in the next subsection.

6.1 Connectivity Evaluation

To evaluate the network connectivity maintenance, first we know that the link (connectivity) between node i and node j is maintained if the distance between them $0 < \|q_i - q_j\| \leq r$. Otherwise this link is considered broken. Then for graph connectivity: a dynamic graph $G(\vartheta, E)$ is connected at time t if there exists a path between any two vertices.

To analyze the connectivity of the network we define a connectivity matrix $[c_{ij}(t)]$ as follows:

$$[c_{ij}(t)] = \begin{cases} 1, \ if \ \ j \in N_i^{\alpha}(t), i \neq j \\ 0, \ if \ \ j \notin N_i^{\alpha}(t), i \neq j \end{cases}$$

and $c_{ii} = 0$.

Since the rank of the Laplacian [2] of a connected graph $[c_{ij}(t)]$ of order n is at most $(n-1)$ or $rank([c_{ij}(t)]) \leq (n-1)$, the relative connectivity of a network at time t is defined as $C(t) = \frac{1}{n-1} rank([c_{ij}(t)])$.

If $0 \leq C(t) < 1$ the network is broken, and if $C(t) = 1$ the network is connected. Based on this metric we can evaluate the network connectivity in our proposed flocking control algorithms.

6.2 Adaptive Flocking Results in Cluttered Environments

The parameters used in the simulation and experiment of the adaptive flocking are specified as follows:

- Parameters of flocking in simulation: number of agents = 50 (randomly distributed in the box of 100x100 size); $a = b = 5$; the active range $r = 8.5$; the desired distance $d = 7$; $\varepsilon = 0.1$ for the σ-norm; $h = 0.2$ for the bump functions $(\phi_\alpha(z), \phi_\alpha^{'}(z))$; $h = 0.9$ for the bump function $(\phi_\beta(z))$.

Parameters of target movement for simulation: The target moves in the line trajectory: $q_t = [100 + 130t,\ t]^T$.

- Parameters of flocking in experiment:

number of agents = 7 (randomly distributed); $a = b = 5$; $d = 1100mm$; the scaling factor $k_c = 1.2$; the active range $r = k_c * d$; $\varepsilon = 0.1$ for the σ-norm; $h = 0.2$ for the bump functions $(\phi_\alpha(z), \phi_\alpha^{'}(z))$; $h = 0.9$ for the bump function $(\phi_\beta(z))$.

Parameters of target movement for experiment: The virtual target moves in the line trajectory: $q_t = [230 + t,\ -3000 + 130t]^T$.

- Experimental setup: In this experiment we use 7 Rovio robots [15] that have omni-directional motion capability. Basically, these robots can freely move in 6 directions. The dynamic model of the Rovio robot can be approximated by Equation (1). However, the localization accuracy of the Rovio robot is low, and the robot does not has any sensing device to sense the pose (position and velocity) of its neighbors or the obstacles. Hence we use a VICON motion capture system setup [16] in our lab (Figure 4) that includes 12 infrared cameras to track moving objects. This tracking system can provide the location and velocity of each moving object with high accuracy.

Figures 5 represents the results of moving target (red/dark line) tracking in the line trajectory using the existing flocking control algorithm (21). Figure 6 represents the results of moving target tracking in the line trajectory using the adaptive flocking control algorithm (34). Figure 7 shows the results of velocity matching among agents (a, a'), connectivity (b, b') and error positions between the CoM (black/darker line) and the target (tracking performance) (c, c') of both flocking control algorithms (34) and (22), respectively. To compare these algorithms we use the same initial state (position and velocity) of mobile agents. By comparing these figures we see that by applying the adaptive flocking control algorithm (34) the connectivity, similar formation and tracking performance are maintained when the network passes through the narrow space between two obstacles (two red/dark circles) while the existing flocking control algorithm (22) could not handle these

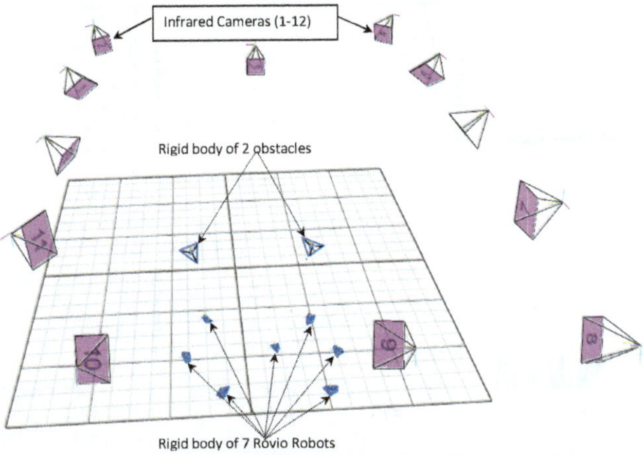

Fig. 4 Experimental setup.

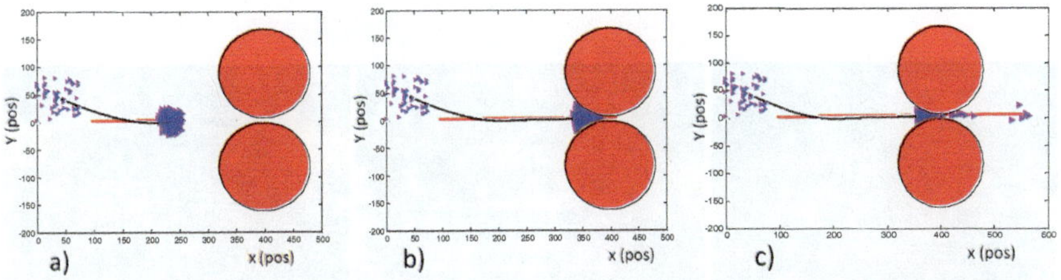

Fig. 5 Snapshots of the mobile agent network (a) when the mobile agents form a network, (b) when the mobile agents avoid obstacles, (c) when the mobile agents get stuck in the narrow space between two obstacles. These results are obtained by using algorithm (22).

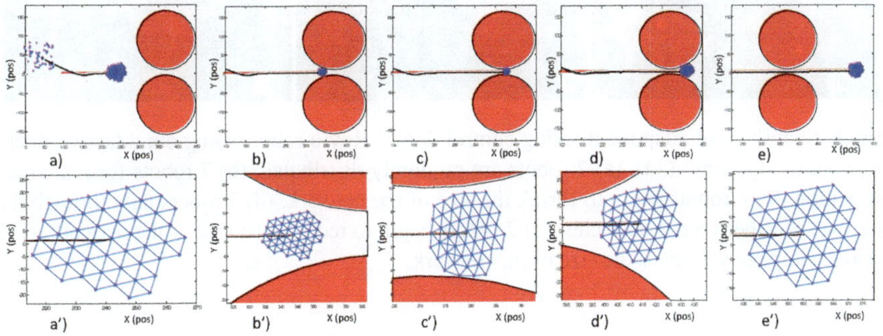

Fig. 6 Snapshots of the mobile agent network (a) when the mobile agents form a network, (b, c) when the mobile agent network shrinks to avoid obstacles, (d) when the mobile agents successfully passed through the narrow space between two obstacles, (e) when the mobile agents recover the original size. (a', b', c', d', e') are closer look of (a, b, c, d, e), respectively. These results are obtained by using our adaptive flocking control algorithm (34).

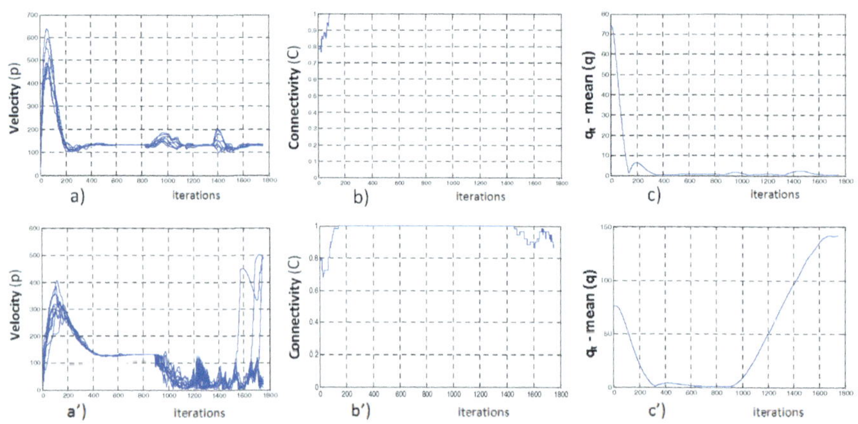

Fig. 7 Velocity matching among agents, connectivity, and error of positions between the CoM and the moving target in (a, b, c), respectively using our adaptive flocking control algorithm (34), (a', b', c') using the algorithm (22).

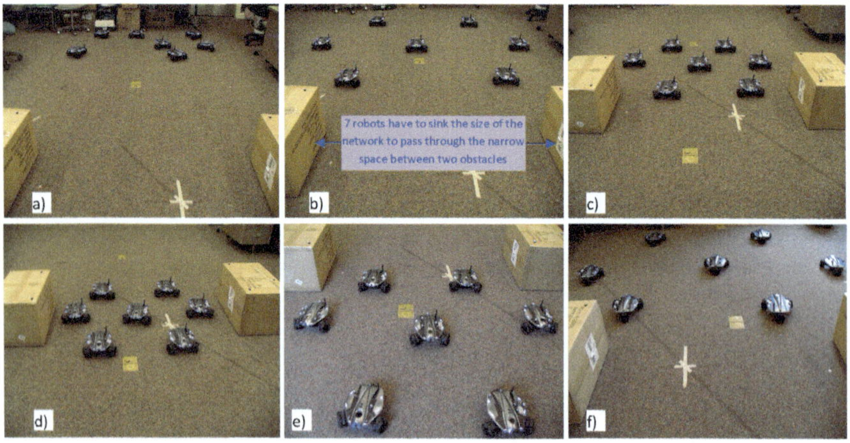

Fig. 8 Snapshots of adaptive flocking control with 7 Rovio robots using our adaptive flocking control algorithm (34). (a) 7 robots are randomly distributed. (b) 7 robots form a lattice formation. (c) 7 robots begin to shrink the size of the network. (d) 7 robots pass through the narrow space between 2 obstacles. (e) 7 robots begin to recover the size of the network. (f) 7 robots completely recover the size of the network.

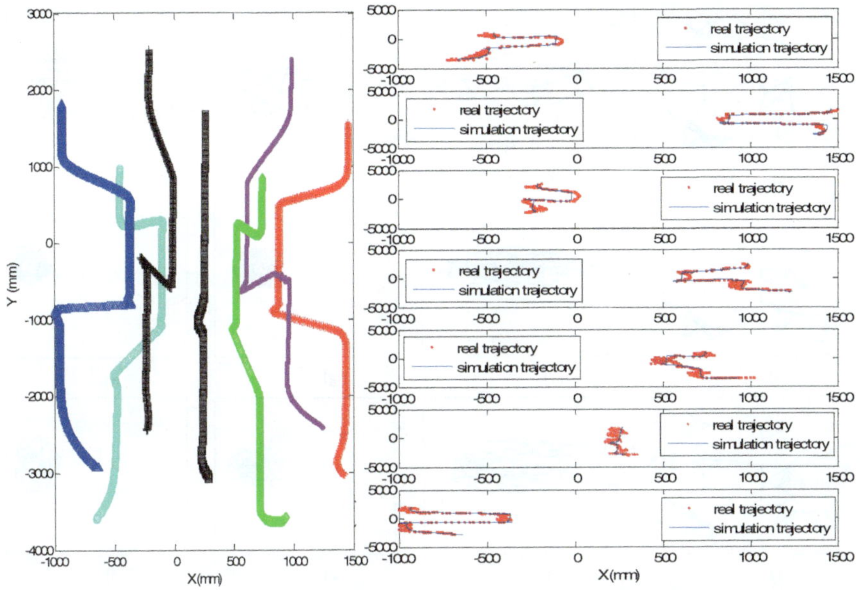

Fig. 9 Trajectories of 7 robots are obtained by using the adaptive flocking control algorithm (34).

problems. In Figures 6 when the network enters the small gap between two obstacles its size is shrunk gradually in order to pass this space, then the network size grows back gradually when it passed. Therefore the connectivity and similar formation are maintained.

Figure 8 shows the snapshots (a to f) of the experiment result for 7 Rovio robots using our adaptive flocking algorithm (34). The results look similar with the simulation result in Figure 6. Figure 9 (Left) shows the trajectories of 7 robots in simulation, and Figure 9 (Right) compares the trajectories of 7 robots in both simulation and experiment.

6.3 Flocking Results in Noisy Environments

The parameters used in this simulation are specified as follows:

- Parameters of flocking: number of agents = 50 (randomly distributed in the square area of 120 x 120 size); $a = b = 5$; the active range $r = 19$; $\varepsilon = 0.1$ for the σ-norm; $h = 0.2$ for the bump functions ($\phi_\alpha^{new}(z), \phi_\alpha(z)$); $h = 0.9$ for the bump function ($\phi_\beta(z)$). The desired distance for the algorithms (22) and *Multi-CoM-Cohesion*, $d = 16$. For the *Multi-CoM-Shrink* algorithm, $r_w = 3.4$, hence $d = r - 2r_w = 19 - 2 \times 3.4 = 12.2$.

- Parameters of target movement:

Case 1: The target moves in a sine wave trajectory: $q_t = [50 + 50t, 295 - 50sin(t)]^T$ with $0 \le t \le 6$.

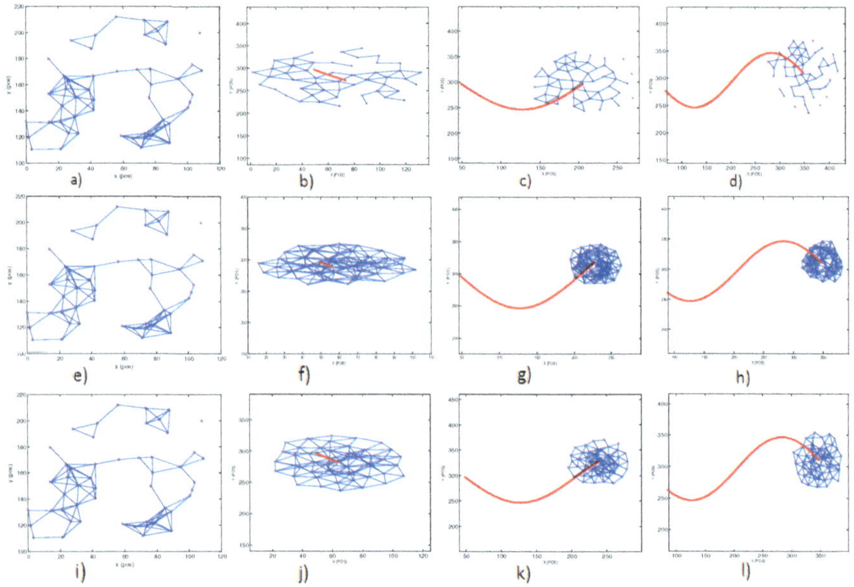

Fig. 10 Snapshots of agents when they are randomly distributed (a, e, i), and when they form a network and track a target (red/dark line) moving in a sine wave trajectory (b, c, d; f, g, h; j, k, l), where (a, b, c, d) are for the algorithm (22), (e, f, g, h) are for the *Multi-CoM-Shrink* algorithm, and (i, j, k, l) are for the *Multi-CoM-Cohesion* algorithm.

Case 2: The target moves in a circle trajectory: $q_t = [310 - 160cos(t),\ 255 + 160sin(t)]^T$ with $0 \leq t \leq 4$.

- The noise used in the simulation is Gaussian with zero mean and a variance of 1.

Figures 10 and 11 show the results of of the moving target (red/dark line) tracking in the sine wave and circle trajectories, respectively in noisy environments for three algorithms, (22), *Multi-CoM-Shrink* and *Multi-CoM-Cohesion*. Especially, Figures 10(a, b, c, d) and 11(a, b, c, d) are for the flocking control algorithm (22). Figures 10(e, f, g, h) and 11(e, f, g, h) are for the proposed flocking control algorithm *Multi-CoM-Shrink*. Figures 10(i, j, k, l) and 11(i, j, k, l) are for the proposed flocking control algorithm *Multi-CoM-Cohesion* .

To compare our proposed flocking control algorithms, *Multi-CoM-Shrink* and *Multi-CoM-Cohesion* with the existing flocking algorithm (22), we use the same initial state (position and velocity) of the mobile agents. Figure 12 shows the results of the tracking performance and the connectivity, respectively: (a, c) are for the flocking control algorithm (22), (b, d) are for the *Multi-CoM-Shrink* flocking control algorithm, and (e, f) are for the *Multi-CoM-Cohesion* flocking control algorithm. Comparing the results in these figures we clearly see that:

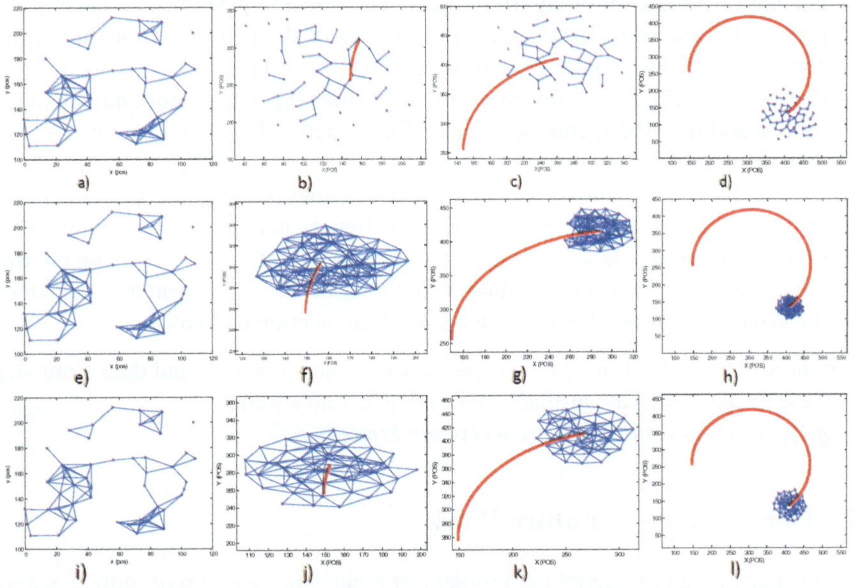

Fig. 11 Snapshots of agents when they are randomly distributed (a, e, i), and when they form a network and track a target (red/dark line) moving in a circle trajectory (b, c, d; f, g, h; j, k, l), where (a, b, c, d) are for the algorithm (22), (e, f, g, h) are for the *Multi-CoM-Shrink* algorithm, and (i, j, k, l) are for the *Multi-CoM-Cohesion* algorithm.

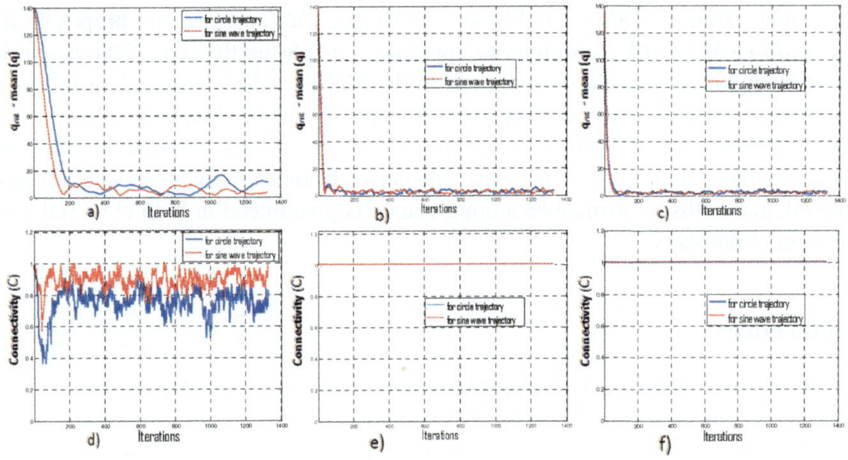

Fig. 12 The tracking performance results (error between the CoM and target positions): (a) is for the algorithm(22), (b) is for the *Multi-CoM-Shrink* algorithm, and (c) is for the *Multi-CoM-Cohesion* algorithm. The connectivity is evaluated by the $C(t)$ value: (d) is for the algorithm (22), (e) is for the *Multi-CoM-Shrink* algorithm, and (f) is for the *Multi-CoM-Cohesion* algorithm.

- For the flocking control algorithm (22): The tracking performance has big errors, and it makes the target out of the center of the network. In addition, the connectivity is lost, or the network is broken ($C(t) < 1$).
- For the *Multi-CoM-Cohesion* algorithm: The tracking performance has small errors. In addition, the agents can quickly form a network (only five iterations) and then maintain connectivity ($C(t) = 1$).
- For the *Multi-CoM-Shrink* algorithm: The tracking performance also has small errors, and the connectivity is maintained after six iterations. However, the size of the network is smaller than that of the *Multi-CoM-Cohesion* flocking control algorithm, and each agent has more neighbors because each agent tries to reduce the distance to its neighbor in order to keep connection to them.

For more details about these results please see some video files and their summary of these results which are available at our ASCC Lab's website.
http://ascc.okstate.edu/projectshung.html

7 Conclusion and Future Work

In this paper, we considered the problem of controlling a group of mobile agents to track a target in cluttered and noisy environments, respectively. First, an adaptive flocking control algorithm is designed to enable mobile agents to track and observe the moving target more effectively in cluttered environments while maintaining their similar formation and connectivity. This means that all mobile agents in the network can surround the target closely which will allow them to observe the target easily for recognition purposes. Second, in noisy environments, two flocking control algorithms, *Multi-CoM-Shrink* and *Multi-CoM-Cohesion*, are proposed. In the *Multi-CoM-Shrink* algorithm our approach is to shrink the size of the network by reducing the distance among agents. In the *Multi-CoM-Cohesion* algorithm our approach integrates local position and velocity cohesion feedbacks in oder to deal with the noise. The stability of the *Multi-CoM-Cohesion* algorithm is investigated based on the Lyapunov approach. Also, the network connectivity preservation is improved, and collision avoidance among agents is guaranteed in both cluttered and noisy environments.

References

1. Tanner, H.G., Jadbabai, A., Pappas, G.J.: Flocking in fixed and switching networks. IEEE Transactions on Automatic Control 52(5), 863–868 (2007)
2. Olfati-Saber, R.: Flocking for multi-agent dynamic systems: Algorithms and theory. IEEE Transactions on Automatic Control 51(3), 401–420 (2006)
3. Reynolds, C.: Flocks, birds, and schools: A distributed behavioral model. In: Computer Graphics, ACM SIGGRAPH 1987 Conference Proceedings, Anaheim California, vol. 21(4), pp. 25–34 (1987)
4. Levine, H., Rappel, W.J., Cohen, I.: Self-organization in systems of self-propelled particles. Phys. Review. E. 63, 017101–017104 (2000)

5. Mogilner, A., Edelstein-Keshet, L., Bent, L., Spiros, A.: Mutual interactions, potentials, and individual distance in a social aggregation. J. Math. Biol 47, 353–389 (2003)
6. Couzin, I.D., Krause, J., James, R., Ruxton, G.D., Franks, N.R.: Collective memory and spatial sorting in animal groups. J. Theor. Biol 218, 1–11 (2002)
7. Su, H., Wang, X., Lin, Z.: Flocking of multi–agents with a virtual leader. IEEE Transactions on Automatic Control 54(2), 293–307 (2009)
8. Olfati-Saber, R.: Distributed tracking for mobile sensor networks with information driven mobility. In: Proceedings of the 2007 American Control Conference, pp. 4606–4612 (2007)
9. La, H.M., Sheng, W.: Flocking control of a mobile sensor network to track and observe a moving target. In: Proceedings of the 2009 IEEE International Conference on Robotics and Automation (2009)
10. Ogren, P., Fiorelli, E., Leonard, N.E.: Cooperative control of mobile sensor networks: Adaptive gradient climbing in a distributed environment. IEEE Transactions on Automatic Control 49(8), 1292–1302 (2006)
11. Olfati-Saber, R., Murray, R.M.: Consensus problems in networks of agents with switching topology and time–delays. IEEE Transactions on Automatic Control 49(9), 1520–1533 (2004)
12. Olfati-Saber, R., Alex Fax, J., Murray, R.M.: Consensus and cooperative in networked multi-agent systems. Proceedings of the IEEE 95(1), 215–233 (2007)
13. Leonard, N.E., Fiorelli, E.: Virtual leaders, artificial potentials, and coordinated control of groups. In: Proceedings of the 40th IEEE Conference on Decision and Control, pp. 2968–2973 (2001)
14. Liu, Y., Passino, K.M.: Stable social foraging swarms in a noisy environment. IEEE Transactions on Automatic Control 49(1), 30–44 (2004)
15. Rovio robot, `http://www.wowwee.com/en/support/rovio`
16. VICON motion system, `http://www.vicon.com/`

Genetic Stigmergy

Joshua Brandoff and Hiroki Sayama

Stigmergy has long been studied and recognized as an effective system for self-organization among social insects. Through the use of chemical agents known as pheromones, insect colonies are capable of complex collective behavior often beyond the scope of an individual agent. In an effort to develop human-made systems with the same robustness, scientists have created artificial analogues of pheromone-based stigmergy, but these systems often suffer from scalability and complexity issues due to the problems associated with mimicking the physics of pheromone diffusion. In this chapter, an alternative stigmergic framework called 'Genetic Stigmergy' is introduced. Using this framework, agents can indirectly share entire behavioral algorithms instead of pheromone traces that are limited in information content. The genetic constructs used in this framework allow for new avenues of research, including real-time evolution and adaptation of agents to complex environments. Experiments are performed using genetic stigmergy as an indirect communication framework for a simulated swarm of robots tasked with mapping an unknown environment. The robots are able to share their behavioral genes through environmentally distributed Radio-Frequency Identification cards. It was found that robots using a schema encouraging them to adopt lesser used behavioral genes (corresponding with novelty in exploration strategies) can generally cover more of an environment than agents who randomly switch their genes, warranting further research to develop its potential.

1 Background: Stigmergy in Natural and Social Systems

Originally described by Pierre Huber in 1810 (Holldobler and Wilson, 2009) and named by Pierre-Paul Grassé in 1959 (White, 2005), stigmergy is a system of

Joshua Brandoff
Museum of Mathematics
e-mail: brandoff@momath.org

Hiroki Sayama
Collective Dynamics of Complex Systems Research Group
Binghamton University, State University of New York
e-mail: sayama@binghamton.edu

Y. Meng and Y. Jin (Eds.): Bio-Inspired Self-Organizing Robotic Systems, SCI 355, pp. 81–103.
springerlink.com

coordination whereby collective action is achieved through indirect interactions between agents via modifications to their local environment. Unlike many presently engineered human-made systems, stigmergic systems are able to self-organize through simple local interactions and without the guidance of a central coordinator. Stigmergy relies upon a number of interacting feedback loops that define how agents and their environment change as a result of interaction (Figure 1).

Each agent possesses internal and external states, with the former invisible to the perception of other agents. Agents are able to perceive and modify their environment through a (usually) small number of sensors and actuators. Guiding the agent is a controller program that modifies an agent's actions depending on the sensed local environment and the agent's internal state. In addition, the controller program itself may be guided by a separate program that changes the agent's interaction dynamics themselves as a function of time or other internal information.

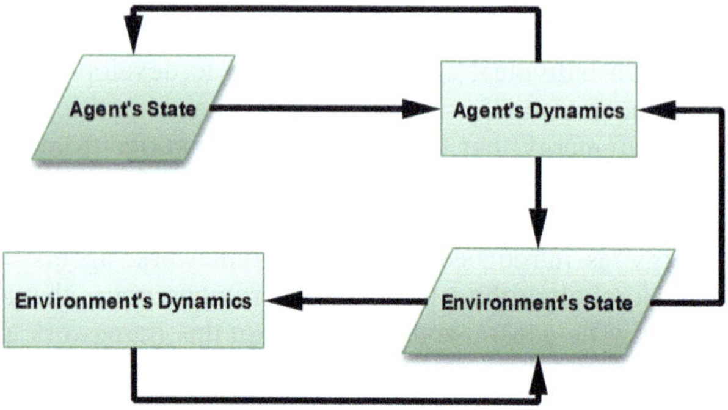

Fig. 1 Flow-chart describing the feedback loops in a stigmergic process (based on figure in Parunak (2006a), p. 164).

In nature, stigmergy is most visible in social insects, which have graced the Earth for at least 50 million years. More than 90% of the signals used in communication by these insects are through chemicals called pheromones, which can trigger various behaviors in other insects of the same species depending on their type and intensity (Holldobler and Wilson, 2009). Through the process of natural selection, these insects gained the ability to create simple "algorithms" that can use pheromone traces to collectively achieve beneficial actions. These collective actions are "satisficial" rather than optimal in nature. The term "satisficing" is defined as the achievement of an adequate or satisfactory outcome rather than the best possible outcome (Simon, 1956). In nature, it is usually impossible to acquire the amount of information necessary to achieve a globally optimal solution (i.e. finding the best food source in the *entire* forest as opposed to one that is "good enough"). Social insects have evolved stigmergic communication to find the most efficient way to complete a task rather than the best way.

The stunning amount of organization possible through pheromonal stigmergy means colonies of social insects can act as a type of *superorganism*, a term often used by evolutionary biologist E.O. Wilson (Holldobler and Wilson, 2009) to describe the emergence of complex collective behavior at a higher scale than that of an individual organism. Up until Grassé's formal introduction of stigmergy, this kind of collective behavior was thought impossible without the guidance of a central controller. It has since attracted the interest of scientists who wish to learn how local stimuli like pheromones are "organized in space and time to ensure the emergence of a coherent adaptive structure and to explain how social insects could act independently yet respond to stimuli provided through the common medium of the environment of the colony" (White, 2005).

While stigmergy is apparent in organisms as diverse as bacteria, slime-molds, and fish (White, 2005), ants have emerged as a primary species of study. Ants have a set of internal "algorithms" that allow them to modify their local environments based on what is immediately apparent (mostly through the use of pheromones) (Holldobler and Wilson, 2009), such as the dead bodies of their kin. Ant species *Lasius niger* and *Pheidole pallidula* are known for building cemeteries through the use of pheromonal stigmergy. If dead ants are initially scattered randomly throughout an environment, their living relatives will "smell" them and start clustering them together (Dorigo, Bonabeau & Theraulaz, 2000). These clusters emerge because of positive and negative feedback loops, which are intrinsic parts of any stigmergic process (Holland & Melhuish, 1999; White, 2005). An ant will tend to put more bodies where bodies already exist because the collective "smell" of larger clusters attracts the living ants. Such clustering indicates that small differences in the initial concentration of pheromones can be amplified over time. Another example of this is seen with ant foraging (Holldobler and Wilson, 2009). While initially the search for food sources is somewhat random, pheromone "trails" left by ants returning from good sources will be reinforced as other ants join to take their share. The stronger the pheromone trail gets, the more ants follow it until the pathways to less plentiful resources diminish and disappear.

Like ants, termites can use similar "winner-takes-all" stigmergic processes to build nests. Termites may initially deposit pheromone-impregnated soil pellets randomly, but the probability of depositing another mud ball in a given location increases with the sensed presence of other mud balls and associated pheromones (Backers, Holland & Deneubourg, 1994; Dorigo, Bonabeau & Theraulaz, 2000). Eventually, mud columns emerge that are further altered through stigmergic processes resulting from the interaction of various concentrations of chemical pheromones, water vapor and carbon dioxide.

Wasps and bees use a combination of pheromones and vision to build complex nests out of hexagonal modules. They can recognize elements of nest construction in process and then, using a small number of internal rules, augment the existing "construction site" in a given way. For instance, Theraulaz et. al. found that with nest building in bees, the probability of adding a cell to a three-wall site is about ten times higher than the case of a two-wall site (Theraulaz & Bonabeau, 1999). After one insect leaves, another can come take its place and make another adjustment using the same internal algorithm with a slightly different local environment. Through the collective interaction of hundreds or thousands of wasps or bees, a full nest structure can emerge.

Stigmergic processes are by no means limited to social insects; examples of stigmergy exist in the human world as well. Holland and Melhuish describe a simple example where several drivers are attempting to negotiate a muddy track. If one car finds the mud in an area on the track too deep, his deep trail marks will act as a sign that alerts other conscientious drivers to avoid that area (Holland & Melhuish, 1999). More refined examples include social networking services like Wikipedia and YouTube, where consumer posts change the "environment of interest" around a piece of media which may, in turn, attract the attention of other users (Parunak, 2006a). Presently, scientists and engineers are continuing to design and develop engineered systems that mimic the dynamics of successful stigmergic systems seen in nature. By focusing on the collective actions of relatively identical classes of agents, they hope to remove the need for processes to be performed by highly specialized (and costly) agents and increase robustness to system failure in more mission-critical applications.

2 Related Work on Artificial Stigmergy

In recent years, stigmergic frameworks have been applied to everything from the routing of data in mobile telecom (Roth & Wicker, 2003) and Peer-to-Peer networks (Mamei and Zambonelli, 2005), to data mining (Ramos & Abraham, 2004) and even the development of military swarm robots (White, 2005). The dynamics of these systems are often closely modeled after the physics of pheromone dispersal seen in termites and ants. Agents in these systems can deposit different types of "virtual pheromones" in their local environments which can be physical, simulated, or even network constructs (White & Salehi-Abari, 2008) where pheromone concentrations can be assigned to nodes or edges. Just like real pheromones, these virtual analogues can be programmed to decay over time. Agents themselves can be programmed to deposit these pheromones at varying rates and increase or decrease their sensitivities depending on the nature of the application or desired interaction between agents (Parunak, 2006b).

Entire classes of *ant algorithms* have been created to use simulated pheromones and ant-like agents to solve distributed optimization and control problems such as vehicle routing, network routing and graph coloring. Ant algorithms are especially adept at addressing Traveling Salesman problems (TSP) where an agent has to find a closed tour of minimal length while hitting every city or node in a network. Ant System (AS), Ant Colony Optimization (ACO) (Dorigo, Bonabeau & Theraulaz, 2000) and Ant-Based Control (ABC) (White, 2005) are just a few of the many types of ant algorithms created to address TSP where virtual ants leave an artificial pheromone trail on the edges that they have crossed once they finish a tour. These pheromones increase the likelihood that other ants will follow the trail and find a destination. Pheromone evaporation is employed to lessen the influence of initial trails (when there is no existing pheromone to influence decision-making) and to allow the system to forget trails that prove ineffective. In one application where an ABC scheme is used for routing calls in a telephone network, "older" virtual agents are even programmed to leave less pheromone over time if it takes them longer to get to their destination (White, 2005). The group size and pheromone dispersion must also be programmed carefully to prevent an overwhelming amount of pheromone to be deposited along paths. These algorithms often produce more optimal paths in TSP-systems than those found using general-purpose algorithms like evolutionary computation or simulated annealing.

Other stigmergic frameworks, such as Ulieru et. al's functional Stigmergic Medical Diagnostic System (SMDS) (Ulieru & Unland, 2006) can be applied to problems that don't fall in the same class as TSP. SMDS is designed to get more accurate medical diagnoses through collective intelligence, rather than relying on the limited or biased perception of one agent (or doctor). First, a request for diagnosis is placed on a virtual blackboard environment where different virtual "diagnosing agents" (specialized for certain classes of ailments) can decide to make an attempt at classifying the problem if it is in their sphere of expertise. If one agent positively comes to a conclusion, its decision is registered in a tree-like format on the blackboard. Other agents, with more specific expertise in that diagnosis class, can then come, examine the existing tree and see if they should tag onto the diagnosis hierarchy (if their own pheromone type is similar enough to the one on a given branch of the tree). The finished diagnosis tree can then be used for more correct medical care.

In the physical realm, swarm robotics has been a prominent test bed for stigmergic frameworks since indirect communication can help ameliorate problems with interference. With swarm robotics, many small agents with limited processing capabilities can interact to achieve beneficial collective behavior. While attempts have been made to use real pheromones in these systems (Wagner, Lindenbaum & Bruckstein, 1999), much research involves the use of virtual pheromones distributed in a physical medium, such as Radio-Frequency Identification (RFID) cards or tags. Robots can read and write information to these objects and the information they create can be read or changed later by other robots or humans (Mamei, Quaglieri & Zambonelli, 2006; Mamei & Zambonelli, 2007). Most RFID-robotic research focuses on using the cards as a means of localizing objects in the environment (Kim & Chong, 2007, Mamei, Quaglieri & Zambonelli, 2006; Mamei & Zambonelli, 2005; Mamei & Zambonelli, 2007; Milella, Cicirelli & Distante, 2008; Patil et. al., 2008) or tracking the location or pose of the robot itself (Bekkali, Sanson & Matsumoto, 2007; Chen et. al., 2007; Howard, Parker & Sukhatme, 2006; Lee & Lee, 2006; Roussos et. al., 2007) or some combination thereof using SLAM (Simultaneous Localization and Mapping) techniques (Kleiner & Dornhege, 2007; Kleiner, Prediger & Nebel, 2006), where passive RFID tags are used by robots to build a "map" of a volatile environment and use it to orient themselves or find human victims (Carbone, Finzi & Orlandini, 2008). In other mapping applications, virtual pheromones are used to prevent trajectory overlap by individual robots (Mamei & Zambonelli, 2007) in an attempt to increase performance. Much of this work focuses on decreasing localization error through statistical techniques like Kalman filtering (Bekkali, Sanson & Matsumoto, 2007), fuzzy inference techniques (Milela, Cicirelli & Distante, 2008) or even through the use of multiple directional RFID antennas (Kim & Chong, 2007).

Other collective robotics applications, such as construction, have a decreased emphasis on pheromone manipulation but still make use of the spatial sorting and clustering seen in the building of termite nests and bee hives (Holland & Melhuish, 1999). A physical nest building implementation was designed where robots were programmed to grip thin circular "pucks" and drop them into clusters (Backers, Holland & Deneubourg, 1994). In this instance, robots essentially ignore each other and only focus on manipulation of local pucks. Interactions in such construction environments can be made more complex by giving the building materials themselves the ability to "talk back" to the robots that are handling them (Werfel & Nagpal, 2006). This is potentially useful in situations where the system must be guided towards a specific structural layout. For other situations where certain *classes* of structures are more desirably than others, researchers such as Bonabeau et. al. attempt to use genetic algorithms to understand which agent instructions produce "better" structures (based on a pre-defined fitness) and what those instructions have in common (Bonabeau et. al., 2000).

As has already been established, the primary benefit of a pheromone-based stigmergic framework is robustness. If individual agents fail, their "traces" or local information will often still be left behind in the environment and not immediately lost (White, 2005), giving the system time to adapt. In addition, no matter how large or dynamic an environment gets, because agents only interact locally they are not overwhelmed (Parunak, 2006a). No agent necessarily needs a global picture because they can work very effectively in parallel to produce a collective behavior (Ramos & Merelo, 2004).

As reviewed above, the robustness of pheromone-based stigmergy in nature has encouraged many researchers to design analogous frameworks in man-made systems. Unfortunately, many of these researchers fall victim to the biomimicry version of "not being able to see the forest for the trees". Efforts to artificially mimic the physics of pheromone diffusion has led to new classes of problems needing to be solved, such as error minimization (Herianto, Sakakibara, & Kurabayashi, 2007; Parunak, 2006b) and the management of "autocatalytic snowball effects" (Dorigo, Bonabeau & Theraulaz, 2000), where, due to runaway feedback processes, virtual pheromones concentrate or diffuse too quickly for proper behaviors or structures to emerge. The cost of true-to-nature artificial analogues of pheromonal stigmergy may be the very robustness they were designed to sustain. If the scientific community instead takes a step back and uses nature as a guide instead of a blueprint, it can open the door for more creative stigmergic frameworks. Thus, researchers may be better served by focusing less on stigmergy as it exists in nature and more on stigmergy "as it could be."

Stigmergy "as it could be" means developing stigmergic frameworks that are inspired by, but do not currently exist in, nature. It is an attempt to reap the benefits associated with natural stigmergy without replicating its constraints. Natural stigmergy evolved in the context of the natural world, not in the world of artificially created systems. Thus, it is reasonable to conclude that the most robust or effective forms of stigmergy for a given human-made system may only conceptually resemble their natural cousins.

3 Proposed Framework

In this chapter, we propose one possible man-made alternative that combines elements of natural stigmergy with the information constructs used in genetic evolution. The goal of this hybridized framework—called "*genetic stigmergy*"—is to encapsulate behaviors in a fully portable, gene-based fashion that frees them from the identity of an individual agent. Such a framework allows for a degree of collective adaptability impossible in natural stigmergy, artificial or otherwise, and thus its potential deserves to be explored.

We define genetic stigmergy as an indirect communication framework where agent behavioral algorithms—represented as collections of virtual "genes"—can be shared, in part or whole, via an external medium (Figure 2). The discretization of algorithms into spatially distributable genes provides a uniform "currency" that agents can use to quickly swap in behaviors that are found to be locally adaptive by other agents. This "hot swapping" allows for real-time optimization of collective behavior without prior knowledge of an environment. Genetic stigmergy differs from pheromonal stigmergy in that the information exchange is not limited by the paradigm of chemical physics and the problems associated with its mimicry. In addition, the genetic information exchanged is more complex than a simple trace or marker and lends itself very easily to evolutionary manipulation.

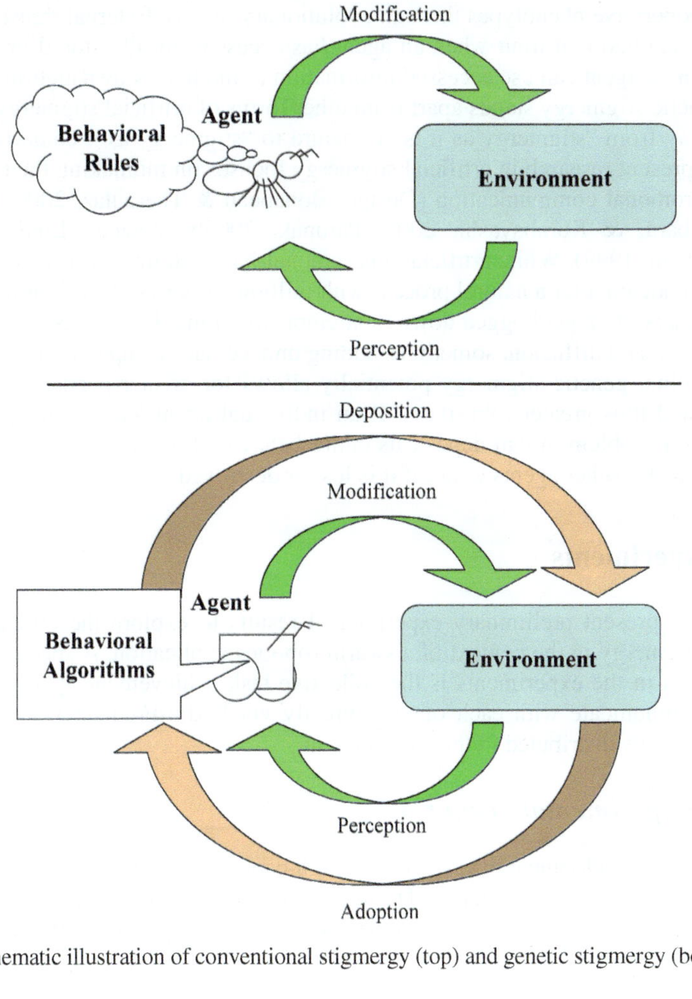

Fig. 2 Schematic illustration of conventional stigmergy (top) and genetic stigmergy (bottom).

An agent may write its own genes to the external medium (heredity), potentially with some minor changes added to them at small probability (allowing for variation). When another agent accesses genes from the medium, it may adopt the genes as its own code (allowing for selection) at another probability that may depend on the "openness" of the previous genes as well as the quality of the new genes written in the medium. Through artificial analogs of heredity, variation and selection, it is possible to include evolutionary processes in a genetic stigmergy framework through such techniques as genetic algorithms or evolutionary programming.

The framework may also be implemented in such a fashion that no restrictions will be imposed on evolvable agent behavior. In such an open-ended system, it is expected that "selfish" individual behaviors that are good at spreading within a population but inconsistent with collective interest may emerge and thereby reduce the collective performance of the population. The removal of non-cooperative phenotypes may be achieved by specifically programming the protocols to remove them from the swarm, or by implicitly suppressing the spread of non-cooperative phenotypes through evolutionary means. External thresholds may also be applied that limit when an agent has access to locally-stored information, or when an agent can use accessed information to modify its own algorithms.

Genetic stigmergy stands apart from other forms of artificial stigmergy by moving away from "stigmergy as it is" in nature to "stigmergy as it could be." Much of the present research in artificial stigmergy focuses on mimicking the mechanics of pheromonal communication (Dorigo, Bonabeau & Theraulaz, 2000; Herianto, Sakakibara, & Kurabayashi, 2007; Parunak, 2006b; Wagner, Lindenbaum & Bruckstein, 1999). While artificial, pheromonal-based stigmergy has the benefit of being modeled after a natural process with millions of years of evolution behind it, researchers often get bogged down in attempts to mimic the physics of pheromone deposition and diffusion, sometimes adding unnecessary complexity to the system. In addition, genetic stigmergy potentially allows for greater persistence of agent states and thus greater robustness. If an individual agent learns a unique way of solving a problem, it can deposit its entire behavioral algorithm (or a representation of it) for other agents to use if it is lost or destroyed.

4 Experiments

Here we present preliminary experimental results to explore the efficacy of genetic stigmergy in the context of a swarm robotics application. A specific scenario we adopt in the experiments is the collective task achievement by swarm robots that communicate with each other indirectly via Radio-Frequency Identification (RFID) cards distributed in the environment.

4.1 Experimental Scenario

The recent development of economical, high-capacity RFID cards has opened up a new opportunity for stigmergy. Through these cards, robotic agents can dynamically exchange complex logical information, such as a genetic code that controls their behavioral rules. Dynamic, real-time modification of agents' behavioral "genes" may increase the adaptability of a swarm to a complex system, which is useful for tasks such as collective exploration of an unknown environment. Certain behaviors may be more adaptive in various areas of an environment (i.e. better at navigating the area more quickly). Using genetic stigmergy, robotic agents do not have to communicate with each other directly and would not need complicated algorithms to manage the physics of pheromone diffusion. RFID cards can be distributed throughout an environment for robots to record their genetic codes.

The goal of the experiments is to demonstrate that the genetic stigmergy framework will provide improvements in a swarm's mapping performance beyond those arising from the random switching of genes. Robots are assumed to carry a multi-locus chromosome where genes at each locus control a robot's reaction to different types of external stimuli. Multiple alleles of the genes at each locus allow for much greater diversity in the robot's response to a specific type of stimuli than was possible in the single-gene experiments reported earlier (Brandoff & Sayama, 2009). RFID cards densely distributed over a space are able to record the frequency of genes deposited to them. Different implementations of genetic stigmergy are tested where robots are encouraged to adopt ("Majority Seeking") or avoid ("Minority Seeking") frequently used genes or randomize their chromosome. In addition, the robustness of genetic stigmergy is demonstrated by exploring the different implementations in environments of varying complexity and through the usage of "accessibility windows" that control when a robot can access the genetic information on a card.

4.2 Simulation Platform

For the experiments, a flexible simulation platform is designed using the Python programming language to quickly and effectively examine the swarm's exploration behavior in environments of varying complexity. The generated virtual environment is composed of OPEN-ROBOTS, open-source RFID-capable mobile robots designed by Abe Howell's Robotics (Howell, 2008), as well as RFID cards and a to-scale environment within which the robots interact (Figure 3). By using the Python-based platform on a multiple-CPU system, four simulations can be run at up to 3600x real-time with different parameter settings.

Robots and obstacles are represented as circular constructs with appropriate radii (7.25 cm for robots, approximately corresponding with their physical counterparts). When visual confirmation is needed to confirm that an algorithm is working properly, the freely available VPython package is incorporated into the simulator. For aesthetic purposes, obstacles and robots are given arbitrary height values (Figure 3). A robot will register the existence of an obstacle if its "sensor circumference" (a radius of 18.5 cm beyond its virtual embodiment) overlaps with the circumference of a given obstacle or a robot or passes beyond the boundary of the environment. Collisions with RFID cards are calculated by testing to see if the robot's center is within the perimeter of a given card.

4.3 Experimental Setup

Because the environments of the simulations are meant to mimic homes or factories with dense, non-overlapping grids of RFID cards in the floor, approximate coverage is determined by counting the total number of RFID cards the swarm interacts with. For all experiments, a swarm of 10 robots are initialized with randomized initial positions in the lower left corner of a square (also see Figure 3), 54 square-meter environment filled with a grid of 6,840 RFID cards (approximately 130 cards per square-meter or 12 cards per square-foot).

Every environment also contains a total of 10 square-meters of obstacles. How this allotment is divided is controlled by a control parameter that sets the maximum possible radius for a given obstacle. The radius of a newly generated obstacle is determined by choosing a number from a uniform distribution between zero and the "max radius" parameter: "R". If R is low, many more obstacles are needed to reach the allotted overall obstacle area. If R is high, fewer obstacles are needed to reach this allotment. Experiments are performed using R values ranging from 1.0 meter to 2.0 meters in increments of 0.2 meters. Some examples are shown in Figure 4.

All objects are placed in open areas inside the environment and do not overlap with other objects are barriers. Robots must be spaced such that their "sensor circumference" does not overlap with those of other robots. These random initializations are to prevent any anomalies that would arise from any given static environment.

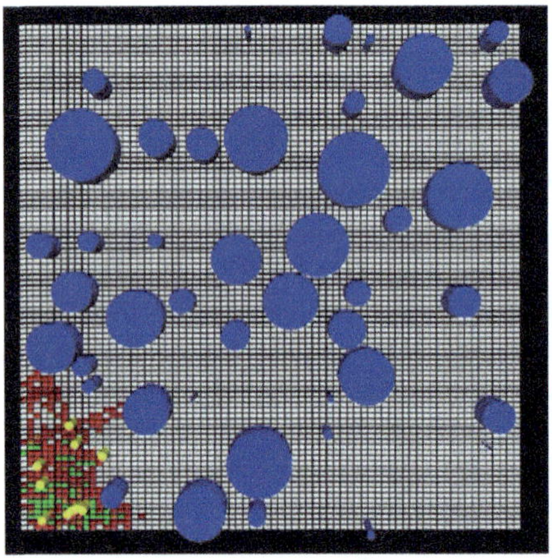

Fig. 3 An "aerial view" of 10 swarm robots (in yellow) exploring a randomly generated virtual environment with a dense grid of RFID cards embedded in the floor. The blue cylinders are obstacles and RFID cards change color depending on their current state.

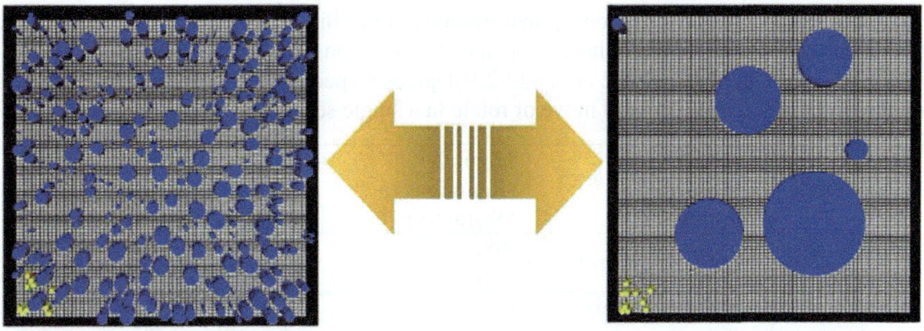

Fig. 4 Examples of environments with R values between 0.2 meters (left) and 2 meters (right). Environments with smaller R values are considered complex to navigate, while environments with larger R values are considered simple.

Robots carry a chromosome with three loci, each containing a gene that controls the robot's reaction to environmental stimuli (see Table 1). The first locus contains genes that control how a robot reacts when it senses a potential collision with another robot. The second locus contains genes that control how a robot reacts to potential collisions with a static obstacle or the environmental boundaries. The third locus controls how a robot behaves when it does not sense any obstacles or other robots in its vicinity. Alleles can be combined in 128 different ways, giving the robot much more algorithmic flexibility than in the single-gene experiments (Brandoff & Sayama, 2009). At the beginning of each simulation, all ten of the robots are initiated with chromosome (1,1,0). The resulting phenotype causes the robots to rotate away from other robots and obstacles, but to otherwise move forward. This chromosome encourages the robots to spread out across the environment instead of staying clumped up in the corner.

The RFID grid acts as a distributed counter system that records the number of times cards have been hit by robots and what genes each robot was carrying. Every time a robot is within range of an RFID card, the RFID card will increment its 'hit' counter and counters corresponding to the individual alleles the robot is carrying. Over time, the RFID card will generate a tally of locally used allele frequencies. Whether a robot can access the information on a card depends upon accessibility windows, another control parameter. These windows are implemented to prevent a robot from adopting new genes too rapidly and essentially "jittering" in place by instantaneously switching behaviors. As indicated by Figure 5, a robot can only modify its chromosome based on the RFID card's information when the card's hit counter is within a "hit window" defined by the experimenter. Depending on the experiment, a window can open after between 1 and 9 hits and can close after between 2 and 10 hits. Figure 6 describes the interaction algorithm between robots and RFID cards. While robots can only access information within a given window, they can "reset" the cards global hit counter to zero after the accessibility window has closed.

Table 1 Breakdown of the chromosomal structure controlling each robot's behavior and the potential genes that control behaviors in specific situations. Note that robots move and rotate in discrete increments, 9.2 cm and 92.1 degrees respectively (the maximum distance a physical OPEN-ROBOT could move or rotate in a single second).

Allele	Locus 1 (Reactions to robots)	Locus 2 (Reactions to Obstacles)	Locus 3 (Reactions to Open Space)
0	Rotate towards robot	Rotate towards obstacle	Move forwards once
1	Rotate away from robot	Rotate away from obstacle	Move backwards once
2	Back away from robot	Back away from obstacle	Move forwards twice and rotate left
3	Randomly choose between rotating towards, away or backing away from robot	Randomly choose between rotating towards, away or backing away from obstacle	Move forwards twice and rotate right
4			Move backwards twice and rotate left
5			Move backwards twice and rotate right
6			Randomly choose between rotating left, right or moving forwards once
7			Randomly choose between rotating left, right or moving backwards once

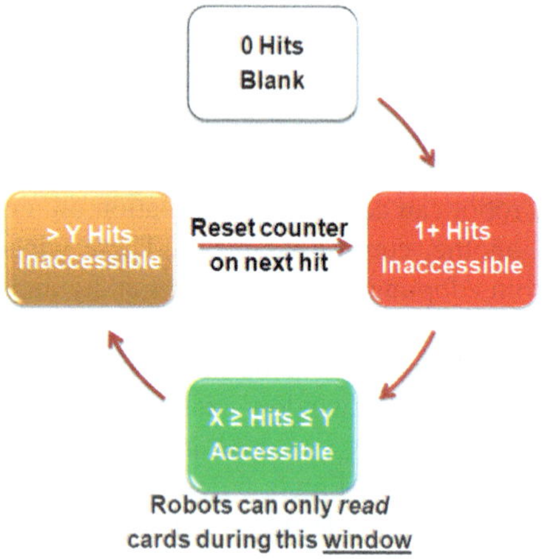

Fig. 5 Access to RFID cards is controlled by virtual "hit counters" stored on the card. Depending on the number of hits, an RFID card can cycle between one of several different accessible or inaccessible states (and colors).

To determine the most efficacious implementation of the genetic stigmergy framework, four robot-RFID card interaction paradigms are tested: “Minority Seeking”, “Majority Seeking”, “Randomization” and “No Threshold”. The “Minority” and “No Threshold” paradigms both encourage robots to adopt the genes least frequently recorded to a card, but the “Minority” paradigm restricts card access to a given accessibility window. “Majority Seeking” and “Randomization” also use accessibility windows but “Majority Seeking” encourages robots to adopt the genes *most* frequently recorded to a card, while “Randomization” forces the robots to randomize their genes (and ignore the information on a card). The effectiveness of each paradigm is tested by performing a series of 30 Monte Carlo simulations for each experimental setting. Simulations are performed at 3600x real-time, so a simulated hour-long trial takes less than one real second.

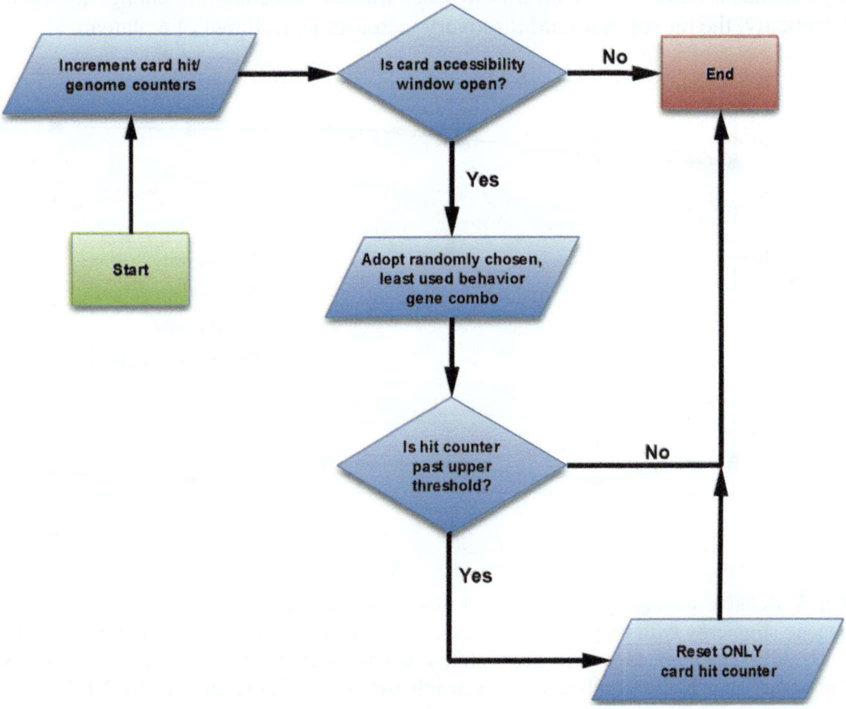

Fig. 6 Flowchart describing interaction between a robot and an RFID card in a single encounter event under the “Minority Seeking” interaction paradigm, where a robot adopts the least used combination of behavior genes based on the information available on the card. Here, the “accessibility window” refers to the card hit thresholds within which a robot can access the genetic information stored on the RFID card.

4.4 Results

Results are summarized in Figure 7, in which the performance of the “Minority Seeking”, “Majority Seeking”, “Randomization” and “No Threshold” paradigms are compared using a fixed accessibility window of 5 to 10 card hits for values of R from 1.0 meters to 2.0 meters. The "Majority Seeking" and "No Threshold" paradigms produce dismal performance, while the "Minority Seeking" and "Randomization" paradigms achieve the highest average performance. Table 2 shows that, with the exception of R = 1.6 meters, the "Minority" paradigm is the best performing implementation (though their performance differences did not reach a statistically significant level for most cases).

Closer examination of the "Minority Seeking" paradigm allows for a greater understanding of how it influences robot behaviors over the course of a simulation and compares to the "Randomization" paradigm. Figures 8(a)-(c) provide tentative evidence that the swarm is collectively reacting to plateaus in the acquisition of "newly found cards" by increasing the rate at which they diversify their behavior and then decreasing it when the plateaus are overcome. In plot (a), for example, card discovery plateaus around t = 40 s until t=50 when the rate of behavioral change increases. Eventually, the rate of new card discovery increases as well, out of a plateau.

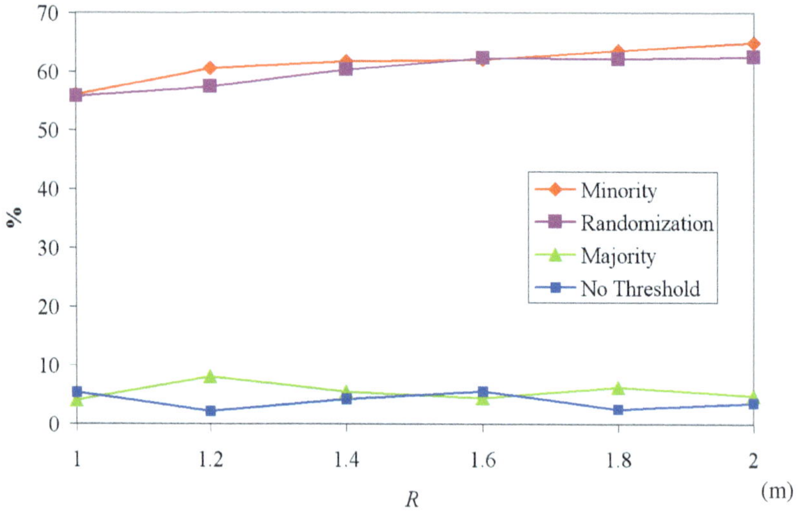

Fig. 7 Relative average performance (% floor coverage) of different implementations of genetic stigmergy with varying obstacle size R. Each value is an average of 30 independent runs. All implementations use accessibility windows of 5-10 hits, except for the "No Threshold" which allows robots to immediately use genetic information on the RFID cards.

Table 2 Average percentage of cards covered for each R. The highest coverage values for each R are in bold red lettering. The p-values of one-sided mean difference tests between coverage results of the Minority Seeking and Randomization conditions are also provided.

R **(m)**	**1.0**	**1.2**	**1.4**	**1.6**	**1.8**	**2.0**
Minority Seeking	**56.0**	**60.5**	**61.6**	61.9	**63.5**	**64.8**
Randomization	55.7	57.3	60.2	**62.2**	62.0	62.4
One-sided p-value	*.435*	***.042****	*.160*	*.420*	*.213*	*.084*
Majority	3.9	8.0	5.4	4.3	6.1	4.7
No Threshold	5.3	2.1	4.1	5.5	2.4	3.5

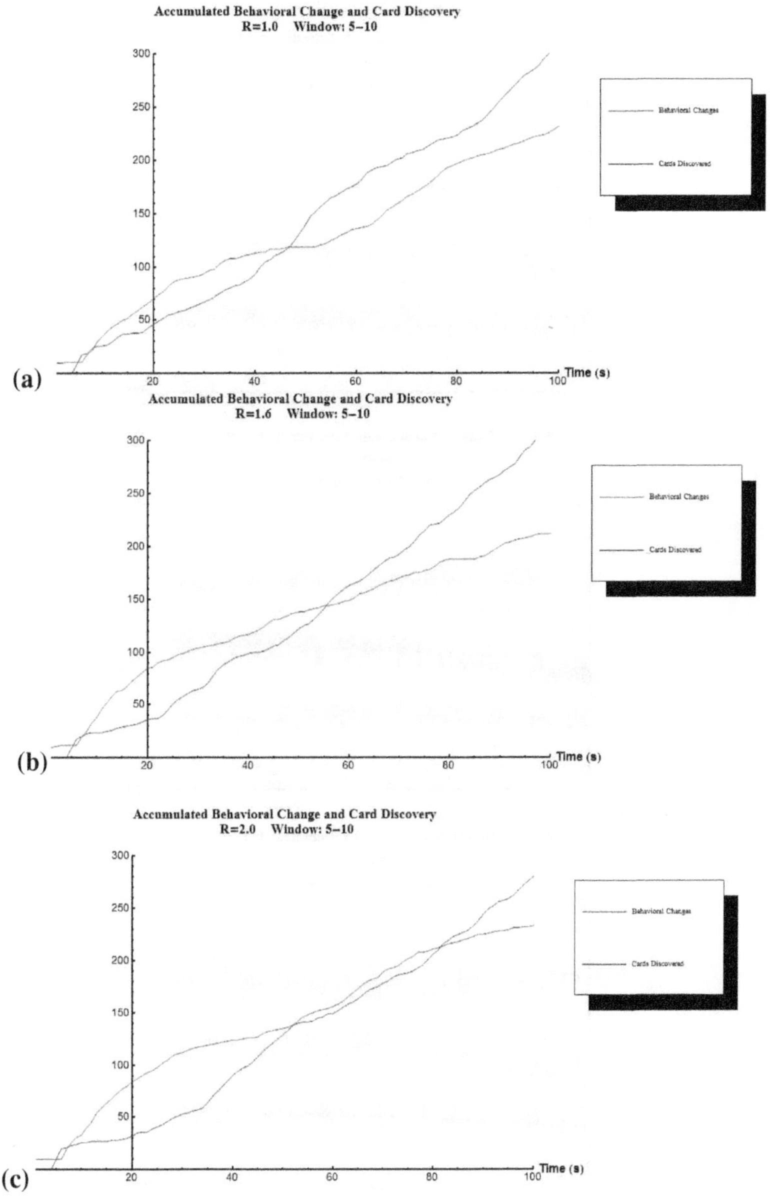

Fig. 8 Time-series plots of the cumulative number of changes in robot behaviors and the cumulative number of cards discovered by robots over 100 seconds at accessibility window 5-10 and R = 1.0 (a), 1.6 (b) and 2.0 (c) meters. Plateaus in accumulation are visible at various time scales where the rate of card discovery or behavioral change slows down. The "Behavioral Changes" curve is blue and the "Cards Discovered" curve is purple.

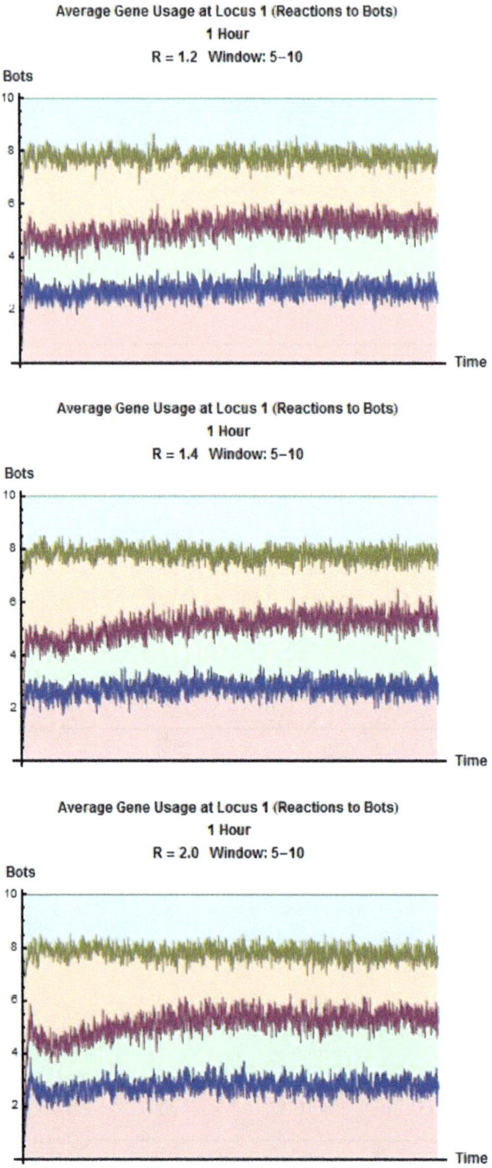

Fig. 9 Average gene usage for all robots at locus 1 (controlling reactions to other robots) for entire simulation (1 hour). Selected R values 1.2, 1.4 and 2.0 are used with accessibility windows of 5-10. Colors correspond to the four different possible alleles such that red indicates "rotate towards", green indicates "rotate away", orange indicates "back away" and blue indicates "random reaction".

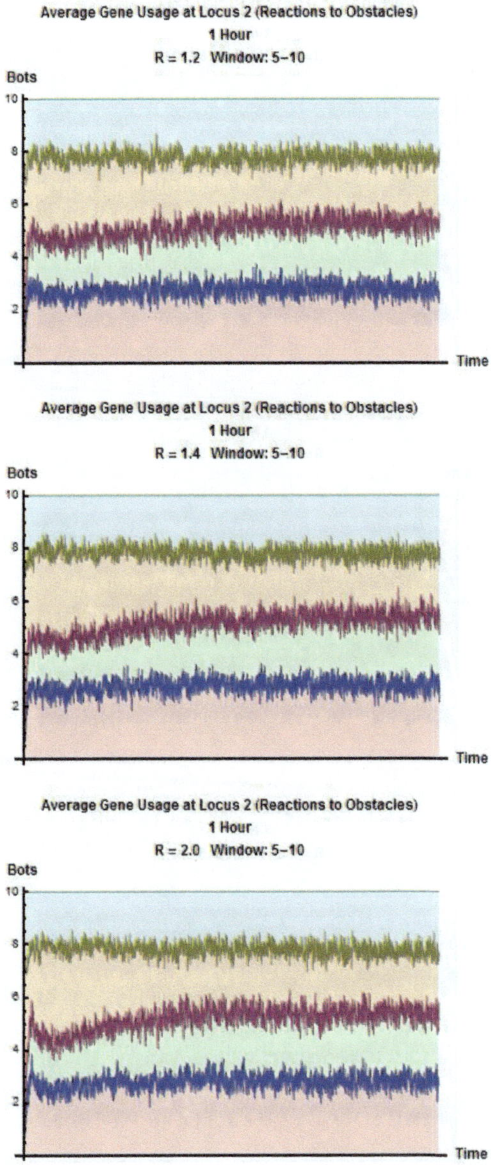

Fig. 10 Average gene usage for all robots at locus 2 (controlling reactions to static obstacles) for entire simulation (1 hour). Selected R values 1.2, 1.4 and 2.0 are used with accessibility windows of 5-10. Colors correspond to the four different possible alleles such that red indicates "rotate towards", green indicates "rotate away", orange indicates "back away" and blue indicates "random reaction".

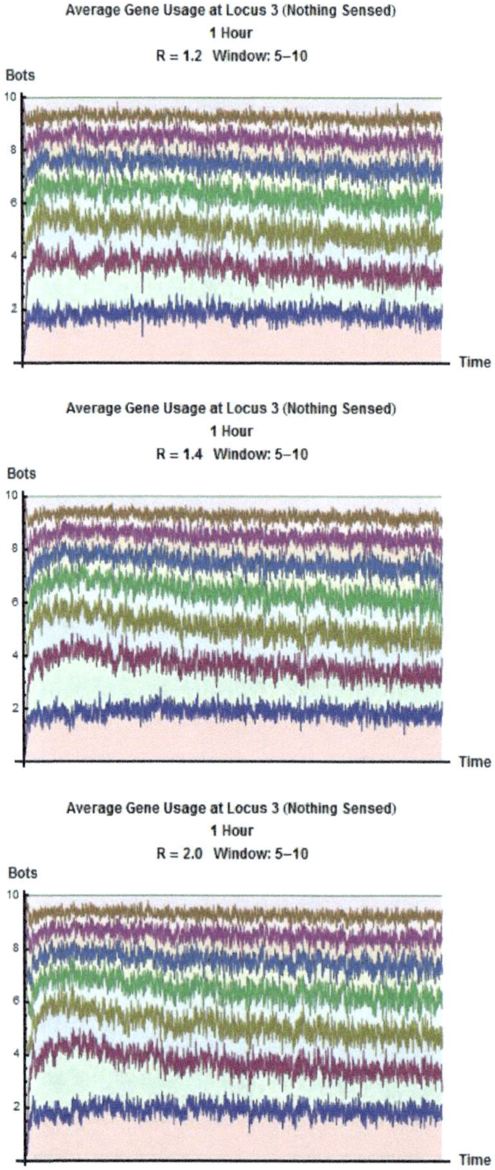

Fig. 11 Average gene usage for all robots at locus 3 (controlling motions when no obstacles/robots) for the entire simulation (1 hour). Selected R values 1.2, 1.4 and 2.0 are used with accessibility windows of 5-10. Colors correspond to the eight different possible alleles such that red, green, blue, cyan, yellow, orange, pink, and purple correspond to "move forwards", "move backwards", "move forwards twice and rotate left", "move forwards twice and rotate right", "move backwards twice and rotate left", "move backwards twice and rotate right", "random rotate or forwards motion", and "random rotate or backwards motion" respectively (see from bottom up).

Figures 9, 10 and 11 all show the average gene usage for all robots at each locus averaged at each second of the overall simulation. The values R = 1.2 and 2.0 meters are used because at these values, the "Minority Seeking" paradigm outperforms the "Randomization" paradigm by the largest margins (3.2% and 2.4% respectively). The value R = 1.4 meters is used as an intermediary value for continuity. The initialization of all robots with the (1,1,0) chromosome (causing them to rotate away from other robots and obstacles and move outwards), appears to skew the initial allele distribution on all loci. The patterns of allele distribution at each locus act as signatures that help differentiate the "Minority Seeking" paradigm from the "Randomization" paradigm, where the alleles are *always* uniformly distributed regardless of initial conditions.

5 Discussion

Among the four paradigms tested in the experiments, the "Minority-Seeking" paradigm encourages local diversity of genetic material, while the "Majority-Seeking" paradigm encourages a "winner-takes-all" course of events. The "Majority-Seeking" paradigm produces poor coverage results on average, while the "Minority-Seeking" paradigm and "Randomization paradigm" produce similarly high coverage rates.

Figure 8 demonstrates that the "Minority Seeking" paradigm (unlike the "Randomization" paradigm) allows robots to collectively react to plateaus in new card discovery by more rapidly modifying their genes. However, while the plateaus described in the results are promising, further experimentation is necessary to determine the existence and strength of any casual relationship present. Also, due to the variance in robot positions, environmental complexity and environmental layout, the scale of plateaus may vary widely over time and space. This may make causal confirmation more difficult.

Figures 9, 10 and 11 help explain why the "Minority Seeking" paradigm outperforms the "Randomization" paradigm. At loci 1 and 2 (Figures 9 and 10), which control how robots react to other robots and obstacles, there appears to be an initial bulge of gene usage associated with the "back away" behavior. This may indicate that the "back away" behavior is useful in finding new RFID cards in the beginning of a simulation (preventing robots from avoiding the gene). After some time, the distribution settles so that all genes are more-or-less equally represented among the swarm at any given time. At locus 3 (Figure 11), the "move forwards" and "move backwards" genes are initially represented somewhat more than other alleles. Later, the allele distribution also settles, but the "move forwards" gene maintains a relatively larger presence in the swarm.

In the "Randomization" paradigm, the haphazard shuffling of genes would, on average, produce an allele distribution such that all genes were represented equally. In the "Minority Seeking" paradigm, this even distribution only appears that the end of the simulations. Even then, the distribution is often slightly skewed in favor of effective genes (such as "move forwards"). The fact that the "Minority Seeking" paradigm allows for adaptive genes to be "over-represented" in the allele distribution is likely the reason why it slightly outperforms the "Randomization" paradigm.

While the genetic stigmergy framework appears promising, several potential flaws or confounding errors in the experimental design must be considered. In the experiments, the order in which robots acknowledge sensor readings indicating obstacles may affect their overall motion. The construction of the simulated environments themselves may pose an issue. The fact that many RFID cards are hidden under obstacles or locked away in permanently inaccessible regions (in the highly complex environments) may skew coverage results. Also, the uniformity of the obstacles (all where circular) and consistently square environment shape may not be fully representative of environments in the real world.

In an effort to more convincingly demonstrate the potential of genetic stigmergy, the framework must be thoroughly examined in the context of other simpler and more complex techniques to determine its true usefulness in practical applications. More advanced techniques may affect better performance at the cost of increased time and effort spent developing individual learning processes. However, this cost-benefit compromise may change depending on the environment and local constraints placed on the system. In addition, a more robust testing of genetic stigmergy under a unified experimental framework is necessary. Future experiments should run multiple Monte Carlo simulations for all algorithm paradigms, varying the density of RFID cards (from very sparse to highly dense) and varying environmental sizes with different arrangements and shapes of obstacles. Only by thoroughly testing each parameter in a consistent fashion can it be determined which parameter(s) is/are most important in the successful implementation of genetic stigmergy. Also, aside from space filling, it is likely there are other engineering applications where genetic stigmergy can more clearly differentiate itself from other techniques. Other application possibilities should be thoroughly explored to determine what is genetic stigmergy's "killer application".

Potential improvements to the system include time-stamping the genetic information deposited to RFID cards or "smarter" RFID cards that can exchange information locally with other RFID cards to help coordinate robots (similar to Werfel & Nagpal, 2006, where blocks and robots can communicate). In addition, allowing for multiple simultaneous accessibility windows in different areas of the map depending on local need (as in Dorigo, Bonabeau & Theraulaz, 2000) may help if, and only if, the swarm can autonomously determine the proper thresholds on the fly. Allowing for genetic evolution within the robotic swarm may help in this regard. Through the introduction of such operators as mutation, novelty can be introduced to the system that helps robots discover more adaptive behavior or ways of interacting with the environment.

Individualized manipulation of accessibility windows through the use of evolutionary operators may be the most important factor in significantly improving the performance of genetic stigmergy. The importance of thresholds has already been thoroughly examined by Dorigo and Bonabeau (Dorigo, Bonabeau & Theraulaz, 2000), who note that simple threshold models have limitations due to their fixed nature and are only valid over short-time scales. In the longer term (perhaps as a function of time), accessibility windows should change and differentiate to allow for agent specialization. Future work will determine if such mutability of thresholds is feasible or realistic.

6 Conclusion

Genetic stigmergy holds great promise as an alternative to pheromonal-based artificial stigmergy for the achievement of collective action through self-organization. Unlike pheromonal-based stigmergy, genetic stigmergy is not mired in unnecessary complications due to attempts to mimic chemical diffusion. Preliminary experimentation indicates that genetic stigmergy may be an effective tool in such fields as swarm robotics, but much theoretical work remains to be done to demonstrate this framework's robustness in robotics and elsewhere. In addition, the implementation of behavior meta-rules to control agent access to local information appears necessary to direct a swarm's emergent behavior to useful ends. These interaction restrictions are even more important in a system where agents can evolve their behaviors in real-time. Future work and further experimentation will address these issues and help develop genetic stigmergy into a viable platform for decentralized communication.

To acquire the underlying code for the simulations in this chapter, please contact Joshua Brandoff at josh.brandoff@gmail.com.

References

Howell, A.: Abe Howell's Robotics (2008), `http://www.abotics.com`

Beckers, R., Holland, O.E., Deneubourg, J.L.: From local actions to global tasks: stigmergy and collective robotics. In: Artificial Life IV, pp. 181–189. MIT Press, Cambridge (1994)

Bekkali, A., Sanson, H., Matsumoto, M.: RFID indoor positioning based on probabilistic RFID map and Kalman filtering. In: WIMOB 2007: Proceedings of the Third IEEE International Conference on Wireless and Mobile Computing, Networking and Communications, p. 21. IEEE, Washington, DC, USA (2007)

Bonabeau, E., Silvain, G., Snyers, D., Kuntz, P., Theraulaz, G.: Three- dimensional architectures grown by simple "stigmergic" agents. BioSystems 56, 13–32 (2000)

Brandoff, J., Sayama, H.: Cultural transmission in robotic swarms through RFID cards. In: Proceedings of the Second IEEE Symposium on Artificial Life (IEEE- CI-ALIFE 2009), pp. 171–178. IEEE, Nashville (2009)

Carbone, A., Finzi, A., Orlandini, A.: Model-based control architecture for atten-tive robots in rescue scenarios. Autonomous Robots 24(1), 87–120 (2008)

Chen, P.Y., Chen, W.T., Wu, C.H., Tseng, Y.C., Huang, C.F.: A group tour guide system with RFIDs and wireless sensor networks. In: IPSN 2007: Proceedings of the 6th International Conference on Information Processing in Sensor Networks, pp. 561–562. ACM, New York (2007)

Dorigo, M., Bonabeau, E., Theraulaz, G.: Ant algorithms and stigmergy. Future Generation Computer Systems 16(9), 851–871 (2000)

Herianto, S.T., Kurabayashi, D.: Artificial pheromone system using RFID for navigation of autonomous robots. Journal of Bionic Engineering 4(4), 245–253 (2007)

Holland, O., Melhuish, C.: Stigmergy, self-organisation, and sorting in collective robotics. Artificial Life 5(2), 173–202 (1999)

Holldobler, B., Wilson, E.O.: Superorganism: The Beauty, Elegance and Strange-ness of Insect Societies. W.W. Norton & Company, Inc., New York (2009)

Howard, A., Parker, L.E., Sukhatme, G.S.: Experiments with a large heterogene-ous mobile robot team: exploration, mapping, deployment and detection. International Journal of Robotics Research 25(5-6), 431–447 (2006)

Kim, M., Chong, N.Y.: RFID-based mobile robot guidance to a stationary target. Mechatronics 17(4–5), 217–229 (2007)

Kleiner, A., Dornhege, C.: Real-time localization and elevation mapping within urban search and rescue scenarios: field reports. Journal of Field Robotics 24(8-9), 723–745 (2007)

Kleiner, A., Prediger, J., Nebel, B.: RFID technology-based exploration and SLAM for search and rescue. In: Proceedings of the IEEE/RSJ International Conference on Intelligent Robots and Systems (IROS), Beijing, China, pp. 4054–4059.

Lee, H.J., Lee, M.C.: Localization of mobile robot based on radio frequency identification devices. In: SICE-ICASE International Joint Conference, pp. 5934–5939 (2006)

Li, J., Poulton, G., James, G.: Agent-Based Distributed Energy Management. In: Orgun, M.A., Thornton, J. (eds.) AI 2007. LNCS (LNAI), vol. 4830, pp. 569–578. Springer, Heidelberg (2007)

Mamei, M., Quaglieri, R., Zambonelli, F.: Making tuple spaces physical with RFID tags. In: SAC 2006: Proceedings of the ACM Symposium on Applied Computing, pp. 434–439. ACM, New York (2006)

Mamei, M., Zambonelli, F.: Physical deployment of digital pheromones through RFID technology. In: AAMAS 2005: Proceedings of the Fourth International Joint Conference on Autonomous Agents and Multiagent Systems, pp. 1353–1354. ACM, New York (2005)

Mamei, M., Zambonelli, F.: Pervasive pheromone-based interaction with RFID tags. ACM Transactions on Autonomous and Adaptive Systems 2(2), 4 (2007)

Milella, A., Cicirelli, G., Distante, A.: RFID-assisted mobile robot system for mapping and surveillance of indoor environments. Industrial Robot 35(2), 143–152 (2008)

Parunak, H.V.D.: A survey of environments and mechanisms for human-human stigmergy. In: Environments for Multi-Agent Systems II, vol. 3830, pp. 163–186 (2006a)

Parunak, H.V.D.: Evolving swarm agents in real time. In: Genetic Programming Theory and Practice III, ch. 2, pp. 15–32 (2006b)

Patil, A., Munson, J., Wood, D., Cole, A.: Bluebot: Asset tracking via robotic lo-cation crawling. Computer Communications 31(6), 1067–1077 (2008)

Ramos, V., Abraham, A.: Evolving a stigmergic self-organized data-mining (2004), `http://arxiv.org/abs/cs.AI/0403001`

Ramos, V., Merelo, J.J.: Self-organized stigmergic document maps: environment as a mechanism for context learning (2004), http://arxiv.org/abs/cs.AI/0412075

Roth, M., Wicker, S.: Termite: ad-hoc networking with stigmergy. In: Global Telecommunications Conference, GLOBECOM 2003, vol. (5), pp. 2937–2941. IEEE, Los Alamitos (2003)

Roussos, G., Papadogkonas, D., Taylor, J., Airantzis, D., Levene, M., Zoumboulakis, M.: Shared memories: a trail-based coordination server for robot teams. In: RoboComm 2007: Proceedings of the 1st International Conference on Robot Communication and Coordination, pp. 1–4. IEEE Press, Piscataway (2007)

Simon, H.A.: Rational choice and the structure of the environment. Psychological Review 63(2), 129–138 (1956)

Theraulaz, G., Bonabeau, E.: A brief history of stigmergy. Artificial Life 5(2), 97–116 (1999)

Ulieru, M., Unland, R.: A Stigmergic approach to medical diagnostics. In: MAS*BIOMED 2006: Second International Workshop on Multi-Agent Systems for Medicine, Computational Biology and Bioinformatics, pp. 87–103. AAMAS, Hakodate (2006)

Wagner, I.A., Lindenbaum, M., Bruckstein, A.M.: Distributed covering by ant-robots using evaporating traces. IEEE Transactions on Robotics and Automation 15, 918–933 (1999)

Werfel, J., Nagpal, R.: Extended stigmergy in collective construction. IEEE Intelligent Systems 21(2), 20–28 (2006)

White, T.: Expert assessment of stigmergy: a report for the Department of National Defence. Technical report A382144. Department of Computer Science. Carleton University, Ottawa, ON, Canada (2005)

White, T., Salehi-Abari, A.: A swarm-based crossover operator for genetic programming. In: Proceedings of the 10th Annual Conference on Genetic and Evolutionary Computation. ACM Press, New York (2008)

From Ants to Robots and Back: How Robotics Can Contribute to the Study of Collective Animal Behavior

Simon Garnier

Abstract. Swarm robotics has developed partly from biological discoveries that have been made on the organization of animal societies during the last thirty years. In this article, I review some of the ways robotics contributes in return to the study of collective animal behavior. I argue that robotics can bring significant improvements in this field, from a technical, conceptual and educational point of view. I base my discussion on five observations I have made while collaborating with computer scientists: robots require a complete specification; robots are physical entities; robots implement new technologies; robots can be inadvertent sources of biological inspiration; and robots are "cool" gadgets.

1 Introduction

Swarm robotics is a scientific discipline that emerged during the 90's at the intersection between two research fields: collective robotics and swarm intelligence. From collective robotics, swarm robotics gets the challenges. Its main goal is to design control algorithms to coordinate the activity of several robots simultaneously. Ideally, this coordination should lead to the achievement of a global task that a single robot could not perform alone, at least in a reasonable amount of time [16, 31, 85]. From swarm intelligence, swarm robotics gets the coordination principles that can serve as a basis to design the aforementioned algorithms. These principles emphasize local communications, distributed control and self-organization to generate collective behaviors that can be very complex, or solve problems that are far beyond the cognitive abilities of the individual robots [9, 53].

Historically, swarm intelligence, and hence swarm robotics, is born from and partly fed by insightful studies about the organization of animal groups, and in

Simon Garnier
Department of Ecology and Evolutionary Biology, Princeton University,
Princeton, NJ 08544, USA
e-mail: sgarnier@princeton.edu

Y. Meng and Y. Jin (Eds.): Bio-Inspired Self-Organizing Robotic Systems, SCI 355, pp. 105–120.
springerlink.com

particular insects societies [9, 14]. Insect colonies, ranging from a few animals to millions of individuals, display fascinating behaviors that combine efficiency with both flexibility and robustness [14]. From the traffic management on a foraging network [20, 29, 12, 80], to the building of efficient structures [10, 11, 62, 76, 78], along with the dynamic task allocation between workers [6, 75, 24, 43], examples of complex and sophisticated behaviors are numerous and diverse among social insects [8, 14, 26].

From the early works in the 80's to the most recent advances [14, 38], the coordination mechanisms used by social animals, and social insects in particular, have proved to be a valuable source of inspiration to solve various complex problems [9, 32, 33, 49]. Among the most common tasks that can be solved by bio-inspired robotics swarms (see Fig. 1), one can find the aggregation of robots [40] or of objects [5], the collective sorting of items [58], the dispersion of robots [71], the collaborative exploration of an environment [18], the localization of a target [47], the collaborative transport of objects [15], the collective selection of a place [37], the unsupervized allocation of tasks [54] and the coordination of displacements [67]. If today most of the studies on artificial swarm intelligent systems are less closely related to a biological model as it was in the past [57, 72], the major role played by biological discoveries to the emergence and the development of the field is largely acknowledged [9].

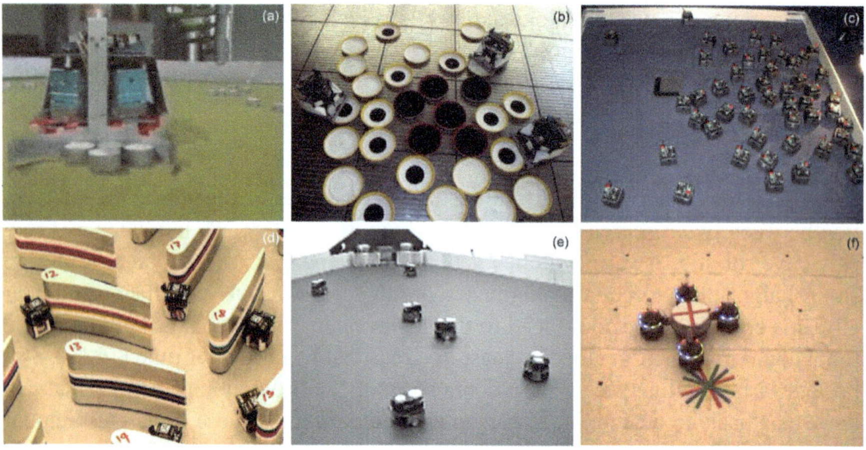

Fig. 1 Examples of swarm behaviors achieved with groups of robots. (a) Aggregation of objects [5]. (b) Sorting of objects [58]. (c) Dispersion of robots [71]. (d) Collaborative exploration of an environment [18]. (e) Localization of a target [47]. (f) Collaborative transport of objects [15].

The inverse statement, that robotics contributed to new discoveries about the functioning of animal societies, is however less common. As a biologist, I worked these last seven years with computer scientists in various laboratories. They most of the time expressed a high interest when I was talking about my biological studies. But speaking of these robotics collaborations with biologists did not always receive the same warm welcome. The criticisms targeted mostly the interest of robotics studies for producing new biological knowledge. They are often seen as over simplified version of animal experiments, or as slow and inconvenient alternatives to mathematical modelling and computer simulations. Robots are perceived as"cool" gadgets, but not as useful tools to study collective behaviors in social animals.

In this article, I discuss several arguments that support the interest of robotics for the study of social organization in biology. These arguments are based on a brief review of the literature on robotics and collective animal behavior, and on my own experience at the frontier between these two fields. My discussion is focused on five particular points that can provide biologists with new tools and new insights to investigate animal sociality. First, robots require a complete specification and, as in animals, their behavior strongly depends on the way they perceive and process information. Second, robots are physical entities that can interact with the real world, and hence with animals. Third, robots implements new technologies that can improve the collection of behavioral data. Fourth, robots can be source of biological questions as they adapt bio-inspired algorithms to their constraints and goals. And finally, robots are indeed "cool" gadgets that attract the attention of people.

2 Why Can Robots Be Useful for the Study of Social Behaviors?

2.1 Robots Require a Complete Specification

The study of collective behaviors in large animal groups makes a significant use of computer simulations and mathematical modelling [14]. Amé et al. [1], for instance, used a mean field modelling approach to explore collective decision making and choice optimality in the cockroach *Blattella germanica*. Couzin and Franks [20] studied traffic organization in the army ant *Eciton burchelli* using individual based computer simulations. More recently, the development of GPU computing drove simulations of complex biological systems to another scale [21], with the possibility to simulate millions of interacting virtual animals, as in Guttal et al. [45] for instance.

These methods are very efficient to understand the link that exists between the activities of individual animals and the global behavior of the group. They mostly rely on a probabilistic description of the animals' individual behaviors. As such they are a simplified description of the characteristics of the animals and of the environment where they live. Besides the obvious gains in performance and mathematical tractability these simplifications introduce, they are also useful to generalize to other species the coordination principles found in a particular one.

One could argue that such a reduced description would, however, skip over the differences between species that are peculiar products of their evolution and of their interactions with the environment. Argentine ants *Linepithema humile* and leaf cutter ant *Atta colombica* both use pheromones to establish trails between their nest and food sources. But while leaf cutter ants stick to a few long lasting trails leading to permanent food sources, Argentine ants rather establish temporary trails that are more adapted to their opportunistic foraging behavior [48]. The general behavior of these two species is the same (laying and following a trail), but slight variations in the pheromone composition, the accuracy to follow the trail [25] or the distribution of resources [23] in the environment are sufficient to generate striking differences. Similarly, it has been shown in bird flock models that changing the way neighbors are taken into account can have a high impact on the stability of the flock. Using topological relationships has been proved to be more efficient in this respect than using geometrical ones [4]. Thus by reducing an animal to a point for instance, and its perceptual apparatus to a circle around him, we somehow put aside the fact that the behavior of an animal is not the product of a simple finite state machine, but a more complex outcome of a perceptual, cognitive and locomotor activity [59, 26].

Today's biology emphasizes an integrative view of nature [13, 51]. The functioning of a cell, an organ or an organism are now observed in relation to higher and lower levels of organization. A cell is integrated in a tissue, which is integrated in an organ, which is integrated in an organism, and so on. And each event that affects one of these levels of organization can potentially cascade to the others. The same integrative view can be apply to the study of animal behavior. In this integrative perspective, robotics could play an important role. Because a robot has to be completely specified to work properly, it forces biologists to think about social behaviors in relation to individual skills of the animal. Several studies have already made use of robots to investigate the link between the individual behavior and the perceptual, cognitive and motor abilities of an animal (see for a review [82, 83, 84]). Ayers and Witting [2], for instance, have designed a robotics model of the American lobster *Homarus americanus*, including both biomechanical and neurological aspects of the animal. As another example, the Psikharpax project [60] aims at developing a robotics rat which sensory-motor equipment and neural control architecture imitate those of real rats. This integrative approach remains to be further extended toward the social level [3], where the exact nature of the social information perceived by an animal is not well understood.

2.2 Robots Are Physical Entities

Robots are not virtual. They are physical entities, and as such they interact with a real environment. This quality, that no simulation or mathematical model will ever have, allows their integration in experiments with real animals. In the 90's and early 2000's, the collaboration between biologists and computer scientists led to several groundbreaking works with human operated robots. For instance, Michelsen *et al.* [61] used an electromechanical device imitating the well known bee waggle dance. They studied the critical parts of this dance that are essential for the bee to communicate the position of a food source. Similar approaches have been used by

Böhlen [7] and Fernandez-Juricic [34] to study social communication in birds, and by Reaney *et al.* [68] to look at courtship signals that females take into account to choose a male in the crab *Uca mjoebergi*.

More recently, the use of reactive robots has opened new perspectives in the study of social behaviors. These new devices are able to adapt their behavior to the behavior of the animals they are interacting with. Varying the rules of interaction in the robot and observing the changes in the behavior of the animals is an interesting solution to systematically explore the social repertoire of the species under study. Moreover, these devices proved to be able to control the behavior of an animal group (see Fig. 2). They can act as reactive repellents (as a sheepdog would) [79] or, camouflaged appropriately, as fake conspecifics that the animals in the group would follow [46].

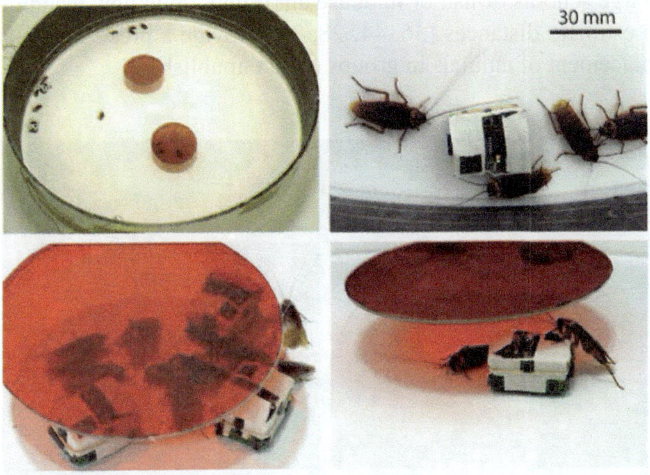

Fig. 2 Example of an experiment mixing animals (here cockroaches *Periplaneta americana*) and robots (here Insbots [74]). Reprinted from [46]

The use of robots in animal experiments is by far the most acknowledged and direct contribution of robotics to the study of social behaviors. In some way, they are modern and sophisticated versions of the lures used, for instance, by Tinbergen from the early 50s [77]. Maybe because of this historical connection, the use of robots as advanced lures has been largely accepted in the behavioral biology community, while pure robotics experiments often fail to convince it of their biological relevance.

2.3 Robots Implements New Technologies

When studying the behavior of an animal, gathering data, processing them and generating models from them can be a long lasting job. When it comes to studying the behavior of multiple individuals at once, and to identifying all the interactions between them, the task may seem overwhelming. But as claimed by Balch *et al.* [3], computer scientists can offer their help on these tasks. In particular, they have at their disposal technical advances likely to facilitate the collection of the large sets of data required in the study of collective animal behavior.

Among them, computer vision algorithms have led to the development of various softwares that automate several tasks in recording behavior. Two of these tasks that are particularly time consuming are animal tracking and behavior labeling. Several solutions already existed to track the movement of one individual or to recognize its behaviors [81, 64, 65]. But the development of collective robotics led them to a new level, with the ability to track dozens of animals at the same time and to automatically labels all the concurrent interactions between them [19, 3] (see Fig. 3). Even in conditions where an accurate tracking is not possible (as, for instance, in large schools of fish or in dense ant trails), optic flow techniques used in robotics to estimate distances [36, 44, 28, 70] provide a precise estimation of the global displacement of animals in groups (data unpublished).

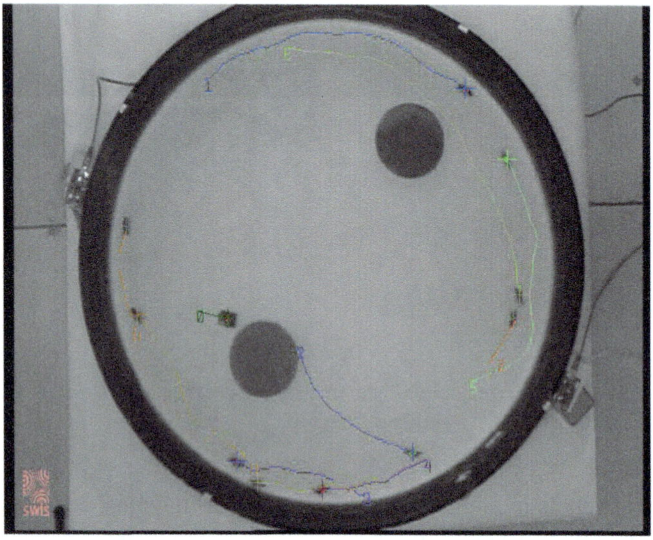

Fig. 3 Example of a versatile tracking software, Swistrack [19], that can track both robots and animals in the same experiment. Source: http://disal.epfl.ch/research/current/leurre/movies/collectivedecision.avi

Another example of devices to collect data about animal behavior comes from the introduction of Radio Frequency Identification (RFID) tags in robotics. These tags allow the development of multi-robot coordination systems where they are used to detect and store robot traffic data, as ants would lay and follow a pheromone trail in the wild [56, 55]. This technology is now slowly showing its potential in the study of collective animal behaviors. Attached to the back of insects, RFID tags allow biologists to precisely detect when an individual enter or exit an area in an experimental environment. This technique has so far been successfully applied to the study of social behaviors in ants [69], bumblebees [66] and paper wasps [73].

Biologists have mastered the art of observing and describing nature in details. But they often lack information about recent technological advances that may simplify their work. Thanks to their high technological competences, computer scientists are therefore likely to fill this gap and to provide new tools to answer questions about social behaviors in animals.

2.4 Robots Can Be Sources of Biological Questions

Natural systems are prolific sources of inspiration for computer scientists. Animal societies are no exception and there exists numerous examples of artificial systems inspired from collective animal behaviors [32, 9, 33]. Among them, ant colony optimization algorithms [27], particle swarm optimization algorithms [50] and bee algorithms [49] are probably the most famous. In each of these cases, engineers started by finding an equivalence between a question they wanted to solve and the computational properties of a biological system. Then, they adapted the mechanism at the origin of these biological properties to the constraints of the artificial system they were working on. During this adaptation phase, they introduced modifications to the original mechanism that may be seen as inadvertent questions addressed to biologists.

One example of these inadvertent questions can be found in the Ant Colony Optimization (ACO) framework. This optimization system, used to solve the travelling salesman problem for instance or the routing of information in communication networks, is inspired from the route selection mechanism discovered in ants. Virtual ants move across a network of interconnected nodes and search a route between two or more of these nodes. Each time an ant finds one of the possible routes, it lays along its path a virtual pheromone that fades with time and whose intensity is inversely proportional to the travelling time required to connect the nodes. When a virtual ant travels within the network and reaches a node, it chooses the next segment to follow in the network according to two parameters: the amount of pheromone on each segment and a heuristic weight that represents the intrinsic desirability of each segment. The higher the values for a segment, the more likely the ant is to choose it. While the amount of pheromone drives the collective choice of the colony through an amplification process, the heuristic weighting has been introduced in the algorithm to reduce the probability for the colony to select a loop. This addition greatly improves the efficiency of ant algorithms in selecting the shortest paths between two or more nodes in a network [27].

This practical consideration raises a very interesting issue in biology: do real ants use an equivalent of this heuristic component when moving within their transport networks, namely their underground nests or their foraging trails? We investigated this question in a recent work with Argentine ants [39, 42]. We observed the

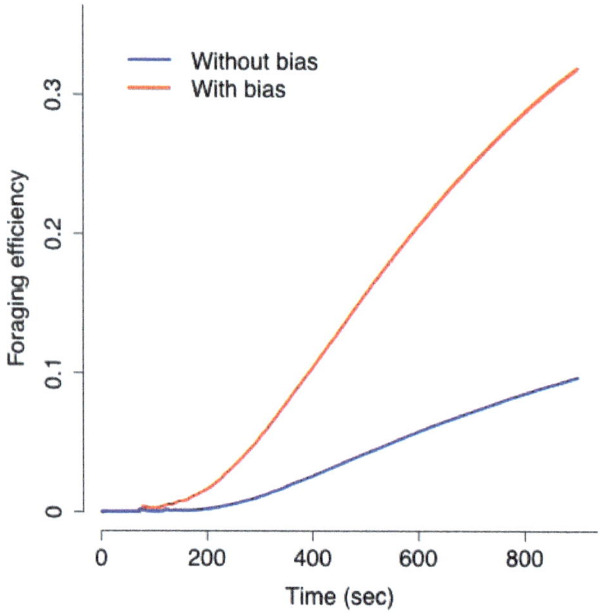

Fig. 4 Simulation results of ants foraging in a network of pheromone trails whose bifurcation geometry is mostly symmetrical during foodbound trips and mostly asymmetrical during nestbound trips. When ant choice at a bifurcation is biased by the bifurcation geometry, the overall foraging efficiency grows three times faster than in the absence of a behavioral bias. This efficiency gain is explained by the lower probability to select a loop in the network, a result similar to the one obtained with the introduction of the heuristic value in ACO algorithms. Adapted from [39]

individual behavior of ants while crossing symmetrical and asymmetrical bifurcations in gallery networks. In the absence of orientation cues, ants crossing a symmetrical bifurcation selected equally either branch that followed the bifurcation. On the other hand, 2/3 of the ants reaching an asymmetrical bifurcation chose the branch that deviates the least from their current heading. We studied with computer simulations the consequences of this latter bias on the pheromone-based collective path selection ability of ants. We simulated colonies of ants foraging within a trail network, the bifurcations of which mimicked those found in natural trail networks (mostly symmetrical during foodbound trips and mostly asymmetrical during nestbound trips). The simulation results show that the foraging efficiency of the colony is more than three times higher (see Fig. 4) after only 15 minutes when the choice of the ants was biased at asymmetrical bifurcations. This result can be mainly explained by the smallest probability for biased ants to select a loop in the network, as predicted by the ACO framework.

This example is not strictly speaking coming from the robotics field. But it illustrates nicely how current research on artificial multi-agent systems, where collective robotics pertains, can loop back to biology. It should encourage biologists to monitor more carefully this field where unexpected questions can emerge about the nature of collective animal behaviors.

2.5 *Robots Are "Cool" Gadgets*

Popularizing science discoveries and concepts is among the everyday tasks of a scientist. Spreading scientific information can be done, for instance, in the form of a lecture given at a university, an article in a general public review, an appearance in a radio or television show, or a demonstration in a science fair or museum. The latter is by far my favorite, since it allows the scientists and the audience to interact directly. However, for a scientist working with animals, this type of event can quickly turn into a nightmare. Ethical and practical questions inherent to the use of animals often limit the demonstration to the display of a poster or a video. Furthermore animals are not machines that execute tasks on command and, as a consequence, demonstration may be restricted to one or two replications a day.

On the contrary, robots are not subject to ethical restrictions and, provided that their battery is charged, they can work at any time of the day. Moreover, they share with animals the ability to trigger empathy in human observers, as attested by the increasing development of so-called "robot pets". They are therefore an ideal platform to explain concepts in animal behavior, when animals themselves cannot be used.

With this purpose in mind (among others), I designed with my colleagues an experimental platform that could demonstrate with robots the mechanism used by ants to select a route to a resource [41]. These insects lay and follow a pheromone trail on their way to or from a resource [48]. Initially, several trails can be formed and amplified by the successive deposits of ants travelling along them. A competition between the amplification processes on each trail often leads to the selection of one of them by the colony, in general the shorter one [14]. This mechanism of selection can be hard to demonstrate to an audience, mostly because ant pheromones are invisible. In our robotics platform, we therefore replace this chemical signal with a visual one, each pheromone drop being simulated by a spot of blue light projected on the ground (see Fig. 5). These spots could accumulate and evaporate, as pheromone drops would, and eventually form a trail. Our robots were equipped with two light sensors that mimicked ant antennae and could follow the trail by simply turning toward the sensor receiving the more light. With this setup, we could reproduce several classical and recent results of the literature (data unpublished), as the selection of the shortest route among two available [22], the ability to redirect a part of the traffic on another route when the main one is overcrowded [29, 30], or the role of the geometry of network bifurcations on the foraging efficiency [39]. More importantly, we could also perform several explanatory demonstrations of these phenomenons to officials, primary and high school classes and TV crews visiting the laboratory [52].

Fig. 5 Robots Alice [17] following a light trail. Adapted from [41]

If the interest of using such robotics experiments to generate new knowledge can be discussed (strictly speaking, our experiments confirmed previously published results), the attractive power that robots exert on human beings, and in particular on children, makes them interesting tools to explain the mechanisms underlying collective behaviors in animals. They are therefore excellent ambassadors to promote this field of research.

3 Conclusions

For thirty years, biologists and computer scientists have investigated coordination mechanisms that allow individuals to perform collectively a task that could not be achieved by a single one. This question is of particular importance for biologists, since it concerns the origins and the functioning of sociality in animals and human beings. For computer scientists, these mechanisms are involved in numerous complex problems ranging from optimization to distributed control. In several occasions, innovative solutions to these problems came from biological observations of animal societies. In this article, I tried to show how computer science, and robotics in particular, can reciprocally be useful to biologists that study collective animal behaviors.

Through the development of new tools, computer scientists can facilitate the collection of biological data (section 2.3) or propose original solutions to interact with animals (section 2.2). But their contribution to the study of collective animal behaviors can be more than a practical and technical help. They can also participate to the development of new ideas and concepts in biology. Robots are integrated platforms, and as such their behavior is the product of their perceptive, cognitive and motor capabilities, in relation with their environment. Therefore, they are interesting systems to combine the different levels of organization that shape collective behaviors (section 2.1). Moreover, improvements that engineers bring to bio-inspired solutions can echo through out the biological world (section 2.4). Finally, the entertaining and engaging character of robots, associated with the total control of their behavior, transform them in interesting instruments for educational purposes.

As it has occurred in the past, it is highly probable that biologists and computer scientists will collaborate in the future. However, the nature of these collaborations may be different. In the 90's, when a biologist and a computer scientist were talking about collective behaviors, their objective was most likely to develop new applications and algorithms from biological observations. Today, such a collaboration has an increasing probability to be dedicated to a biological problem instead, as shown by the success of recent projects [37, 46, 35, 63].

Acknowledgments

The author would like to thank Jennifer K. Peterson and Alexandre Campo for kindly reviewing this manuscript. Simon Garnier is currently at Princeton University were he is supported by a Fyssen grant and a Searle Scholar grant (on the behalf of Professor Iain Couzin). His work cited in this paper was performed while he was at the University of Toulouse (France) and was partly supported by an European community grant given to the Leurre project under the Information Society Technologies Programme (1998-2002), contract FET-OPEN-IST-2001-35506 of the Future and Emerging Technologies arm, by the Programme Cognitique from the French Ministry of Scientific Research and by a research grant from the French Ministry of Education, Research and Technology.

References

1. Amé, J.M., Halloy, J., Rivault, C., Detrain, C., Deneubourg, J.L.: Collegial decision making based on social amplification leads to optimal group formation. Proceedings of the National Academy of Sciences of the United States of America 103(15), 5835–5840 (2006), doi:10.1073/pnas.0507877103
2. Ayers, J., Witting, J.: Biomimetic approaches to the control of underwater walking machines. Philosophical transactions. Series A, Mathematical, physical, and engineering sciences 365(1850), 273–295 (2007), doi:10.1098/rsta.2006.1910
3. Balch, T., Dellaert, F., Feldman, A., Guillory, A., Isbell, C., Khan, Z., Pratt, S., Stein, A., Wilde, H.: How Multirobot Systems Research Will Accelerate Our Understanding of Social Animal Behavior. Proceedings of the IEEE 94(7), 1445–1463 (2006), doi:10.1109/JPROC.2006.876969

4. Ballerini, M., Cabibbo, N., Candelier, R., Cavagna, A., Cisbani, E., Giardina, I., Lecomte, V., Orlandi, A., Parisi, G., Procaccini, A., Viale, M., Zdravkovic, V.: Interaction ruling animal collective behavior depends on topological rather than metric distance: evidence from a field study. Proceedings of the National Academy of Sciences of the United States of America 105(4), 1232–1237 (2008), doi:10.1073/pnas.0711437105
5. Beckers, R., Holland, O.E., Deneubourg, J.L.: From local actions to global tasks: Stigmergy and collective robotics, pp. 181–189. MIT Press, Cambridge (1994)
6. Beshers, S.N., Fewell, J.H.: Models of division of labor in social insects. Annual Review of Entomology 46(1), 413–440 (2001), doi:10.1146/annurev.ento.46.1.413
7. Böhlen, M.: A robot in a cage. In: Proceedings of IEEE International Symposium on Computational Intelligence in Robotics and Automation. Monterey, CA (1999)
8. Bonabeau, E.: Self-organization in social insects. Trends in Ecology & Evolution 12(5), 188–193 (1997), doi:10.1016/S0169-5347(97)01048-3
9. Bonabeau, E., Dorigo, M., Theraulaz, G.: Swarm intelligence: from natural to artificial systems. Oxford University Press, USA (1999)
10. Buhl, J., Deneubourg, J.L., Grimal, A., Theraulaz, G.: Self-organized digging activity in ant colonies. Behavioral Ecology and Sociobiology 58(1), 9–17 (2005), doi:10.1007/s00265-004-0906-2
11. Buhl, J., Gautrais, J., Solé, R.V., Kuntz, P., Valverde, S., Deneubourg, J.L., Theraulaz, G.: Efficiency and robustness in ant networks of galleries. The European Physical Journal B 42(1), 123–129 (2004), doi:10.1140/epjb/e2004-00364-9
12. Burd, M.: Ecological consequences of traffic organisation in ant societies. Physica A: Statistical Mechanics and its Applications 372(1), 124–131 (2006), doi:10.1016/j.physa.2006.05.004
13. Cain, C.J., Conte, D.A., Garcia-Ojeda, M.E., Daglio, L.G., Johnson, L., Lau, E.H., Manilay, J.O., Phillips, J.B., Rogers, N.S., Stolberg, N.S., Swift, H.F., Dawson, M.N.: INTEGRATIVE BIOLOGY: What Systems Biology Is (Not, Yet). Science 320(5879), 1013a–1014a (2008), doi:10.1126/science.1157405
14. Camazine, S.: Self-organization in biological systems. Princeton University Press, Princeton (2001)
15. Campo, A., Nouyan, S., Birattari, M., Groß, R., Dorigo, M.: Negotiation of Goal Direction for Cooperative Transport, pp. 191–202. Springer, Heidelberg (2006), doi:10.1007/11839088_17
16. Cao, Y.U., Fukunaga, A.S., Kahng, A.B.: Cooperative mobile robotics: Antecedents and directions. Autonomous Robots 4(1), 7–27 (1997)
17. Caprari, G., Siegwart, R.: Mobile micro-robots ready to use: Alice. In: IEEE/RSJ International Conference on Intelligent Robots and Systems, pp. 3295–3300. IEEE, Los Alamitos (2005)
18. Correll, N., Martinoli, A.: Robust Distributed Coverage using a Swarm of Miniature Robots. In: Proceedings 2007 IEEE International Conference on Robotics and Automation, pp. 379–384. IEEE, Los Alamitos (2007), doi:10.1109/ROBOT.2007.363816
19. Correll, N., Sempo, G., De Meneses, Y., Halloy, J., Deneubourg, J.I., Martinoli, A.: SwisTrack: A Tracking Tool for Multi-Unit Robotic and Biological Systems. In: 2006 IEEE/RSJ International Conference on Intelligent Robots and Systems, pp. 2185–2191 (2006), doi:10.1109/IROS.2006.282558
20. Couzin, I.D., Franks, N.R.: Self-organized lane formation and optimized traffic flow in army ants. Proceedings of The Royal Society Biological sciences 270(1511), 139–146 (2003), doi:10.1098/rspb.2002.2210
21. Dematté, L., Prandi, D.: GPU computing for systems biology. Briefings in Bioinformatics 11(3), 323–333 (2010), doi:10.1093/bib/bbq006

22. Deneubourg, J.L., Goss, S.: Collective patterns and decision making. Ethology, ecology and Evolution 1(4), 295–311 (1989)
23. Deneubourg, J.L., Goss, S., Franks, N., Pasteels, J.M.: The blind leading the blind: Modeling chemically mediated army ant raid patterns. Journal of Insect Behavior 2(5), 719–725 (1989), doi:10.1007/BF01065789
24. Deneubourg, J.L., Goss, S., Pasteels, J.M., Fresneau, D., Lachaud, J.P.: Self-organization mechanisms in ant societies (II): Learning in foraging and division of labor. Experientia Supplementum 54, 177–196 (1987)
25. Deneubourg, J.L., Pasteels, J.M., Verhaeghe, J.C.: Probabilistic behaviour in ants: A strategy of errors? Journal of Theoretical Biology 105(2), 259–271 (1983), doi:10.1016/S0022-5193(83)80007-1
26. Detrain, C., Deneubourg, J.L.: Self-organized structures in a superorganism: do ants "behave" like molecules? 3(3), 162–187 (2006), doi:10.1016/j.plrev.2006.07.001
27. Dorigo, M., Stützle, T.: Ant Colony Optimization. MIT Press, Cambridge (2004)
28. Duchon, A., Warren, W., Kaelbling, L.: Ecological Robotics: Controlling Behavior with Optic Flow. In: Proceedings of the seventeenth annual conference of the Cognitive Science Society, p. 164. Lawrence Erlbaum, Pittsburgh (1995)
29. Dussutour, A., Fourcassié, V., Helbing, D., Deneubourg, J.L.: Optimal traffic organization in ants under crowded conditions. Nature 428(6978), 70–73 (2004), doi:10.1038/nature02344.1
30. Dussutour, A., Nicolis, S., Deneubourg, J.L., Fourcassié, V.: Collective decisions in ants when foraging under crowded conditions. Behavioral Ecology and Sociobiology 61(1), 17–30 (2006), doi:10.1007/s00265-006-0233-x
31. Dutta, I., Bogobowicz, A.D., Gu, J.J.: Collective robotics - a survey of control and communication techniques. In: Proceedings International Conference on Intelligent Mechatronics and Automation, pp. 505–510 (2004)
32. Eberhart, R.C., Shi, Y., Kennedy, J.: Swarm Intelligence. Morgan Kaufmann, San Francisco (2001)
33. Engelbrecht, A.: Fundamentals of computational swarm intelligence. Wiley, New York (2005)
34. Fernandez-Juricic, E., Gilak, N., McDonald, J.C., Pithia, P., Valcarcel, A.: A dynamic method to study the transmission of social foraging information in flocks using robots. Animal Behaviour 71, 901–911 (2006), doi:10.1016/j.anbehav.2005.09.008
35. Floreano, D., Mitri, S., Magnenat, S., Keller, L.: Evolutionary Conditions for the Emergence of Communication in Robots. Current Biology 17(6), 514–519 (2007)
36. Franceschini, N., Pichon, J.M., Blanes, C., Brady, J.M.: From Insect Vision to Robot Vision [and Discussion]. Philosophical Transactions of the Royal Society B: Biological Sciences 337(1281), 283–294 (1992), doi:10.1098/rstb.1992.0106
37. Garnier, S., Gautrais, J., Asadpour, M., Jost, C., Theraulaz, G.: Self-Organized Aggregation Triggers Collective Decision Making in a Group of Cockroach-Like Robots. Adaptive Behavior 17(2), 109–133 (2009), doi:10.1177/1059712309103430
38. Garnier, S., Gautrais, J., Theraulaz, G.: The biological principles of swarm intelligence. Swarm Intelligence 1(1), 3–31 (2007), doi:10.1007/s11721-007-0004-y
39. Garnier, S., Guérécheau, A., Combe, M., Fourcassié, V., Theraulaz, G.: Path selection and foraging efficiency in Argentine ant transport networks. Behavioral Ecology and Sociobiology 63(8), 1167–1179 (2009), doi:10.1007/s00265-009-0741-6
40. Garnier, S., Jost, C., Gautrais, J., Asadpour, M., Caprari, G., Jeanson, R., Grimal, A., Theraulaz, G.: The embodiment of cockroach aggregation behavior in a group of microrobots. Artificial Life 14(4), 387–408 (2008)

41. Garnier, S., Tâche, F., Combe, M., Grimal, A., Theraulaz, G.: Alice in Pheromone Land: An Experimental Setup for the Study of Ant-like Robots. In: IEEE Swarm Intelligence Symposium, SIS 2007, pp. 37–44 (2007)
42. Gerbier, G., Garnier, S., Rieu, C., Theraulaz, G., Fourcassié, V.: Are ants sensitive to the geometry of tunnel bifurcation? Animal Cognition 11(4), 637–642 (2008), doi:10.1007/s10071-008-0153-4
43. Gordon, D.M.: The organization of work in social insect colonies. Nature 380(6570), 121–124 (1996), doi:10.1038/380121a0
44. Green, W.E., Oh, P.Y., Sevcik, K., Barrows, G.: Autonomous Landing for Indoor Flying Robots Using Optic Flow. In: ASME International Mechanical Engineering Congress, vol. 2, pp. 1347–1352. ASME, Washington (2003)
45. Guttal, V., Couzın, I.D.: Social interactions, information use, and the evolution of collective migration. Proceedings of the National Academy of Sciences of the United States of America, 1–6 (2010) (in press), doi:10.1073/pnas.1006874107
46. Halloy, J., Sempo, G., Caprari, G., Rivault, C., Asadpour, M., Tâche, F., Said, I., Durier, V., Canonge, S., Amé, J.M., Detrain, C., Correll, N., Martinoli, A., Mondada, F., Siegwart, R., Deneubourg, J.L.: Social Integration of Robots into Groups of Cockroaches to Control Self-Organized Choices. Science 318(5853), 1155–1158 (2007), doi:10.1126/science.1144259
47. Hayes, A., Martinoli, A., Goodman, R.: Distributed odor source localization. IEEE Sensors Journal 2(3), 260–271 (2002), doi:10.1109/JSEN.2002.800682
48. Hölldobler, B., Wilson, E.O.: The ants. Belknap Press of Harvard University Press, Cambridge (1990)
49. Karaboga, D., Akay, B.: A survey: algorithms simulating bee swarm intelligence. 31(1-4), 61–85 (2009), doi:10.1007/s10462-009-9127-4
50. Kennedy, J., Eberhart, R.: Particle swarm optimization. In: Proceedings of ICNN 1995 - International Conference on Neural Networks, pp. 1942–1948. IEEE, Piscataway (1948), doi:10.1109/ICNN.1995.488968
51. Kitano, H.: Systems biology: a brief overview. Science 295(5560), 1662–1664 (2002), doi:10.1126/science.1069492
52. Kneser, J.: Collective Minds: The Intelligence of Swarms (2009)
53. Kube, C.R., Zhang, H.: Collective Robotics: From Social Insects to Robots. Adaptive Behavior 2(2), 189–218 (1993), doi:10.1177/105971239300200204
54. Labella, T.H., Dorigo, M., Deneubourg, J.L.: Division of Labor in a Group of Robots Inspired by Ants' Foraging Behavior. ACM Transactions on Autonomous and Adaptive Systems 1(1), 4–25 (2006)
55. Mamei, M., Zambonelli, F.: Spreading pheromones in everyday environments through RFID technology. In: 2nd IEEE Symposium on Swarm Intelligence, pp. 281–288. Citeseer (2005)
56. Mamei, M., Zambonelli, F.: Pervasive pheromone-based interaction with RFID tags. ACM Transactions on Autonomous and Adaptive Systems 2(2), 4–es (2007), doi:10.1145/1242060.1242061
57. Martinoli, A.: Collective Complexity out of Individual Simplicity. Artificial Life 7(3), 315–319 (2001)
58. Melhuish, C., Wilson, M., Sendova-Franks, A.: Patch sorting: Multi-object clustering using minimalist robots, pp. 543–552. Springer, Heidelberg (2001), doi:10.1007/3-540-44811-X_62
59. Menzel, R., Giurfa, M.: Cognitive architecture of a mini-brain: the honeybee. 5(2), 62–71 (2001)

60. Meyer, J., Guillot, A., Girard, B., Khamassi, M., Pirim, P., Berthoz, A.: The Psikharpax project: towards building an artificial rat. Robotics and Autonomous Systems 50(4), 211–223 (2005), doi:10.1016/j.robot.2004.09.018
61. Michelsen, A., Andersen, B.B., Storm, J., Kirchner, W.H., Lindauer, M.: How honeybees perceive communication dances, studied by means of a mechanical model. Behavioral Ecology and Sociobiology 30(3), 143–150 (1992), doi:10.1007/BF00166696
62. Mikheyev, A.S., Tschinkel, W.R.: Nest architecture of the ant Formica pallidefulva: structure, costs and rules of excavation. Insectes Sociaux 51(1), 30–36 (2004), doi:10.1007/s00040-003-0703-3
63. Mitri, S., Floreano, D., Keller, L.: The evolution of information suppression in communicating robots with conflicting interests. Proceedings of the National Academy of Sciences of the United States of America 106(37), 15786–15790 (2009), doi:10.1073/pnas.0903152106
64. Noldus, L.P.J.J., Spink, A.J., Tegelenbosch, R.A.: EthoVision: a versatile video tracking system for automation of behavioral experiments. Behavior Research Methods, Instruments, & Computers 33(3), 398–414 (2001)
65. Noldus, L.P.J.J., Trienes, R.J.H., Hendriksen, A.H.M., Jansen, H., Jansen, R.G.: The Observer Video-Pro: New software for the collection, management, and presentation of time-structured data from videotapes and digital media files. Behavior Research Methods, Instruments, & Computers 32(1), 197–206 (2000)
66. Ohashi, K., D'Souza, D., Thomson, J.: An automated system for tracking and identifying individual nectar foragers at multiple feeders. Behavioral Ecology and Sociobiology 64(5), 891–897 (2010)
67. Pugh, J., Martinoli, A.: Small-scale robot formation movement using a simple on-board relative positioning system. In: Proceedings of the International Symposium on Experimental Robotics (2006)
68. Reaney, L.T., Sims, R.A., Sims, S.W.M., Jennions, M.D., Backwell, P.R.Y.: Experiments with robots explain synchronized courtship in fiddler crabs. Current Biology 18(2), R62–R63 (2008)
69. Robinson, E.J.H., Richardson, T.O., Sendova-Franks, A.B., Feinerman, O., Franks, N.R.: Radio tagging reveals the roles of corpulence, experience and social information in ant decision making. Behavioral Ecology and Sociobiology 63(5), 627–636 (2008), doi:10.1007/s00265-008-0696-z
70. Ruffier, F., Franceschini, N.: Optic flow regulation: the key to aircraft automatic guidance. Robotics and Autonomous Systems 50(4), 177–194 (2005), doi:10.1016/j.robot.2004.09.016
71. Schwager, M., McLurkin, J., Rus, D.: Distributed coverage control with sensory feedback for networked robots, p. 49. The MIT Press, Cambridge (2007)
72. Sharkey, A.J.C.: Swarm robotics and minimalism. Connection Science 19(3), 245–260 (2007), doi:10.1080/09540090701584970
73. Sumner, S., Lucas, E., Barker, J., Isaac, N.: Radio-Tagging Technology Reveals Extreme Nest-Drifting Behavior in a Eusocial Insect. Current Biology 17(2), 140–145 (2007), doi:10.1016/j.cub.2006.11.064
74. Tâche, F., Asadpour, M., Caprari, G., Karlen, W., Siegwart, R.: Perception and behavior of InsBot: Robot-Animal interaction issues. In: IEEE International Conference on Robotics and Biomimetics - ROBIO, pp. 517–522. IEEE, Los Alamitos (2005), doi:10.1109/ROBIO.2005.246321
75. Theraulaz, G., Bonabeau, E., Deneubourg, J.L.: Response threshold reinforcements and division of labour in insect societies. Proceedings of the Royal Society B: Biological Sciences 265(1393), 327–332 (1998), doi:10.1098/rspb.1998.0299

76. Theraulaz, G., Gautrais, J., Camazine, S., Deneubourg, J.L.: The formation of spatial patterns in social insects: from simple behaviours to complex structures. Philosophical transactions. Series A, Mathematical, physical, and engineering sciences 361(1807), 1263–1282 (2003), doi:10.1098/rsta.2003.1198
77. Tinbergen, N., Perdeck, A.: On the Stimulus Situation Releasing the Begging Response in the Newly Hatched Herring Gull Chick (Larus Argentatus Argentatus Pont.). Behaviour 3(1), 1–39 (1951), doi:10.1163/156853951X00197
78. Tschinkel, W.R.: Subterranean ant nests: trace fossils past and future? Palaeogeography, Palaeoclimatology, Palaeoecology 192(1-4), 321–333 (2003), doi:10.1016/S0031-0182(02)00690-9
79. Vaughan, R., Stumpter, N., Henderson, J., Frost, A., Cameron, S.: Experiments in automatic flock control. Robotics and Autonomous Systems 31, 109–117 (2000)
80. Vittori, K., Talbot, G., Gautrais, J., Fourcassié, V., Araújo, A.F.R., Theraulaz, G.: Path efficiency of ant foraging trails in an artificial network. Journal of Theoretical Biology 239(4), 507–515 (2006), doi:10.1016/j.jtbi.2005.08.017
81. Vrinten, D.H., Hamers, F.F.T.: 'CatWalk' automated quantitative gait analysis as a novel method to assess mechanical allodynia in the rat; a comparison with von Frey testing. Pain 102(1-2), 203–209 (2003)
82. Webb, B.: What does robotics offer animal behaviour? Animal Behaviour 60, 545–558 (2000), doi:10.1006/anbe.2000.1514
83. Webb, B.: Can robots make good models of biological behaviour? Behavioral and Brain Sciences 24(06), 1033–1050 (2001), doi:10.1017/S0140525X01000127
84. Webb, B.: Using robots to understand animal behavior, vol. 38, ch.1, pp. 1–58. Academic Press, London (2008), doi:10.1016/S(X)65-3454(08)00001-6
85. Yan, Y., Tang, Z.: Control Architecture for Autonomous Multi-Robot System: Survey and Analysis. In: International Conference on Intelligent Computation Technology and Automation, USA, vol. 4, pp. 376–379. IEEE Computer Society Press, Los Alamitos (2009)

Part II: Self-Reconfigurable Modular Robots

On Self-Optimized Self-Assembling of Heterogeneous Multi-robot Organisms

Serge Kernbach, Benjamin Girault, and Olga Kernbach

Abstract. This chapter is devoted to a bio-inspired self-assembling of heterogeneous robot modules into specific topological configurations. The approach involves several algorithmic inspirations from biological regulatory networks for achieving environmental dependability and considers constraint-based optimization techniques for finding optimal connections between heterogeneous modules. Scalability and locality of sensor information are addressed.

1 Introduction

Self-assembling is an important process, where a disordered set of existing components aggregates into a well-ordered structure by following simple assembling rules [1]. Such a process takes place on macro- or micro- levels without external guidance by utilizing several self-organizing mechanisms. A great source of inspiration for self-assembling algorithms is a molecular self-assembling, which appears in forms of e.g. crystals, colloids, or self-assembled monolayers [2]. Macroscopic self-assembling is primarily related to a robot research, where robot modules aggregate into complex topological structures to achieve a flexible functionality [3].

A substantial difference between micro- and macro- self-assembling consists in the size and capabilities of components as well as in the appearing problems and challenges. On the micro-level such components are molecules utilizing chemical bounding forces, whereas on the macro-level, components are robot modules, capable of docking with each other [4]. These modules can have very simple form [5], or possess several cognitive features, such as sensing, objects detection/recognition, actuation, communication and others [6]. Microscopic and

Serge Kernbach · Benjamin Girault · Olga Kernbach
Institute of Parallel and Distributed Systems, University of Stuttgart,
Universitätstr. 38, 70569 Stuttgart, Germany
e-mail: {serge.kernbach,giraulbn,
olga.kernbach}@ipvs.uni-stuttgart.de

Y. Meng and Y. Jin (Eds.): Bio-Inspired Self-Organizing Robotic Systems, SCI 355, pp. 123–141.
springerlink.com

macroscopic self-assembling targets also different objectives. When self-assembling on the micro-level is normally a large-scale phenomenon appearing as a uniform structural pattern [7], on the macro-level we are interested primarily in creating dedicated low-scale structures with a desired functionality. Typically, these structures are aggregated robots, which demonstrate locomotive functionality in a legged, clambering or rolling form [8].

Despite differences, micro- and macro- self-assembling shares also several common problems. One of them is a recruitment of new elements for assembling, where we can find several bio- and chemo- inspired works [9] leading to an efficient recruitment strategy [10]. However, the major effort on both levels is related to a guided self-assembling [11]: formation into a desired final form and, as a consequence, a need of multiple optimization steps, required to obtain such a form. Artificial programmability and self-optimization can be addressed in several ways: we used here some ideas from gene regulatory networks [12] and its multiple algorithmic inspirations, e.g. [13], [14]. Another interesting scientific challenge is the distribution of self-regulating mechanisms and integration of multiple constraints, appearing during the assembling process.

This chapter is based on the previous works [15], [16] and extends them towards two-steps optimization, which happen during the expression of high-level topological descriptions into a concrete configuration and an assembling of modules into this configuration. Sec. 2 gives a common pictures of self-assembling procedure and discusses main optimization steps. Sec. 3 introduces several constraints, which appear during robot-robot assembling, and Sec. 4 considers constraint-optimization approach. In Sec. 5 we focus on specific tasks such as grouping, or scaling of topologies. Finally, Sec. 6 demonstrates some results and Sec. 7 concludes this work.

2 General Self-Assembling Scenario

Problems of robot self-assembling and self-disassembling are well-known in reconfigurable robotics, see e.g. [17] or [18]. Here several high- and low- dimensional approaches [15],[19] are distinguished. To be more strong in definitions, we define self-assembling as a process, where robot modules R_i establish multiple bilateral connections $R_k : R_p$ (denoted as docking between modules R_k and R_p, see Fig. 1(a)), which step by step lead finally to an appearance of the topology Φ.

Each topology Φ has some macroscopic functionality; more exactly, each particular connection $R_k : R_p$ introduces a degree of freedom (DoF) φ_i. As shown in Fig. 1(b), each connection between modules introduces only one DoF; by a combination of all DoF, an organism is capable of a collective movement. We express interactions between all φ_i as a macroscopic functionality F of an artificial organism. Figs. 1(c) and 1(d) provide two examples of such rotating and wheeled macroscopic functionalities. The relationship between Φ and F is complex and depends not only on the involved DoFs, but also on the environment. We discuss these issues in the second comment below.

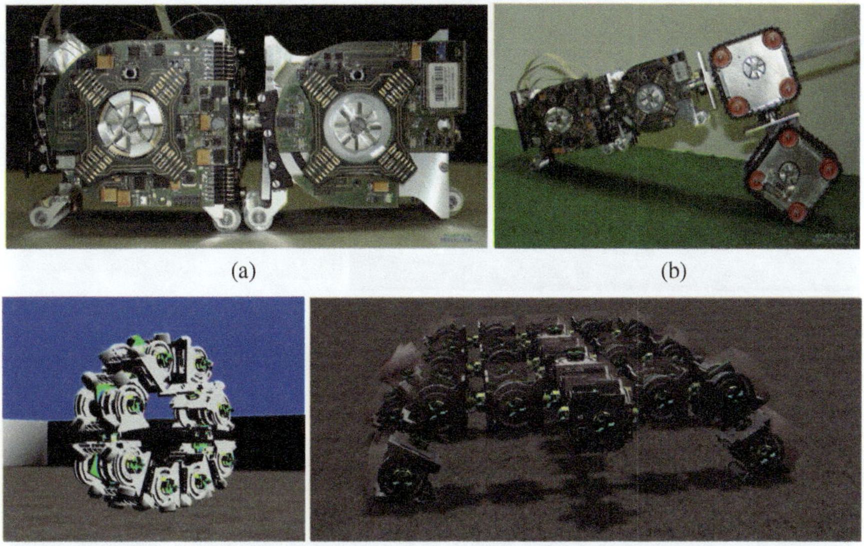

(a) (b) (c) (d)

Fig. 1 **(a)** Examples of a bilateral connection $R_k : R_p$; **(b)** Several 1DoF connections produce a macroscopic locomotion; **(c,d)** Two examples of macroscopic functionalities, achieved by many robot modules.

The process of self-assembling can be split on different steps, which are sketched in Fig. 2. Firstly, different types of robot modules R_i and other objects (such as energy "cubes") are placed on the arena, see Fig. 2(a). In this stage all robots exist in the so-called swarm mode, i.e. as independent robots. On the second step, they select from possible existing topologies a small subset Φ^S, which is optimal for given environment conditions. Such modules, which are needed for these topologies, group in some area of arena, see Fig. 2(b). This grouping approach is described in Sec. 5.1. After this first optimization, robots perform the second optimization, where a set of Φ^S is reduced to one Φ. Robots take into account the number of modules in the aggregation site, their capabilities and availability for the desired functionality F. Finally, the chosen modules queue up in a right spatial order for docking, Fig. 2(c), and dock into an organism Φ, see Fig. 2(d). Here, robots utilize different recruitment approaches, e.g. [10], to add a robot into an existing topology. After docking, all robot modules exist in the so-called organism mode, i.e. they are co-dependent on each other.

The general scenario, shown in Fig. 2, indicates only the main stages during the real robot self-assembling. We have to make two comments for this scenario.

1. On-line/Off-line issue. The first comment is related to the way of how to obtain the final topology Φ. This approach originates from [15] and consists in the following idea. The set of pre-generated building blocks for patterns $\{\Phi\}$ can be maximized so that to cover the most functionally useful behaviors in different predictable environmental situations. For example, this can be performed by off-line

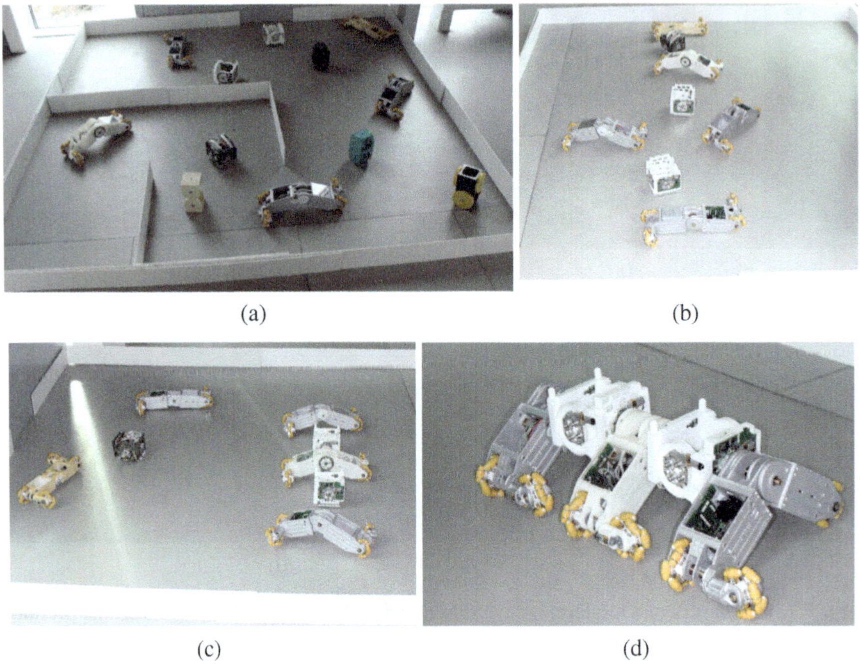

Fig. 2 Different envisaged steps of the self-assembling scenario. (a) Initial stage – all robots are irregularly distributed on the arena; (b) Approaching of the selection modules for docking; (c) Ordering of modules in 2D disconnected form; (d) Docking of modules on 2D grid into an assembled organism and collective actuation into 3D state.

evolutionary simulation, by utilizing bio-inspired mechanisms from insect or animals, and in general is done *in advance*. These patterns can structurally be perturbed by a few modules *on-line* depending on the environment. This perturbation creates some deviation in the expected functionality and behavior, which can be handled by different on-line adaptive mechanisms. This finally reduces the set $\{\Phi\}$ to some Φ_1, Φ_2, ..., Φ_m, which makes sense in a concrete environment. We will denote this smaller subset as Φ^S. Thus, the problem of finding a specific solution is narrowed down to the problem of optimizing a deviation from one of the pre-generated patterns, for which all controlling mechanisms exist. Since a linear optimization is very fast, for example, the linear sum assignment problem is of $O(n^3)$ complexity [20], this approach can be run on-board and on-line.

2. Topology, Functionality and Environment. In general, the degrees of freedom φ_i between robots $R_k : R_p$ depend on both R_k and R_p, i.e., we can encounter the situation when both R_k, R_p are relevant, one of them is relevant and none of them are relevant, see more in [16]. Moreover, a common functionality F of an organism is determined by interactions between all φ_i. It is hardly possible to say in advance which functionality can be useful or not for a particular environment. There are two possibilities to explore a usefulness of functionalities: performing on/off-boar simulation and by trial-and-error approach with different topologies in real environment. Both have advantages and drawbacks and are used for finding Φ. Essential scientific challenge here represents a programmability of self-assembling and its dependability on environment (constraints of robots). We need to find a balance between desirability (design goals) of topologies and capabilities to adapt to fluctuations of environment.

Thus, the self-assembling requires two-steps optimization:

- Environment-dependable generation of topologies from $\{\Phi\}$ to Φ^S. We can denote this process as expression of topology from some general building-blocks-like description (by analogy with gene expression process) into a set of topologies with desired functional properties. Example of such approach is discussed in [15], where we defined a set of operators and basic elements for a generation of topologies.
- On the second step, Φ^S is optimized taking into account a *current* situation in terms of the number of robots in the assembling area, availability for docking, structure of local environment, assembling dynamics (for example amount of collisions) and other conditions. As a result, robots firstly produce a final topology Φ, and secondly start a docking approach.

These two steps are demonstrated in Fig. 3. The whole approach consists of three functional parts ($F1-F3$) and three descriptive parts ($D1-D3$). Each functional part represents in fact a controller, running on-board of a robot module. It takes as input a corresponding description, executes all necessary activities and generates an output description. After this, the control over the robot(s) is passed to the next controller. Thus, descriptions $D1-D3$ are interfaces between controllers and controllers themselves work to some extent as operators over $D1-D3$. The main ideas of $F1$ and $F3$ controllers are already represented in [15] and [10], further we concentrate on the $F2$ optimization controller and introduce it in more detail.

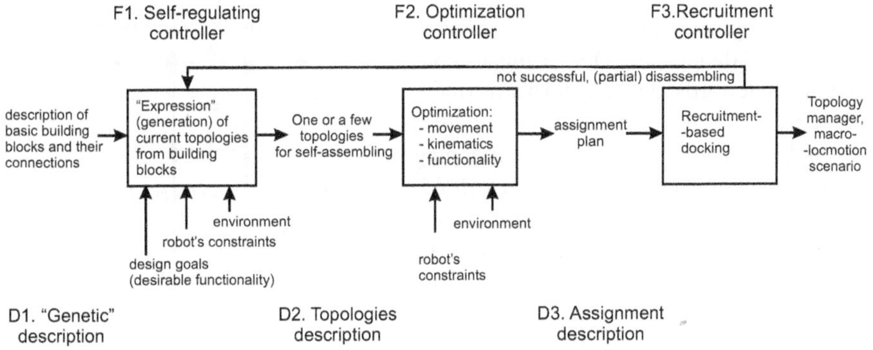

Fig. 3 General sketch of the self-assembling scenario.

3 Optimization Controller: Transition from Φ^S into Φ and the Role of Constraints

The self-assembling on the stage of optimization controller aims at several tasks: grouping of robots, identification and collection of constraints, selection of topologies, which satisfy these constraints, and finally a generation of one concrete topology Φ from a set of possible topologies Φ^S. This is done in several ways. Firstly, Φ^S is compared with the current situation. For example, when other robots are already building some structure, this structure has a higher priority in the decision process. When there are no such structures, a robot compares current positions of known robots in order to decide which pattern is the most suitable for this configuration or which pattern is the most suitable for the given surface. When a pattern is selected, a robot solves the constrained assignment problem in order to determine a position in this pattern, where it goes to dock.

Secondly, when a robot decides about one of the patterns in Φ^S, it is not constrained by this pattern, a robot can perturb the pattern ϕ_i by templates. This can happen only in two following cases. In the first case, the pattern, generated for n robots, already has n robots. The $n+1$ robot, joining to this organism, starts building a new scalability core. For example, when an organism with n robots has four legs, the $n+1$ robot can start building additional legs. In the second case, based on observation in the environment, a robot can estimate a need of specific perturbations, e.g. to make legs longer to overstep some obstacles. However, the difficulty is that other robots may not know about this initiative and the whole pattern become desynchronized. Solution of this problem involves more communication between robots, as suggested in [15].

A topology Φ of an organism is represented by the connection placeholder $R_k : R_p$ and DoF functionality in this connection. For example in Fig 4, six modules are shown, this topology is described by five connection placeholders. In each $R_k : R_p : x$, the last "x" denotes the DoF: 1 means connection sidewards, 2 means forwards (there are only two ways to connect two modules R_1 and R_2). The generation of the mapping between robots and connection placeholders represents a classical constrained assignment problem. Choosing a partner R_i for docking, i.e. the generation of $x_l^i \leftrightarrow x_k^j$ is independent of each other. It underlies several requirements: firstly it should satisfy local constraints υ_i, secondly, each $x_l^i \leftrightarrow x_k^j$ pair has an associated local cost, and finally, the whole appearing topology has its own global costs.

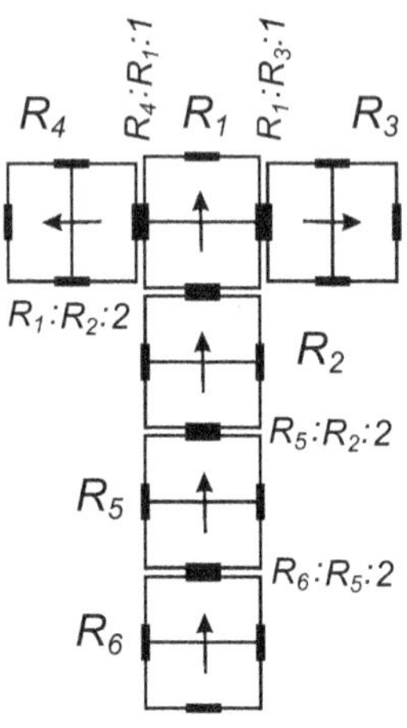

Fig. 4 Topology of an organism, $R_k : R_p : x$ is a connection placeholder.

1. Connectivity Constraints. As mentioned, there are multiple constraints, imposed on connectivity, kinematic properties, heterogeneity and others. Generally, the connectivity means the number of elements, connected to each of modules. For example, the central element of the cross has the connectivity 4 (modules connected from each side). Connectivity constrains the number of connections and can be effectively utilized in a description of topologies. When c_i is the connectivity of the i-element, where i goes from 1 to n (n is the number of robots in the topology; in contrast N is a total number of robots), the topology Φ can be described as $n+1$ set $(c_1, c_2, ..., c_n, c_t)$, c_t is a total number of connections in the topology with n robots. In general case, max. of c_i is equal to the maximal connectivity of the platform. All c_i are re-ordered from c_{max} to c_{min} so that the first element c is always that one, which has a maximal degree of connectivity. The topology Φ can be described as

$$\Phi = \{c_{max}, c_{max-1} ..., c_{min+1}, c_{min}, c_t\}, c_i \in \{1,2,3,4\}. \tag{1}$$

The description, defined by (1) has different topological properties, see more in [16]. Generally, there are basic topologies, which are unique, provided the topology is coherent (coherent topology = no disconnected nodes). To eliminate disconnected topologies, a coherency constraint has to be integrated into *Constraint Satisfaction Problem* (*CSP*) and the *Constraint Optimization Problem* (*COP*) solver. Basic topologies can be perturbed by one or several modules, this increases n and c_t. Such perturbed topologies are not unique. One of possible ways to deal with perturbed topologies is indicated in [15].

2. Kinematic Constraints. Topology Φ defined by (1) creates connections, which are invariant to robot's IDs. To integrate kinematics into topology, Φ should be supplemented with a functional description: it means to involve the desired degrees of freedom φ_i for a particular connection. Since each node has max. four connections (i.e. in general case different φ_i), the functional topology should include all of them. We use the agreement, that when only one φ is specified for a connectivity, it means $\varphi_i = \varphi$. Now we can generalize Φ from (1):

$$\Phi = ((c_{max} : \{\varphi\}_{max}), (c_{max-1} : \{\varphi\}_{max-1}), ..., (c_{min} : \{\varphi\}_{min}), c_t) \tag{2}$$

Thus, φ defines a kinematic constraint imposed on a $R_k : R_p$ link. To be more formal, we introduce several following properties.

Definition 1 (Functional Constraint). A constraint φ is a *functional constraint* if it verifies:

$$\varphi = ((T_1, D_1), (T_2, D_2)) \tag{3}$$

with T_i the type of the module i and D_i the type of the dock of module i participating in the link, module 1 being the module having the constraint and module 2 being the other module.

From the viewpoint of functional constraints, the topology Φ can be then defined as:

$$\Phi = ((c_1, (\varphi_{1,1}, \dots, \varphi_{1,c_1})), \dots, (c_n, (\varphi_{n,1}, \dots, \varphi_{n,c_n}))) \tag{4}$$

The variable (see description of the CSP approach in Sec. 4) whose value is true means that there is a link between the nodes, and if its value is false, then there is no such a link. The cost of the link is the cost of docking for two nodes if no dynamical obstacles were present in the environment. The constraints are the tricky part of this optimization step. First, we want an acyclic connected topology, which means a topology without loops. Such a topology is achieved when the two following properties hold:

Property 1 (Connected Topology Characterization). A topology of n nodes is connected if and only if every subset of $i \leq \lfloor \frac{n}{2} \rfloor$ nodes has at least one link involving one of its node to one node of the complementary subset of nodes in the group.

Property 2 (Acyclic Connected Graph Characterization). A connected topology is acyclic if and only if it has exactly $n-1$ links between two nodes.

Such a topology is indeed a tree. Writing the constraints for the LP solver from those two properties is now easy and straightforward. There will be one constraint per subset of $i \leq \lfloor \frac{n}{2} \rfloor$, and one constraint to check that the topology is acyclic. Thus, the topology generated is ensured to be a single tree.

Finally, we include the functional constraints. To do that we need to split the whole optimization process in two phases: the Constraint Satisfaction and the Constraint Optimization Problems. CSP consists of the running LP solver giving an assignation of the constraints to the nodes. We suppose that this assignation is valid, *i.e.*, the type of node is compatible with the type constraints. Then we add a constraint ensuring that there are enough links satisfying functional constraints. For instance, if the node i needs to dock to at least N_A of type A and N_B of type B, then there will be two more constraints verifying that the number of links to the node i of type A (resp. B) is greater of equal to N_A (resp. N_B).

3. Local costs. Robots have different on-board capabilities to measure distances between robots, their orientation, relative rotation and other parameters [21]. Moreover, as shown in Fig. 5(a), robots can measure distance not only to direct neighbors, but also to any visible object/robot. However, the further away the robots are, the less accurate is the measurement. Moreover, not visible robots, see Fig. 5(b), are not included into the local cost matrices.

The measured distance $S_{i,j}$ between R_i and R_j is one of the local costs for docking: the closer R_i and R_j are to each other, the "cheaper" is their docking $x_l^i \leftrightarrow x_k^j$. Other costs of $x_l^i \leftrightarrow x_k^j$ are the distance cost αS_{l-k}^{i-j}, rotation cost βS_{rot}^i, cost of "being hidden" S_{k-hid}^i, where α, β are coefficients. The cost S_{k-hid}^i is the price for the robot R^i of not-knowing the situation around R^i. For example, when the costs of connection between $x_1^1 \leftrightarrow x_1^2$ are $S_{1-1}^{1-2} = \alpha S_{1-2}^{1-2} + \beta S_{rot}^2 + S_{1-hid}^2$. More generally, local costs can include also any other factors, which determine a value of a particular connection. All local costs between all $x_l^i \leftrightarrow x_k^j$ are collected in the local costs matrix $\underline{\underline{S}}$.

Since morphogenesis is distributed and egocentric, the generations of $x_l^i \leftrightarrow x_k^j$ and $x_k^j \leftrightarrow x_l^i$ are asymmetric, i.e. from the viewpoint of the module R_i a cost of

(a) (b)

Fig. 5 **(a)** Local costs, from the viewpoint of a robot; **(b)** Global view, where some robots are not visible.

connection between R_i and R_j may be different than the connection between R_j and R_i from the viewpoint of the module R_j. This leads to the effect, that the robot R_i knows precisely only its local costs, all other costs can be estimated only roughly.

4. Global costs. Issue of global costs represents a crucial point. In general, it means how good the whole organism fits the environment and can be expressed as velocity of motion, energy consumption, weight and any other global factor for a specific environment. Normally, it requires multiple tests with different configurations and is very expensive or even impossible for on-line estimation in real situations. Therefore the proposal is to use a set of different a-priori-tested topologies, whose global costs for different environments are known. During experiments, depending on availability of tools and sensed environment, robots have to agree in which configuration they should collectively work.

4 Constraint-Based Optimization

The problem of constraint-based optimization can be formulated in two different ways, as described in [15] and [16]. The first approach, shown in Fig. 6(a), considers a classical assignment for each $ID_i \rightarrow R_i$, where ID_i and a robot's ID and R_i is a label of a robot in a topology. In the following table we display horizontally costs of all possible permutations between $ID_i \rightarrow R_j$ and vertically the connections $R_i \rightarrow R_j$ from the Fig. 6(a). The cost matrix (example of costs) takes the following form

	1:2	1:3	1:4	1:5	1:6	2:3	2:4	2:5	2:6	3:4	3:5	3:6	4:5	4:6	5:6
1:2	35	40	80	36	30	41	42	31	32	20	55	23	60	21	32
2:3	35	40	80	36	30	41	42	31	32	20	55	23	60	21	32
2:4	35	40	80	36	30	41	42	31	32	20	55	23	60	21	32
4:5	35	40	80	36	30	41	42	31	32	20	55	23	60	21	32
5:6	35	40	80	36	30	41	42	31	32	20	55	23	60	21	32

(5)

(a) (b)

Fig. 6 Example of the assembling problem. **(a)** Approach based on the assignment for each $ID_i \rightarrow R_i$; **(b)** Approach based in the linear program for the objective function Θ with $s_{i,j}$, shown are connections only to R_1.

where the assignment should satisfy

$$C_{ID_2 \rightarrow R_2} = \sum_{i}^{4} \sum_{j}^{4} S_{i,j} = 3. \qquad (6)$$

The constraint (6) is a degree of connectivity for the main core element R_2. The problem, shown in Fig. 6(a), can be formulated as e.g. quadratic assignment problem (QAP), see e.g.[22] or as constraint-satisfaction problem (CSP) and constraint-optimization problem (COP), see e.g. [23]. Since QAP is a NP hard problem, we will solve this in the CSP+COP way. Solving the assignment problem, taking into account (6) for all elements, we receive the following assignment matrix

	1:2	1:3	1:4	1:5	1:6	2:3	2:4	2:5	2:6	3:4	3:5	3:6	4:5	4:6	5:6
1:2	1	0	0	0	0	0	0	0	0	0	0	0	0	0	0
2:3	0	0	0	0	1	0	0	0	0	0	0	0	0	0	0
2:4	0	0	0	0	0	1	0	0	0	0	0	0	0	0	0
4:5	0	0	0	0	0	0	0	0	0	0	0	1	0	0	0
5:6	0	0	0	0	0	0	0	0	0	0	0	0	0	0	1

(7)

This approach has two essential drawbacks: need of robot's ID and large matrices (5),(7). Thus, this approach was improved in [16] and has the following form. First of all, we need to define the objective function Θ, which is specified by s_i, see Fig. 6(b) (here only R_1 is shown). When n robots are involved into some topology Φ, the variables $\underline{x}$ represent all possible bilateral connections between there robots. The vector of variables has m components:

$$m = \frac{n!}{(n-2)!2!}. \qquad (8)$$

There are several different Θ, in the experiments we used

$$s_i = f_i(R_k : R_p) = D(R_k : R_p) + F(R_k : R_p), k, p = 1, ..., n; k \neq p, i = 1, ..., m \quad (9)$$

where $D(R_k : R_p)$ is a distance between neighbor R_k and R_p, $F(R_k : R_p)$ satisfaction of functional constraints (0 when satisfied or $> maxD(R_k : R_p)$ not satisfied); all of them are estimated only to locally visible robots R_k and R_p. The optimization is formulated as a linear program (LP), which optimize the linear objective function $\Theta = \underline{s}^T \underline{x}$, where $\underline{s}$ is the vector of costs and $\underline{x} = (x_1, x_2, ... x_m)^T$ is a vector of variables, which are bounded by 0 and 1. LP is constrained as follows

$$\underline{\underline{A}}\underline{x} = \underline{b}, \quad x_i \in \{0, ..., 1\}, \quad (10)$$

where $\underline{\underline{A}}$ is a matrix and $\underline{b}$ is a vector of numerical coefficients, which form m linear equations (in general case inequalities). In this form it is known as integer program. Finally, by solving (10), all variables x_i take "0" or "1" so that to optimize $\underline{s}^T \underline{x}$. Now $\underline{\underline{A}}$ and $\underline{b}$ in (10) have to be defined; they reflect the connectivity constraints of the corresponding topology. For the topology shown in Figs. 4 and 6, this leads to the following linear problem

$$\underline{\underline{A}} = \begin{pmatrix} 1\,1\,1\,1\,1\,0\,0\,0\,0\,0\,0\,0\,0\,0\,0 \\ 1\,0\,0\,0\,0\,1\,1\,1\,1\,0\,0\,0\,0\,0\,0 \\ 0\,1\,0\,0\,0\,1\,0\,0\,0\,1\,1\,1\,0\,0\,0 \\ 0\,0\,1\,0\,0\,0\,1\,0\,0\,1\,0\,0\,1\,1\,0 \\ 0\,0\,0\,1\,0\,0\,0\,1\,0\,0\,1\,0\,1\,0\,1 \\ 0\,0\,0\,0\,1\,0\,0\,0\,1\,0\,0\,1\,0\,1\,1 \\ 1\,1\,1\,1\,1\,1\,1\,1\,1\,1\,1\,1\,1\,1\,1 \\ 0\,0\,0\,0\,0\,0\,0\,0\,0\,0\,0\,0\,0\,0\,0 \\ ... \\ 0\,0\,0\,0\,0\,0\,0\,0\,0\,0\,0\,0\,0\,0\,0 \end{pmatrix}, \underline{b} = \begin{pmatrix} c_{max} \\ c_{max-1} \\ c_{max-2} \\ c_{min+2} \\ c_{min+1} \\ c_{min} \\ c_t \\ 0 \\ ... \\ 0 \end{pmatrix}, \quad or \quad \underline{b} = \begin{pmatrix} 3 \\ 2 \\ 2 \\ 1 \\ 1 \\ 1 \\ 5 \\ 0 \\ ... \\ 0 \end{pmatrix}. \quad (11)$$

As mentioned, all c_i are disconnected from robots, i.e. we have to map the set of c_i to all possible combinations between these robots

$$(c_{max}, c_{max-1}, ..., c_{min}) \rightarrow Permutation(R_1, R_2, ..., R_n). \quad (12)$$

Since the number of permutations is equal to $n!$, computational power of the most of microprocessors allows computation for n below 10 closely to real time. This is more than enough for a large diversity of cores, complex topologies are created through scalability. Since variable x_i points to connections between robots, defined by (12), the vector $\underline{b}$ is equal to the set of c_i in the order from c_{max} to c_{min} and the matrix $\underline{\underline{A}}$ creates corresponding placeholders. There are several comments to this approach.

1. All topologies have a list of associated constraints Υ such as a total N of robots or requirement on heterogeneous modules. Moreover, each topology has a list of global costs: consumed energy, velocity of motion on a normal surface, geometry of concave and convex obstacle treatable with this topology or a possible geometry of docking elements.

2. Before start self-assembling, robots check whether N of available robots match the set of possible topologies. For instance, when there are topologies requiring $\{15, 17, 22, 25\}$ robots and there is only 20 available robots, they can self-assemble only into first two topologies.
3. Several topologies require tools or specialized robots. Robots should check availability of these specialized robots and correspondingly limit the set of possible topologies.
4. Self-assembling starts from the core element with the highest degree of connectivity. When such a core element is already built, robots consider the next element with the lower degree of connectivity and so on, until the whole structure is assembled.
5. When the selected topology is already partially assembled, and more free robots arrived to the assembling place, robots can decide instead of disassembling and new assembling to create a new core in the already assembled structure. Prerequisite is that the topology can be scaled up in the number of cores.

5 Grouping and Scaling Approaches

The CSP/COP-based optimization approach, described the previous section, is a distributed and decentralized process that each module in the assembling area performs. It cannot be described as a single task to achieve, the whole approach depends on interactions between modules and particular optimization processes running on-board. In this section we focus on two other tasks, which precede or follow the CSP/COP optimization: grouping and scalability approaches for different cases of $N > n$.

5.1 Grouping Approach

Grouping is required for the case when the number of available robots N is larger than the number of required modules n in a topology. Thus, for $N > n$ we need to choose the modules that will participate in the aggregation. The first idea is to choose the n nearest modules. This strategy works for topologies without any constraints on the type of modules, however can fail in other cases. Consequently, we need firstly to consider the type of modules in the neighborhood, and secondly, if a neighbor belongs to the already finished topology, it has to be removed from the list of available robots. Moreover, robots can have internal constraints that make some actions more difficult than others. For example, non-holonomic robots will not be able to move laterally, straight forward movement is cheaper than moving in any other direction. Hence, distances between A and B, from the point of view of A will be the complexity of A to reach B.

To perform grouping, a simple function selecting iteratively each nearest neighbor can be considered. However, this can lead to a non-optimal solution. To perform the grouping optimization, an application of LP solver is more suited. Each neighbor has assigned a Boolean value by the LP solver: `true` if and only if the module is selected to be part of the group. The objective function is the sums of distances (as described in the previous sections) between the calling module and the selected modules of the neighborhood. The involved constraints are the number of modules selected (equal to the number of modules in the topology) and the type constraints (there is at least as much robot of a given type as required by the topology type constraints).

5.2 *Scaling Approach*

There are two general cases, where we need to consider a scalability of self-assembling. Firstly, we frequently encounter the situation, when the number of available robots N is larger than the number of robots in the topology n. Here there are several strategies, which are considered below. Secondly, the topology with n robots can join to another topology of m robots; in this case the scalability represents the generating problem for $n+m$ topology. Such problem can be solved by morphing algorithms when basic hardware modules are able for this functionality. In our case, we solve the problem of $n+m$ topologies by a disassembling of one topology and an assembling of free robots into another one.

As mentioned, in the first case, there are several possibilities to scale N in the relation to n:

(1) for $N = xn, x = 1,2,3,...$, the topology with n robots can be replicated x times. Each of these new topologies is an independent structure. This is the simplest form of scalability, which can be denoted as the behavioral scalability.

(2) x topologies from the previous case can join into one common structure. This is typically segmented body construction, where n robots within one segment are repeated x times. This is the structural scalability.

(3) the robots from $N \ mod \ n > 0$ cannot create a new topology. These robots are still useful for the already existing topology, as e.g. energy reserve, so these robots can perturb the topology Φ, this is the perturbational scalability.

(4) finally, $N \ mod \ n > 0$ robots are not aggregating with any other structures, they build a "reserve" for e.g. self-repairing.

The experiments performed in [16] and here indicated that a combination of the (1) and (4) strategies is the most useful approach for creating multiple artificial organisms. All results described in Sec. 6 utilize this strategy. For the cases (2) and (3) we need to recalculate kinematic constraints. For this we need to perform hybrid topological-kinematic techniques, which are described e.g. in [24].

6 Implementation and Results

We performed several series of experiments with real robots and in simulation. The implementation on the real platform was intended to test computational properties as well as to estimate the level of distortion in creating the objective function Θ. Since currently there are not enough robots for testing scalability, several experiments are performed in simulation, which is done in AnyLogic. For implementation of

LP solver for CSP, we used lp_solve 5.5 routine (see lpsolve.sourceforge.net) of Mixed Integer Linear Programming solver (C++ version is used for real robots, Java version is used for simulation). Robots use Blackfin double core as the main CPU (in each module) with 64 Mb SDRAM on board. The robot arena was approximately 50x larger than the size of the robot. For measuring the time of experiments, we use the notion of "iteration of the autonomy cycle". This is an internal value of robots and allows estimating the running time more precisely to the steps of the CSP/COP approach (and not to the motion of the robot). Tests are performed with two topologies: "T"-like form shown in Fig. 4 and a snake. These topologies are used also for scalability tests. Additionally to experiments presented in [15], [16], we explored here different strategies for a grouping approach and its impact on a performance, and measured a performance of self-assembling at different scaling parameters.

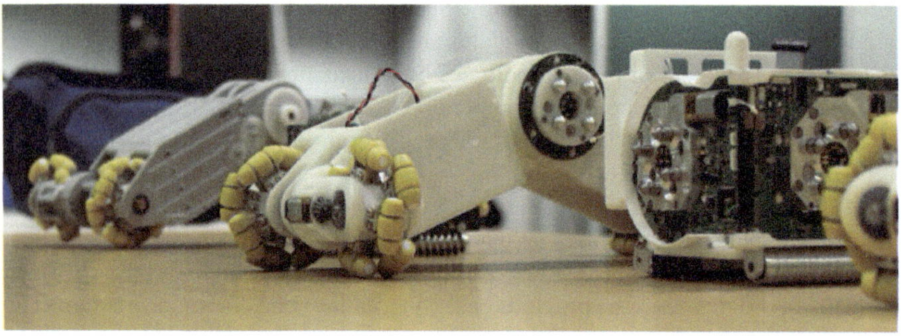

Fig. 7 Ordering phase of the self-assembling approach. Five robots approached each other and moved into right spatial positions for creating a topology shown in Fig. 2(d). During this approach, a massive collision avoidance behavior can be observed.

Grouping approach. As mentioned in Sec. 5.1, grouping is a necessary step before self-assembling. It allows selecting the appropriate n robots from the swarm of N robots for building a topology. Implementing a grouping strategy, we encounter several difficulties. First of all, robots are not always visible to each other, see Fig. 7, i.e., estimation of the most closest robots to a group is in many situations not possible.

In the implemented strategy, we add a robot into a group when it is visible at least to one robot which already included into a group. Assigning to a group is performed by the "first visible, first served" principle. When a robot receives an invitation to join to a group, and can confirm or reject it. When no other invitation is accepted, a robot confirms its intention to join to the group and moves towards a visible inviting robot. When a robot approached the group closely enough, it starts CSP/COP procedure and waits a resolution of the assignment problem. After this, it moves to the right spatial position (the ordering phase).

To evaluate performance of this strategy, we used two other approaches: a candidate for joining a group was selected randomly based on WiFi connection (i.e. no position information), and the ideal case when only the closest robot was invited to join a group (by using global information). In Fig. 8 we plot the performance of the self-assembling with the neighbor-based grouping strategy and with a random grouping. In Figs. 8(a), 8(b) the performance is estimated as the sum of distances s_i between all robots of a group, whereas Figs. 8(c), 8(d) demonstrate the performance as the sum of distances s_i between all robots. Fig. 8(d) shows the ideal case with the closest neighbors. We can see that selection of robots for grouping has an impact on the assembling strategy. In many performed experiments it reduces the approaching time on 30%-50% and makes the ordering phase more early. In ideal case the ordering phase starts almost immediately after the start of experiments.

The second series of experiment was performed to investigate the scalability performance of the self-assembling strategy. The idea of this experiment originates

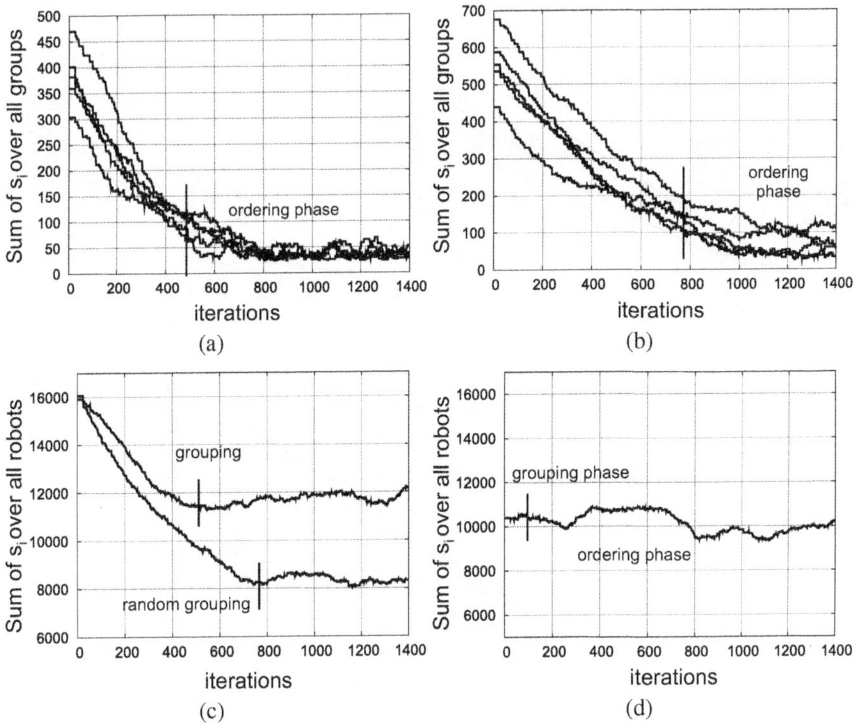

Fig. 8 Different grouping strategies, $n = 5$, $N = 25$. **(a)** Grouping of locally visible robots to the each other, shown is the sum of all distances between the robots in each group; **(b)** Random grouping, all robots are equally distributed on the arena, shown is the sum of all distances between the robots in each group; **(c)** Comparison between both strategies, shown is the sum of s_i over all robots; **(d)** Ideal grouping strategy, shown is the sum of s_i over all robots.

from [16], where we proposed to increase a connectivity of robots. It is achieved by movement of all robots to a specific point on the arena (it was a right, upper corner, which can be approached by using a compass information). When all robots are closely enough to each other, they can perform a faster grouping, creation of cost matrices and ordering. The encountered problem was that robots massively increased a number of collisions and this slowed down the performance. Moreover, the more robots are participated in the experiment the higher was the number of collision, and the slower was the assembling. In Fig. 9 we investigated the number of collisions for different grouping strategies and for different N of robots. We encountered that the collision behavior has two well observable phases: the low-slope phase during the grouping and the high-slope phase during the ordering. The slope of the ordering phase is almost the same at all strategies, and is independent of the number of modules. This can be explained by the Fig. 7, where we can see that the ordering causes strongly local collision avoidance behavior and so is independent of

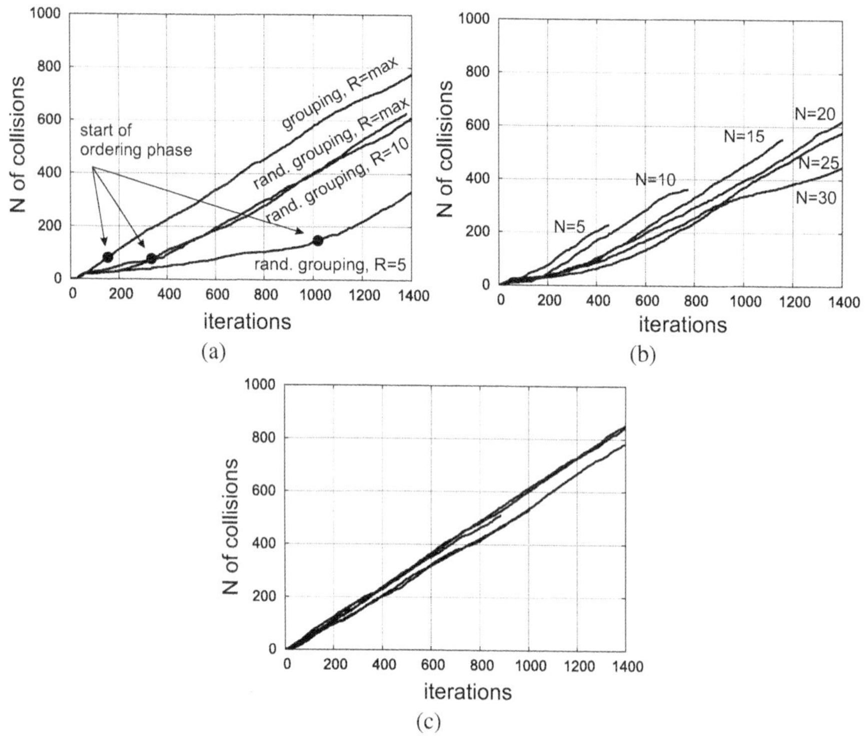

Fig. 9 Scalability performance as a number of collisions over the running time of an experiment, R is the sensing radius, $R = max.$: a robot can observe the whole arena, $R = 5, 10$: a robot can observe a range of 5 and 10 body lengths. **(a)** Scalability in relation to different grouping strategies and different visibility radii; **(b)** The strategy with random grouping and with the minimal sensing radius, shown are N of collisions for different N of robots; **(d)** The strategy with neighbor grouping, shown are N of collisions for different N of robots.

the total number of modules. However, different grouping strategies have different collision performance. As it follows from Fig. 9(c), the neighbor-based grouping has the best scalability properties (it is almost independent of N), whereas as the random-based strategies, especially with limited sensing radius, has a worse performance when increasing the number of robots. This behavior we observed also during earlier experiments. Thus, the best scalability strategy is to increase the level of locality not only for all robots during the aggregation, but also to increase the level of locality for the grouping phase.

7 Conclusion

This chapter is devoted to a self-assembling strategy, which utilizes generating and optimizing approaches, known in biological regulatory networks. We primarily focused here on the optimization controller, which narrows down the set of generated topologies to a particular description of connections between modules and performs local optimization defined by the objective function. This function takes into account connectivity and functional constraints, local and several global costs. Due to flexibility of costs and constraints, this approach is very useful for modules with different geometry and functionality, i.e. for heterogeneous reconfigurable robots.

For experimental part we mainly worked on the grouping strategies and scalability performance for different such strategies and different N. These experiments are additional to the already published results. We estimated that neighbor-based grouping, even not optimal in a global sense, can provide shorter aggregation time due to spatial optimization process. Non-optimality of spatial grouping is substantially limited by the capabilities to detect a neighbor robot. To improve the performance, we suggested increasing the level of locality by pre-aggregating all robots into a specific area of the robot arena. In the previous experiments, this strategy created multiple bottlenecks due to massive collision avoidance behavior among robots. In the improved approach, when a local grouping is selected, the pre-aggregated self-assembling is well scalable to different N, tested for $n = 5$ and N between 5 and 30.

For further works, we would like to verify these tests for the topologies shown in Fig. 2 with three different types of robots. It is also of interest to investigate the influence of environmental conditions for very simple robots (like chemical molecules) and to estimate whether it is possible to transfer results from macroscopic to a microscopic self-assembly.

References

1. Ariga, K., Hill, J.P., Lee, M.V., Vinu, A., Charvet, R., Acharya, S.: Challenges and breakthroughs in recent research on self-assembly. Science and Technology of Advanced Materials 9(1), 014109 (2008)
2. Whitesides, G.M., Kriebel, J.K., Love, J.C.: Molecular engineering of surfaces using self-assembled monolayers. Science Progress 88, 17–48(32) (2005)

3. Levi, P., Kernbach, S. (eds.): Symbiotic Multi-Robot Organisms: Reliability, Adaptability, Evolution. Springer, Heidelberg (2010)
4. Chiang, C.-J., Chirikjian, G.: Modular robot motion planning using similarity metrics. Auton. Robots 10(1), 91–106 (2001)
5. Miyashita, S., Hadorn, M., Hotz, P.E.: Water floating self-assembling agents. In: Nguyen, N.T., Grzech, A., Howlett, R.J., Jain, L.C. (eds.) KES-AMSTA 2007. LNCS (LNAI), vol. 4496, pp. 665–674. Springer, Heidelberg (2007)
6. Castano, A., Shen, W.-M., Will, P.: Conro: Towards deployable robots with inter-robots metamorphic capabilities. Journal of Autonomous Robots 8, 309–324 (2000)
7. Samori, P., Francke, V., Müllen, K., Rabe, J.P.: Self-assembly of a conjugated polymer: From molecular rods to a nanoribbon architecture with molecular dimensions. Chemistry - A European Journal 5, 2312–2317 (1999)
8. Kernbach, S.: Towards application of collective robotics in industrial environment. In: Rigatos, G. (ed.) Industrial Systems: Modelling, Automation and Adaptive Behaviour, pp. 18–49. IGI Global (2010)
9. Yu, C.-H., Haller, K., Ingber, D., Nagpal, R.: Morpho: A self-deformable modular robot inspired by cellular structure. In: IEEE/RSJ International Conference on Intelligent Robots and Systems, IROS 2008, pp. 3571–3578 (2008)
10. Liu, W., Winfield, A.F.T.: Autonomous morphogenesis in self-assembling robots using IR-based sensing and local communications. In: Dorigo, M., Birattari, M., Di Caro, G.A., Doursat, R., Engelbrecht, A.P., Floreano, D., Gambardella, L.M., Groß, R., Şahin, E., Sayama, H., Stützle, T. (eds.) ANTS 2010. LNCS, vol. 6234, pp. 107–118. Springer, Heidelberg (2010)
11. Huie, J.C.: Guided molecular self-assembly: a review of recent efforts. Smart Materials and Structures 12(2), 264 (2003), `http://stacks.iop.org/0964-1726/12/i=2/a=315`
12. Davidson, E.H.: The Regulatory Genome: Gene Regulatory Networks In Development And Evolution. Academic Press, London (2006)
13. Jacob, C., Burleigh, I.: Genetic programming inside a cell. In: Genetic Programming Theory and Practice III, vol. 9, pp. 191–206. Springer, Heidelberg (2005)
14. Banzhaf, W.: On evolutionary design, embodiment, and artificial regulatory networks. In: Iida, F., Pfeifer, R., Steels, L., Kuniyoshi, Y. (eds.) Embodied Artificial Intelligence. LNCS (LNAI), vol. 3139, pp. 284–292. Springer, Heidelberg (2004)
15. Kernbach, S.: From robot swarm to artificial organisms: Self-organization of structures, adaptivity and self-development. In: Levi, P., Kernbach, S. (eds.) Symbiotic Multi-Robot Organisms: Reliability, Adaptability, Evolution, pp. 5–25. Springer, Heidelberg (2010)
16. Kernbach, S.: Heterogeneous self-assembling based on constraint satisfaction problem. In: Martinoli, A., Mondada, F. (eds.) 10th International Symposium on Distributed Autonomous Robotics Systems. Springer Tracts in Advanced Robotics. Springer, Heidelberg (2011)
17. Salemi, B., Shen, W.-M.: Distributed behavior collaboration for self-reconfigurable robots. In: Proc. of the IEEE International Conference on Robotics and Automation (ICRA-2004), New Orleans, USA, pp. 4178–4183 (2004)
18. Lau, H.Y.K., Ko, A.W.Y., Lau, T.L.: The design of a representation and analysis method for modular self-reconfigurable robots. Robot. Comput.-Integr. Manuf. 24(2), 258–269 (2008)
19. Castano, A., Will, P.: Representing and discovering the configuration of CONRO robots. In: Proc. of IEEE Int. Conf. on Robotics and Automation (ICRA-2001), vol. 4, pp. 3503–3509. IEEE, Los Alamitos (2001)

20. Burkard, R., Dell'Amico, M., Martello, S. (eds.): Assignment Problems. Society for Industrial and Applied Mathematics (2009)
21. Kernbach, S., Meister, E., Scholz, O., Humza, R., Liedke, J., Ricotti, L., Jemai, J., Havlik, J., Liu, W.: Evolutionary robotics: The next-generation-platform for on-line and on-board artificial evolution. In: Tyrrell, A. (ed.) Proc. of the IEEE Congress on Evolutionary Computation (IEEE CEC-2009). IEEE Press, Trondheim (2009)
22. Loiola, E., de Abreu, N.M., Boaventura-Netto, P., Hahn, P., Querido, T.: A survey for the quadratic assignment problem. European Journal of Operational Research 176(2), 657–690 (2007)
23. Kornienko, S., Kornienko, O., Priese, J.: Application of multi-agent planning to the assignment problem. Computers in Industry 54(3), 273–290 (2004)
24. Kernbach, S., Meister, E., Schlachter, F., Kernbach, O.: Adaptation and self-adaptation of developmental multi-robot systems. International Journal On Advances in Intelligent Systems 3(1,2), 121–140 (2010)

Morphogenetic Self-Reconfiguration of Modular Robots

Yan Meng and Yaochu Jin

Abstract. It is still a challenging problem to self-reconfigure modular robots to different morphologies to adapt to dynamic environments. To tackle this problem, a new computational framework inspired from biological morphogenesis is suggested in this chapter. First, we introduce two reconfigurable modular robots, Cross-Cube and Cross-Ball, which are developed for various complex pattern reconfigurations using the pro-posed morphogenetic approach. Then, a hierarchical morphogenetic self-reconfiguration model is presented for both Cross-Cube and Cross-Ball. In this hierarchical model, the layer 1 controller is responsible for the adaptive pattern generation based on the current environmental constraints and the task requirements in hands. The layer 2 controller automatically generates target configurations to guides the modules to converge into the target pattern. Both the layer 1 and layer 2 controllers are generic and can in principle be applied to other reconfigurable modular robots. The controller in layer 3 is hardware dependent that mainly deals with the physical constraints of module movements. This hierarchical morphogenetic model is applied to each module of the reconfigurable modular robot in a distributed manner, where each module makes its configuration movements only based on its local sensory information and shares information with its local neighboring modules. Extensive simulation results have demonstrated the feasibility and efficiency of the proposed module design as well as the corresponding hierarchical morphogenetic model for both Cross-Cube and Cross-Ball modular robots to construct various configurations.

Yan Meng
Department of Electrical and Computer Engineering,
Stevens Institute of Technology, Hoboken, NJ 07030, USA
e-mail: yan.meng@stevens.edu

Yaochu Jin
Department of Computing, University of Surrey, Guildford, Surrey, GU2 7XH, UK
e-mail: yaochu.jin@surrey.ac.uk

Y. Meng and Y. Jin (Eds.): Bio-Inspired Self-Organizing Robotic Systems, SCI 355, pp. 143–171.
springerlink.com

1 Introduction

Reconfigurable modular (RM) robots are modular robots with a variable morphology, where they are able to deliberately change their own shape by reorganizing the connectivity of their modules to adapt to new environments, perform new tasks, or recover from damages. Each module is an independent unit that is able to connect it to or disconnect it from other units to form various structures/patterns dynamically. Compared with conventional robotic systems, RM robots are potentially more robust and more adaptive under dynamic environments.

Generally, the mechanical design of RM robots can be classified as chain-based and lattice-based. Chain-based RM robots [11, 34, 29, 26] consist of modules which are connected in serial chains to form line, tree or loop structures, which are scalable and easy in motion planning. However, it is hard for chain-based RM robots to build arbitrary complex 3D patterns. To address this is-sue, lattice-based RM robots [32, 3, 37, 8] have been proposed, where modules are aligned in discrete grid lattices in space and can build up complex 3D patterns. As a result, the control and motion-planning for lattice-based RM robots become more complicated. In addition, some hybrid RM robots have been developed by combining the advantages of both types of RM robots, such as ATRON [13], M-TRAN II [14], M-TRAN III [15], and SUPERBOT [25], which aim to provide more complex and flexible morphologies. However, it is difficult for most available RM robots to build arbitrary complex configurations due to mechanical constraints of hardware modules.

Self-organization and self-repairing are critical for the success of RM robots. Usually, reconfiguration occurs for purposes of locomotion, manipulation, or for the creation of stable static structures. Self-organization of RM robots is especially challenging as it involves coordinating the movement of a possibly large number of modules while enforcing motion constraints imposed by their physical design. Some centralized reconfiguration mechanisms have been proposed for various RM robots [23, 35, 36, 6]. However, RM robots using centralized approaches for predefined patterns are very vulnerable to system failures/malfunctions and non-scalable to more complex and large-scale systems. To address this issue, some decentralized methods [20, 24, 12, 31] have been proposed, where each module in a RM robot makes its own movement decision without a centralized controller. To further improve the self-organization capability, some researchers turn their attentions to biological systems for inspirations. Fukuda et al. [7] first proposed an optimal structure decision method for a distributed CEBOT (cell structured robot) system based on the cell structures for static and mobile-based manipulators. Shen et al. [28] proposed a hormone-inspired adaptive communication protocol and adaptive distributed control protocol to allow modules to use hormone-like messages to accomplish locomotion and self-reconfiguration. An artificial homeostatic hormone systems was proposed in

[27] for the self-reconfiguration of robotic organisms, where the parameterization of the underlying dynamical system was encoded into a "genome" of the organism which is subject to selection through evolutionary algorithms. A cellular automata approach was developed in [30] for self-reconfiguration of a modular robot.

However, most of these approaches need some predefined rules to reconfigure into a limited set of predefined patterns. To our knowledge, very few of the current RM robots can automatically reorganize their structures to adapt to different environments based on their online sensory information. Although self-organization is believed to be the most important feature of RM robots, the ability to adapt their configurations autonomously to environmental changes largely re-mains to be demonstrated. To address this issue, we start to study the morphogenesis procedure in multi-cellular organisms. Morphogenesis is the biological process that causes an organism to develop its shapes. During the morphogenesis, genes in each cell are expressed, resulting in various cellular functions. The expression of the genes is regulated by their own protein products as well as proteins produced by other genes in the same cell or neighboring cells through intracellular and intercellular diffusion, forming a complex gene regulatory network (GRN). This underlying GRN plays critical rules in the organism morphogenesis.

The connection between RM robots and multi-cellular organisms is straightforward. Each unit in a RM robot can be treated as a cell, and there exist analogies in control, communication and physical interactions between cells in a multi-cellular organism and modules in a RM robot. For example, control in both RM robots and multi-cellular organisms are decentralized. In addition, global behaviors of both RM robots and multi-cellular organisms emerge through local interactions of the units, which include mechanic, magnetic and electronic mechanisms in RM robots, and chemical diffusion and cellular physical interactions such as adhesion in multi-cellular organisms.

Therefore, inspired by the embryonic development of multi-cellular organisms [33], morphogenetic self-reconfiguration of two RM robots, namely, Cross-Cube [16, 17] and Cross-Ball [18], are proposed in this chapter using underlying principles of biological morphogenesis. The mechanical designs of both Cross-Cube and Cross-Ball RM robots belong to the hybrid RM robots, which combine the features of both chain-based and lattice-based RM robots. And a hierarchical morphogenetic model is proposed for the self-reconfiguration of both Cross-Cube and Cross-Ball RM robots. Generally speaking, the layer 1 controller is the pat-tern generation layer, which aims to generate appropriate patterns for RM robots to adapt to the current environment and the tasks in hands. Layer 2 controller is a gene regulatory network (GRN) based controller, which automatically produce reconfiguration plans for modules to converge to the target patterns generated from layer 1. This two-layer model is enough for the Cross-Cube RM robots due to the high flexibility of the module movements in Cross-Cube. However, the mechanical and electrical design of Cross-Cube is very complex and hard to be implemented in physical prototype. Therefore, we improve the design of Cross-Cube and develop a new RM robot, called Cross-Ball. Compared with Cross-Cube, the complexity of the mechanical and electrical design of the Cross-Ball is significantly decreased at the cost of losing some flexibility of module movements of Cross-Ball. Therefore, an additional layer 3 controller is required for Cross-Ball to deal with the physical constraints of module movements.

The major contributions of the proposed morphogenetic self-reconfiguration of RM robots are listed as follows. (1) Due to the high flexibility of module movements of both Cross-Cube and Cross Ball RM robots, various complex pat-terns can be constructed using these two RM robots. (2) With the proposed hierarchical morphogenetic model, both RM robots can automatically self-reconfigure their modules to appropriate patterns to adapt to dynamic environments based on onboard sensory information of RM robots. So far, very few of the existing RM robots have such self-reconfiguration capability. (3) The proposed hierarchical morphogenetic model is executed on each module of both RM robots in a distributed manner, where each module makes its own decisions based on its local perceptions by itself and its immediate neighboring modules. In this manner, both RM robots are very robust to system failures or malfunctions and can self-repair themselves automatically. The rest of the chapter is organized as follows. First, a brief introduction of multi-cellular morphogenesis is presented in Section 2. A generic hierarchical morphogenetic model is introduced in Section 3. The hardware designs and corresponding hierarchical morphogenetic models for Cross-Cube and Cross-Ball RM robots are described in Section 4 and Section 5, respectively, as well as the experimental results of each RM robot. Section 6 gives the conclusion and future work.

2 Multi-cellular Morphogenesis

Morphogenesis is one fundamental biological process in developmental biology that guides an organism to develop its body plan. Multi-cellular morphogenesis is under the control of gene regulatory networks (GRN). When a gene is expressed, information stored in the genome is transcribed into mRNA and then translated into proteins. Some of these proteins are transcription factors that can regulate the expression of their own or other genes, thus resulting in a complex network of interacting genes termed as a GRN. To understand the emergent morphology resulting from the interactions of genes in a regulatory network, reconstruction of gene regulatory pathways using a computational model has become popular in systems biology [1]. A large number of computational models for GRNs have been suggested [5, 4], which can largely be divided into discrete models, such as random Boolean networks and Markovian models, and continuous models, such as ordinary differential equations and partial differential equations. Sometimes, GRN models also distinguish them-selves as deterministic models and stochastic models according to their ability to describe stochasticity in gene expression.

3 A Generic Hierarchical Morphogenetic Model

Inspired by the mechanisms of biological morphogenesis, a generic hierarchical morphogenetic model is proposed, which in principle can be applied to various RM robots for self-reconfiguration in dynamic environments. First, the target pat-tern that a RM robot needs to form can be generated automatically based on the current environments and the task requirements, which is the layer 1 controller of the hierarchical model. Then, the layer 2 controller is required to automatically generate self-reconfiguration plans to guide each module in a RM robot to con-verge to the target pattern generated by the layer 1 controller. Then, based on different mechanical design and physical constraints of module movements of RM robots, different motion control methods are needed to move the modules to their destination positions in the target pattern by following the self-reconfiguration plan provided by the layer 2 controller. Fig. 1 shows the block diagram of this hierarchical morphogenetic model. Due to different hardware design and physical constraints of modules, different RM robots may need to develop different hierarchical controllers for each layer, but they all share the same fundamental control architecture described in Fig. 1.

4 Self-Reconfiguration of Cross-Cube RM Robots

4.1 Hardware Design of Cross-Cube

Cross-Cube is a RM robot we developed in a robot simulator using a real time physics engine PhysX [16]. Each module in a Cross-Cube robot is a cubical structure having its own computing and communication resource and actuation capability. Each module can perceive its local environment and communicate with its neighboring modules using on-board sensors. Each Cross-Cube module consists of a core and a shell as shown in Fig. 2. The core is a cube with six universal joints. Their default heading directions are bottom, up, right, left, front, and back, respectively. Each joint can be attached to or detached from the joints of its neighbor modules. The axis of each joint can be actively rotated, extended, and returned to its default direction. Basic module movements include rotating, climbing, and parallel movements. Fig. 2 illustrates a rotating movement of two modules. Parallel movement means that a module moves to a next position which is parallel to its current position. Climbing movement means that a module moves to a diagonal neighboring position. Please refer to [16] for the detailed mechanical design of Cross-Ball and its motion capabilities.

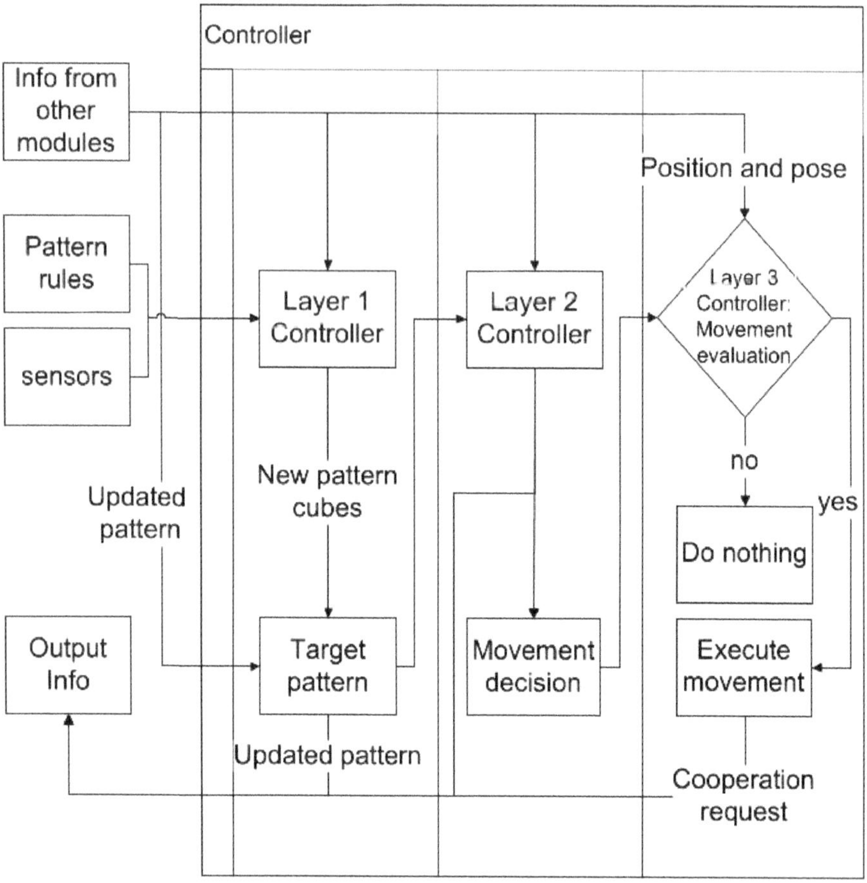

Fig. 1 The flowchart of the proposed hierarchial morphogenetic controller for each module in a RM robot

4.2 A Hybrid Hierarchical Model

Based on the generic hierarchical morphogenetic model described in Fig. 1, a hybrid hierarchical model was proposed for the Cross-Cube RM robot in our previous work [16], where the layer 1 controller is a rule-based method to generate the target pattern. The rules are described in terms of a look-up-table. The layer 2 controller is a GRN-based controller to reconfigure the modules to the target pattern automatically. Since there is no physical constraint in module movements of the Cross-Cube modules, specific motion controller is not needed for Cross-Cube RM robots. Therefore, we only focus on the controllers of the first two layers here.

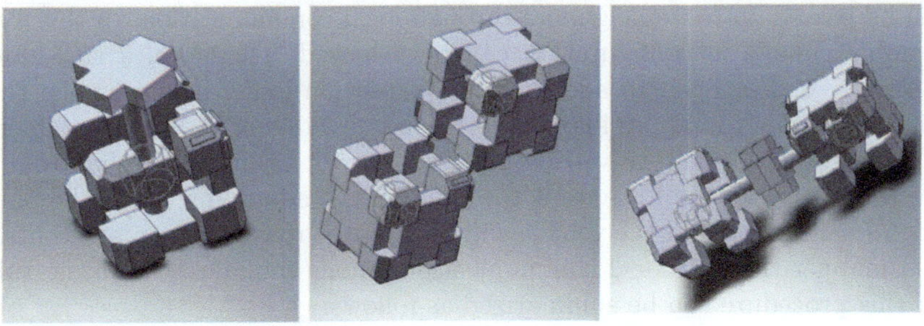

Fig. 2 Mechanical demonstration of the Cross-Cube module. (a) The joints; (b) The locks on the boundaries of the modules. (c) Rotation and extension of the joints of the modules.

4.2.1 A Rule-Based Method for Layer 1 Controller

Initially, a number of basic configurations for different environments can be represented in a look-up-table for a given mission. Table 1 lists one definition of a vehicle pattern using a look-up-table. In the table, x, y, and z are 3D coordinates of grid positions, MG denotes morphogen value and PID stands for position identification. The target pattern is defined by the morphogen values on grid positions.

Table 1 Definition of a vehicle pattern.

Positions (x, y, z, MG, PID)		Joints (PID1, PID2, RD)
(0, 0, 0, 10, 0)	(1, 0, 3, 10, 10)	(0, 1, 0)
(1, 0, 0, 10, 1)	(2, 0, 3, 10, 11)	(2, 3, 1)
(2, 0, 0, 10,2)	(0, 0, 4, 10,12)	(6, 7, 0)
(3, 0, 0, 10, 3)	(1, 0, 4, 10, 13)	(8, 9, 1)
(1, 0, 1, 10, 4)	(2, 0, 4, 10, 14)	(12, 13, 0)
(2, 0, 1, 10, 5)	(3, 0, 4, 10, 15)	(14, 15, 1)
(0, 0, 2, 10, 6)	(0, 0, 1, -1, 16)	
(1, 0, 2, 10, 7)	(3, 0, 1, -1, 17)	
(2, 0, 2, 10, 8)	(0, 0, 3, -1, 18)	
(3, 0, 2, 10, 9)	(3, 0, 3, -1, 19)	

Adaptation to environmental changes is of paramount importance in RM robots. A mechanism is needed to define and modify the target pattern of a RM robot adaptively to respond to environmental changes based on local sensory feedback. For such tasks, it is assumed that each module is equipped with a sensor to detect the distance between the module and obstacle in the environment. Once a module receives such sensory feedback, this information will be passed on to its neighbors through local communication. In this way, a global change in configuration can be achieved. Therefore, a rule-based controller is developed to address this adaptation issue. It is assumed that initially all robot modules know the heading direction of the vehicle pattern. When a robot needs to traverse a path whose width is narrower than that of the robot, the width of the front row will be first adapted to fit in the path. The remaining rows of the vehicle will be adapted row by row in a decentralized manner through local communication among modules. The basic rules for this procedure can be summarized as follows:

- Rule 1: Once a module in the front row detects obstacle(s), it passes this information through local communication to its neighboring modules until all the modules are reset to 'unstable' state for initialization. Refer to the next subsection for the finite state machine of modules.
- Rule 2: If some modules in the first row detect an obstacle, they will make a decision on whether the robot needs to reconfigure itself to avoid the obstacle. If yes, these modules will estimate how many modules need to be removed and this information is passed to other modules in the same row through local communication. Therefore, the morphogen values of these need-to-remove positions are set up as negative values while others as positive values.
- Rule3: After the GRN-based pattern formation controller (will be discussed next) finishes the reorganization of the modules in one row, the states of these modules are set to be 'stable'.
- Rule 4: If a row of a vehicle pattern is filled in by stable modules, these modules set the morphogen values for the position of its next row to be positive.
- Rule 5: The pattern generation procedure stops when all the modules change to the 'stable' state.

This rule-based approach will allow the modular robots to adapt its configurations to the current environmental constraints.

4.2.2 A GRN-Based Method for Layer 2 Controller

Once the target pattern is initialized in a look-up-table and adapted different patterns to respond to the environmental changes based on the above rule-based method, the next step is to configure the Cross-Cube modules to converge to the target pattern. By setting any single module as the origin, all other modules can figure out their relative positions to this origin easily through local communications. Based on the relative positions and the information on the target pattern, each module can produce different types of proteins to attract other modules to fill in its neighboring positions with positive morphogen values, or repel its neighbor modules from positions with negative morphgen values.

Finite State Machine of Modules. The finite state machine of a Cross-Cube module is shown in Fig. 3, where each module has five different states, namely, 'stable', 'unstable', 'attracting', 'repelling', and 'repelled'. The 'stable' state is the final state of the module. The "attracting" state means the module is attracting other modules to fill in some of its neighboring grids. The 'unstable' state means the module can respond to attractions. The 'repelling' state means the module is repelling unwanted neighboring modules away. The 'repelled' state means that the module responds to repelling requests and move away from the current grid.

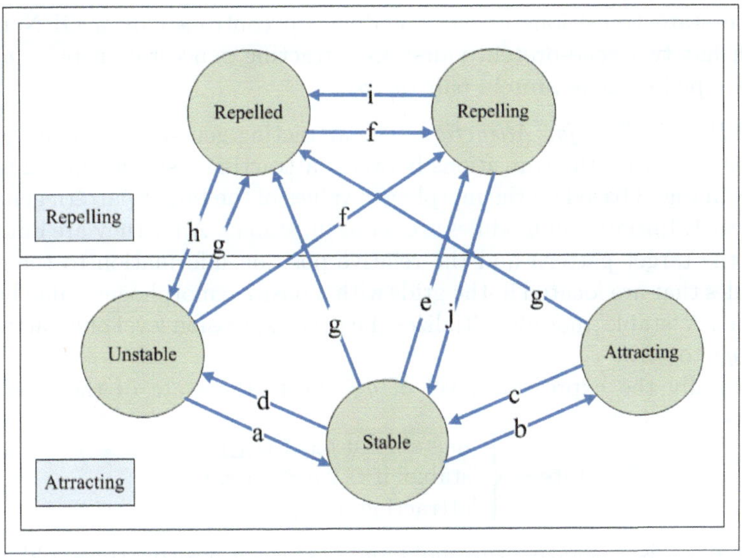

Fig. 3 The finiate state machine of a Cross-Cube module.

When an 'unstable' module arrives at the destination grid, it changes its state to "stable" (arrow *a*). When a 'stable' module has neighboring grids need to be filled, it will change its state to 'attracting' (arrow *b*) to attract 'unstable' modules to fill in those grids. Once those neighboring grids are occupied, the 'attracting' module returns to the 'stable' state (arrow *c*). A 'stable' module may also give up its current position to fill in a grid with higher morphogen value in the pattern by turning its state to 'unstable' (arrow *d*).

A 'repelling' module repels unwanted neighbor modules from their current positions by setting their states to 'repelled'(arrow *g*). Then the 'repelled' module moves away from its current position, and switches its state to 'unstable' (arrow *h*). A module can be triggered to the 'repelling' state in two situations. The first one is when a 'stable' module finds out that some of its neighboring modules need to be removed (with a negative morphogen value), it changes its state to 'repelling' (arrow e)and switches the state of those neighbors to 'repelled' (arrow g). When all the 'repelled' modules have left, the 'repelling' module returns to the 'stable' state (arrow j). The second situation is when a deadlock happens, where a moving module is blocked by its neighboring modules. To resolve this deadlock, the blocked module switches its state to 'repelling' (arrow f), and attempts to change the state of all its neighbors to be 'repelled' (arrow g). This removes some of its neighboring modules to make room for the blocked module to move away. Then the 'repelling' module returns to the 'repelled' state (arrow i). To prevent the deadlock situations, the repellent states always have a higher priority than the attracting states.

The state transitions of each module are controlled by a GRN model, which has two gene-protein pairs: an attracting gene-protein pair (g_A, p_A) and a repelling gene-protein pair (g_p, p_p).

Gene-Protein Pair for Attraction. The attracting gene-protein pair (g_A, p_A) is used to control the transitions between 'attracting', 'stable' and 'unstable' states in Fig. 3 based on the morphogen values of the target pattern generated by layer 1. Initially, all modules are set as 'unstable'. After they are initialized with the target pattern and the relative position information to the origin, modules that are located in the grids with a positive morphogen value become 'stable'. A 'stable' module initializes the gene expression level of its attracting gene g_A to zero.

Basically the expression level of g_A affects the state of the module as follows:

$$\text{state} = \begin{cases} \text{'unstable' if } g_A < G_{A\_L} \\ \text{'stable' if } G_{A\_L} < g_A < G_{A\_H} \\ \text{'attracting' if } g_A > G_{A\_H} \end{cases} \tag{1}$$

where G_{A_L} is a negative threshold and G_{A_H} is a positive threshold.

Each 'stable' module generates attracting protein p_A for all the empty neighboring grids with a positive morphogen value. The local generated p_A and received p_A from other modules will regulate the expression level of g_A. When g_A is higher than G_{A_H}, the state changes to 'attracting', and the attracting protein p_A is diffused to other modules. p_A is defined as

$$p_A^{ij} = k_1 \cdot mp_j \tag{2}$$

where p_A^{ij} is the concentration of attracting protein generated by module i for its neighbor grid j. k_1 is the discount rate of p_A^{ij} over distance, which is a defined as 0.8 in experiments. mp_j is the morphogen value of grid j.

The gene expression level of the attracting gene $g_A(t)$ is defined by:

$$\frac{dg_A^i(t)}{dt} = -k_2 \cdot g_A^i(t) + k_3 \cdot \sum_j^{p_A^{ij}} - k_4 \cdot \sum_k^{p_A^{rec}} \tag{3}$$

where $g_A^i(t)$ is the gene expression level of modular robot i. The first term in Eqn.(3) means that $g_A^i(t)$ decays over time. $\sum_j p_A^{ij}$ represents the sum of all the generated attracting proteins by robot module i for its empty neighboring grids. The more proteins the module generates for its empty neighboring grid, the higher the g_A expression level it will have, which means it will have a higher possibility to change its state from 'stable' to 'attracting'. $\sum_k p_A^{rec}$ is the sum of the protein concentrations received from other modules by robot module i. The module may turn to 'unstable' if other modules have stronger attractions so that $g_A(t)$ is reduced to a value less than G_{A_L}. Unstable modules choose to fill in the attracting grid with the highest P_A from all the received attracting proteins. $k_2, k_3,$ and k_4 are constant coefficients.

Gene-Protein Pair for Repelling. The 'repelling' and 'repelled' states are controlled by the repelling gene-protein pair (g_P, p_P). A 'repelling' module generates p_P and diffuses it to the neighboring modules that need to be repelled. The repellent protein p_P is defined as a positive constant. Each module has a gene g_P whose expression level decides the state transition to "repelled" state. The gene expression level of g_P is initialized as 0 and can be regulated by p_P through Eqn.(4)

$$\frac{dg_P^i(t)}{dt} = -k_5 \cdot g_P^i(t) - k_6 \cdot \sum p_p^{rec} \tag{4}$$

$$\text{state of grid } i = \text{repelled when } g_P^i < -mp_i \tag{5}$$

where $g_P^i(t)$ is the gene expression level of the repellent gene at time t. p_P^{rec} is the concentration of the received repellent protein. mp_i is the morphogen value of the current grid. k_5 and k_6 are constant coefficients. The first item in Eqn.(4) denotes that $g_p(t)$ decays over time. $\sum p_P^{rec}$ indicates the sum of all p_P received by module i from its immediate neighbors. g_P is reduced with more received p_P. Eqn.(4) indicates that modules with a lower morphogen value are more likely to be repelled.

4.2.3 Experimental Results

To evaluate the efficiency and robustness of the proposed hybrid approach to the self-reconfiguration of Cross-Cube RM robots, several case studies have been conducted in a robot simulator. This simulator is used to simulate the behaviors and interaction of Cross-Cube RM robots with a physical world using C++ and the PhysX engine from nVidia (http://en.wikipedia.org/wiki/PhysX). In the following experiments, the system parameters are setup as follows: $k_1 = 0.7$, $k_2 = 1$, $k_3 = 1$, $k_4 = 0.5$, $k_5 = 2$, $G_{A\_L} = -1$, $G_{A\_H} = 1$, $G_{P\_L} = -2$, $C_P^{ij} = 0.7$. Protein concentration decays to 80% of its previous level when it diffuses into a neighbor module.

To evaluate the performance of the GRN-based controller for pattern formation layer, first, we can predefine a fixed target pattern using a look-up table as shown in Table 1. Based on this predefined target pattern, the modules of Cross-Cube need to autonomously configure themselves to form the target pattern using the GRN-based layer 2 controller. A set of snapshots of this pattern formation procedure in the experiment is depicted in Fig. 4. From Fig. 4, we can see that the Cross-Cube can automatically form a given target pattern through self-reconfiguration using the proposed GRN-based controller.

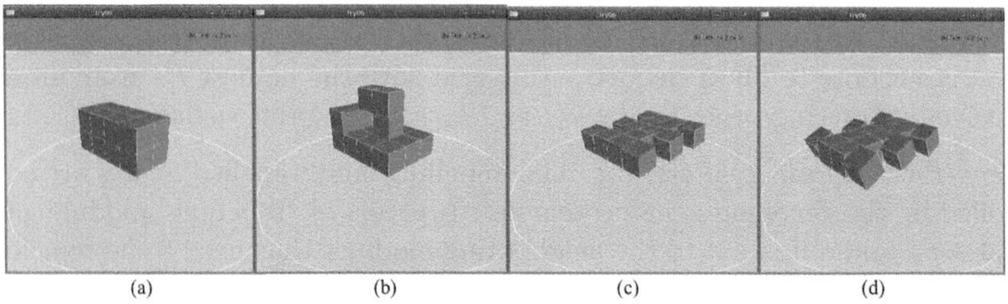

Fig. 4 Autonomous reconfiguration of a Cross-Cube RM robot from a rectangle to a vehicle using the hybrid hierarchical model.

To evaluate the self-repairing performance of the GRN-based layer 2 controller, another experiment is conducted here. First, the look-up table of the target pattern (i.e., a vehicle patter here) for robot modules is predefined. When the vehicle is moving, an "explosion" occurs and some functional modules are blown away. The backup modules then move to fill in the damaged modules. Fig. 5 shows a snapshot of this self-repairing procedure using the proposed hierarchical framework on Cross-Cube modules. This experiment demonstrates that the proposed approach is efficient for self-repair of a modular robot in the presence of some failed modules.

To verify the efficiency and robustness of the rule-based controller for pattern generation, an adaptive vehicle pattern is constructed automatically using the proposed hybrid hierarchical model. During the pattern generation process, the positive morphogen value is set as 10 and the negative morphogen value is -10. A set of snapshots showing the adaptation of the vehicle pattern to environmental changes is provided in Fig. 6. First, the pattern generation controller generates a vehicle pattern based on the width of path it needs to traverse using the rule-based method. As the vehicle moves forward, a narrower path is detected. Consequently, a new vehicle pattern that can fit in this narrower tunnel is generated. Then steps are detected in front of the robot, new target patterns are dynamically generated to allow the robots to climb the steps, and eventually a new vehicle pattern is generated to continue its locomotion task after finishing the climbing. During this procedure, the GRN-based controller for pattern formation layer would automatically reconfigure the modules to form the new target patterns.

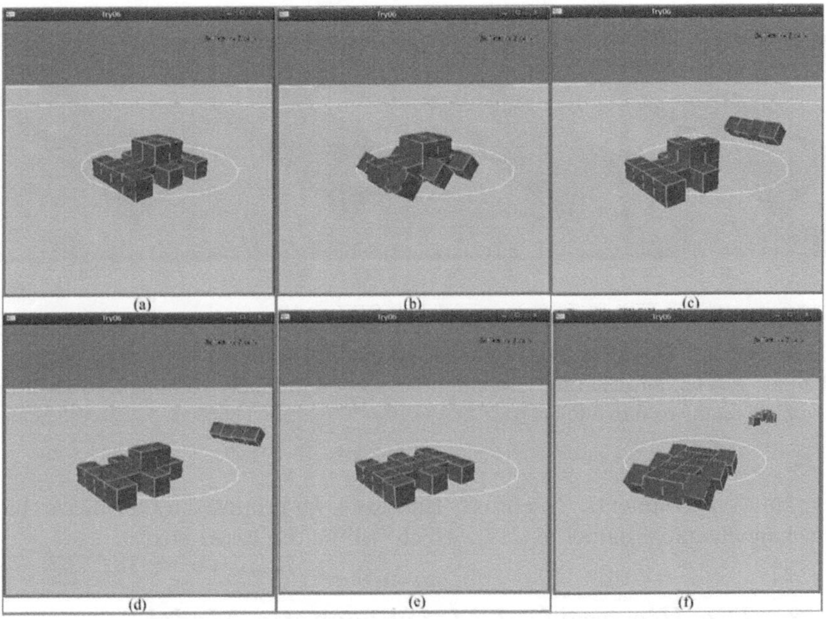

Fig. 5 A set of snapshots of the self-repairing of CrossCube using the GRN-based controller. (a) A vehicle pattern is formed. (b) The vehicle pattern moves forward. (c) Some modules are blown off when the explosion happens. (d) The failed part is filled up by the backup modules. (e) The vehicle is repaired. (f) The repaired vehicle continues moving.

4.2.4 Intermediate Summary

In this hybrid hierarchical model for the self-reconfiguration of Cross-Cube RM robots, layer 1 controller defines the target pattern of the modular robots in a look-up-table and adapts the target pattern using a rule-based method, while layer 2 controller organizes the modules autonomously to achieve the target pattern. The major limitation of this hybrid hierarchical model is that the target pattern has to be predefined in a look-up-table, and the current design of the first layer is a heuristic rule-based method, which can only generate some simple patterns and is hard to generate various patterns for

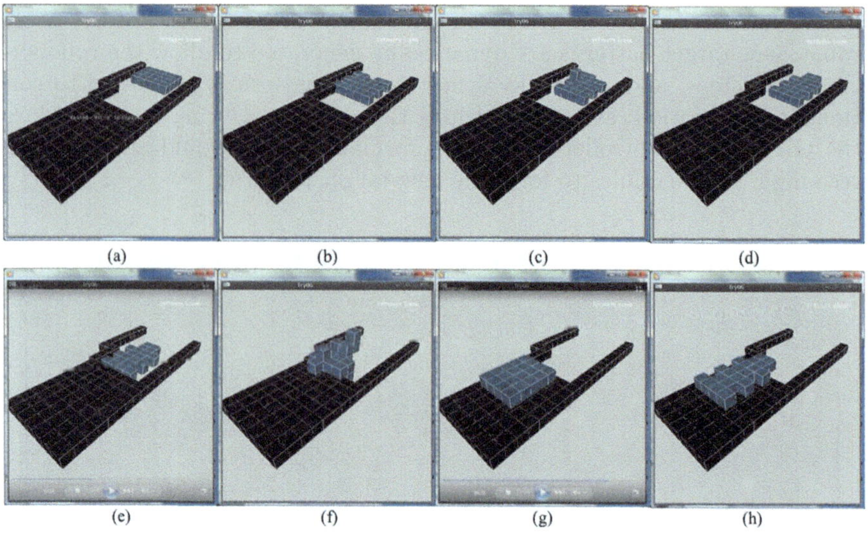

Fig. 6 A set of snapshots demonstrating a series of reconfigurable processes during locomotion and climbing. The robot first adapted its width to the narrow path, then changed its configuration for climbing up a step, and finally reconfigured itself into a vehicle again to move forward.

dynamic environments. To address this issue, we proposed a new hierarchical mechanochemical model in [17], which will be discussed next.

4.3 The Hierarchical Morphogenetic Model

4.3.1 Layer 1 Controller: A Mechanochemical Model

Morphogenesis of multi-cellular systems provides a nice example in which self-organizing pattern formation is realized through cell-cell and cell-environment interactions. One cellular environment is a complex mixture of nonliving material that makes up the extracellular matrix (ECM) [21, 22] . In biological systems, the ECM is the extracellular part of animal tissue secreted by the cells that usually provides structural support to the cells. Meanwhile, the shape of ECM is also affected by the behaviors of the cells through the interactions between ECM and cells. The interaction dynamics between cells and the ECM can be described by a morphogenetic mechanochemical model proposed by Murray et al. [22]. In this model, the simultaneous development of pattern formation and morphogenesis is a closed-loop system where the extracellular matrix (ECM) can be treated as environmental constraints, which is therefore well suited for constructing layer 1 in a hierarchical morphogenetic model. Therefore, a virtual-cell (v-cell) based approach is developed to transfer continuous biological systems to discrete robotic systems. The basic hypothesis and assumptions of this approach are listed as follows:

- There are three basic components in this model: v-cells, ECM, and environment (including task requirements and local environmental constraints). At the beginning, a predefined number of v-cells start to proliferate from a fixed grid. V-cells can proliferate, interact with other v-cells, and diffuse in a 3D space. The ECM and environment provide constraints on the v-cell behaviors.
- Each discrete grid can contain multiple v-cells. The v-cells can only move to the grids that are either occupied by modules, or empty grids that are the modules' immediate neighbors. By following these rules, the v-cells will not move to the grids that are occupied by obstacles or disconnected from the robot system.
- Each grid is associated with one ECM value, which is determined by the density of v-cells of the immediate neighboring grids. If more than one ECM value is generated on the same grid by v-cells from different grids, the final ECM value is the sum of all the generated ECM values.

Based on the above assumptions, a new mechanochemical model for self-organization of modular robots is developed. As shown in Fig. 7, the v-cells (red dots) can move freely in the modules of a modular robot and their neighboring discrete grids, while the ECM and task requirements can produce environmental and task constraints to the behavior of the v-cells.

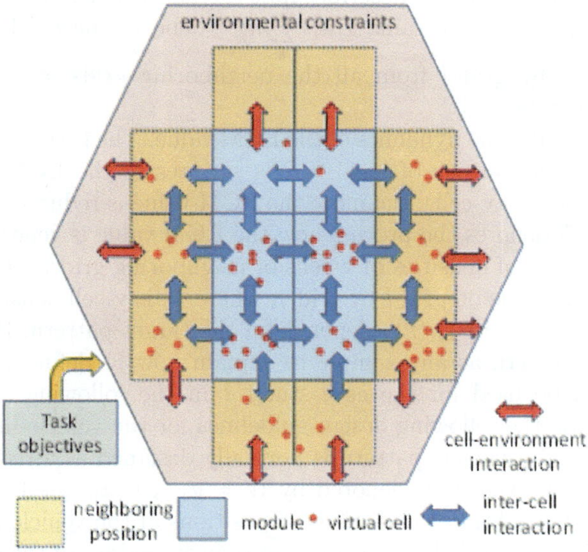

Fig. 7 The virtual-cell based mechanochemical model at a macro-level view.

The density of the v-cells and the ECM value in a grid can be defined as follows:

$$\frac{dn_i}{dt} = r \cdot n_i(N - n_i) - d \cdot K_i \cdot M_i - a \cdot \frac{\rho_i}{n_i + \rho_i} + \sum_k n_k^{rec} \tag{6}$$

$$\frac{d\rho_i}{dt} = -b \cdot \frac{n_i}{n_i + \rho_i} + e \cdot \sum_j f_{ji}(n_j) \tag{7}$$

where n_i and $\rho_i$represent the density of the v-cells and the ECM value in grid i, respectively. The first term in Eqn.(6) denotes the proliferation rate of the v-cells, where N is the predefined maximum number of v-cells allowed in the grid, and $(r \cdot N)$ is the maximum value of linear proliferation rate of the v-cells in the grid. The proliferation rate of the v-cells will be expedited when the local density of v-cells is low, and will be reduced when the density of v-cells approaches to N.

$K_i \cdot M_i$ denotes the random dispersal of the v-cells, which depends on a dispersal control vector K_i and a density gradient vector M_i for grid i. Since v-cells are allowed to move only to their immediate neighbors, v-cells can have up to 6 directions to move to: up, down, left, right, forward and backward. The dispersal control vector K_i is defined as $K_i = \left[k_i^{up}, k_i^{down}, k_i^{left}, k_i^{right}, k_i^{forward}, k_i^{backward}\right]$, where each element of K_i represents the dispersal rate for each direction. A higher dispersal rate means a faster dispersal of v-cells in that direction.

The third item in Eqn.(6) describes how the density of v-cells is decreased by the ECM value on the same grid. $\sum_k n_k^{rec}$ denotes the total densities of v-cells received by grid i from all the neighboring grids. r, d, and a are predefined constants.

Eqn.(7) describes the dynamics of the ECM value. The first item of Eqn.(7) describes how the density of local v-cells suppresses the ECM value. The higher the density of v-cells, the more the ECM value is reduced. The second item in Eqn.(7) denotes the process how the ECM value is generated in grid $i$based on the sum of densities of v-cells of neighboring grids. $f_{ji}(n_j)$ is the generated ECM value on grid i by grid j, where n_j is v-cell density in grid j.

The detailed rules of $f_{ji}(n_j)$ depends on the target pattern. Here, we use a vehicle-like pattern as an example to explain how to define $f_{ji}(n_j)$. This definition will be used for the case study 1 in the following experiments. We first provide the following design guidelines for the vehicle-like patterns. (1)The chassis of a vehicle pattern is normally designed as a rectangle. The total width of the vehicle is denoted by $W = x_{\max} - x_{\min} + 1$, where $x_{\max}$ and $x_{\min}$ are the rightmost and leftmost positions of the vehicle pattern. (2) Wheel modules are needed on the left and right boundary of the chassis, and cannot be immediate neighbors, otherwise they cannot rotate freely. (3) The target pattern should be built from bottom up. Only when the bottom floor is filled up and there are still modules left, the top floor will be filled.

To optimize the chemical pattern formation using the mechanochemical model, the covariance matrix adaption evolution strategy (CMA-ES) [10] is employed for tuning the parameters of $f_{ji}(n_j)$ in the model. CMA-ES is a stochastic, iterative optimization method belonging to the class of evolutionary algorithms. For example, in the vehicle-like pattern, the parameters to be tuned are: the width of the vehicle W and ECM_{up}. Based on different target patterns, different $f_{ji}(n_j)$ is applied. Therefore, different parameters will be tuned by CMA-ES.

The final target pattern is determined by the morphogen value in each grid. V-cells tend to flow to the grids with lower ECM values. Therefore, the morphogen value mp_i of grid i can be defined as the difference between the density of v-cells and the ECM value in each grid as $mp_i = n_i - \rho_i - z_i$, where z_i is a threshold which is determined by $(n_i - \rho_i)$ so that all the grids with positive morphogen values belong to the target pattern, and others have negative morphogen values. A higher value of morphogen value indicates a higher priority for the grid to be filled by a module. This mechanochemical process stops when the number of the grids with positive morphogen values reaches the number of the available robot modules.

Both v-cells and ECM are virtual objects which don't physically exist. They are represented by variables which are stored in the memory of each module and can be exchanged between the neighboring modules through diffusion. For those grids occupied by modules, the density of v-cells and ECM value are estimated by the hosting modules. For those empty grids where no module exists, the density of v-cells and ECM values are maintained by neighboring modules in a distributed way. Please be noted that only those empty grids which are immediate neighbors of the modules will be considered here.

4.3.2 Layer 2 Controller: A GRN-Based Model

As described above, the target pattern of a RM robot is defined by morphogen values associated to the grids in the target pattern. To place the modules in the grids with positive morphogen value and keep grids with negative morphogen value empty, a local coordinate system must be built up. This can be achieved by setting an arbitrary module to be the origin of the coordinate system [9], and thus all other modules can figure out their relative positions to the module on the origin through local communications. With the help of relative positions and the morphogen values of grids in the target pattern, each module can decide: (1) if attracting proteins should be produced to attract other modules to fill in its neighboring grids, or (2) if repelling proteins should be produced to remove neighboring modules, or (3) if the module should respond to the attraction and repellence requests sent out by other modules. Thus, the reconfiguration of the modular robot is controlled through changing the state of the individual modules. The basic model of the layer 2 controller is similar to the one used in the hybrid controller we discussed before. Therefore, we skip this part here.

4.3.3 Experimental Results

To evaluate the efficiency and robustness of the proposed hierarchical morphogenetic model for the self-reconfiguration of Cross-Cube RM robots, several case studies have been conducted in the same robot simulator we discussed before. In the following experiments, the system parameters of the hierarchical model are set up as follows. The parameters of the GRN model in layer 2 are: $k_2 = 0.7$, $k_3 = 1$, $k_4 = 1$, $k_5 = 0.5$, $k_6 = 2$, $G_{A\_L} = -1$, $G_{A\_H} = 1$, and $G_{P\_L} = -2$. The parameters of the model in layer 1 are: $r = 0.005$, $N = 200$, $d = 1$, $a = 0.05$, $b = 20$ and $e = 1$. For case study 3, the CMA-ES method is employed to evolve the pattern parameters in the mechanochemical model to achieve optimal pattern formation.

First, a vehicle pattern in an open space is generated using the proposed hierarchical morphogenetic model. The parameters in $f(c_i)$ are defined as follows: The parameters in $f(c_i)$ are: $x_{\min} = 0$, $x_{\max} = 4$, $ECM_{up} = 100$, $ECM_{wheel} = n * 100$, where n is the density of v-cells in input position. The pattern threshold Z is set to 170. The robots are distributed in a 4x6 rectangle. Since there is no environmental constraints in an open space, a one-floor vehicle pattern will be generated, which matches the guidelines of the vehicle-like pattern design. The ECM valued generated by the v-cells on the left and the right boarders can be used to generate the wheels. The generated ECM value by a wheel slows down the diffusion of the v-cells into the wheel's neighboring grids. As a result, no two wheel modules can be immediate neighbors so that wheels can freely rotate. A set of snapshots of this experiment are shown in Fig. 8.

Second case study is to verify if the hierarchical morphogenetic model can self-reconfigure the Cross-Cube RM robots to adapt to environmental changes. More specifically, we want to show that a Cross-Cube RM can climb stairs by changing its pattern autonomously. The starting grid of the robot is initialized with 4000 v-cells and the robot is distributed in a 3x3 square. First, in order to trigger the robot to move forward, a forward "force" is applied on v-cells. To update the density of v-cells in each grid, this forward "force" can be obtained from $\sum_k n_k^{rec}$ in Eqn.6. Second, we need to define $f_{ji}(n_j)$ for climbing patterns. Since we want the pattern to be able to move forward and climb up the stairs, diffusion direction of v-cells should be forward and up while keeping the width of current pattern. Therefore, if grid j is at the leftmost or rightmost position of the pattern, $f_{ji}(n_j) = 100 * n_j$, where i can be the left or right grid of j. Third, since the robot has to climb up on the stairs, the moving-up action of the v-cells should be encouraged and the moving-down should be restricted. Therefore the dispersal control vector is

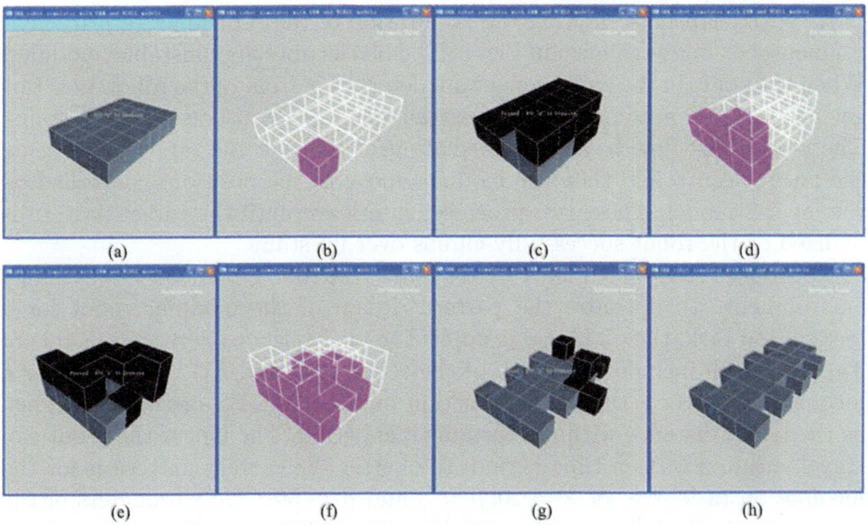

Fig. 8 A set of snapshots of a vehicle pattern generation of a modular robot in an open space.

defined as $k_i^{up} = 1, k_i^{down} = 0, \quad k_i^{left} = 0.5, \quad k_i^{right} = 0.5, \quad k_i^{forward} = 0.7,$ and $k_i^{backward} = 0.7$.

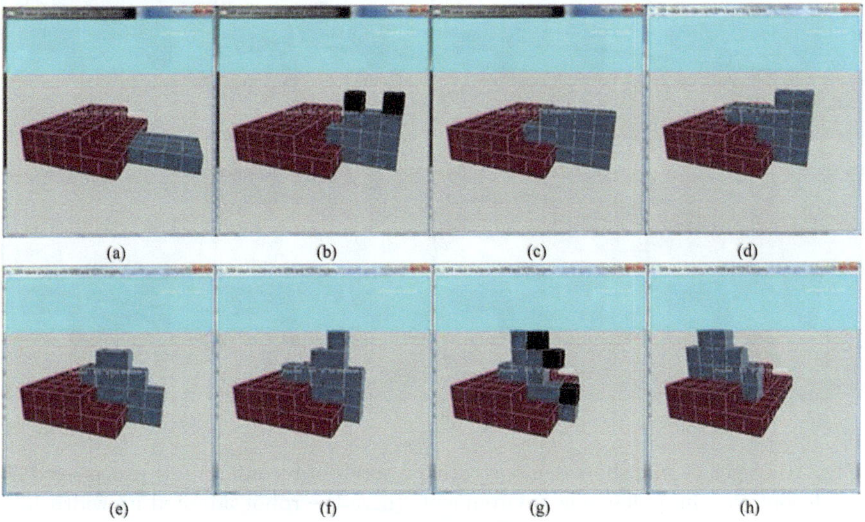

Fig. 9 A set of snapshots demonstrating the reconfigurable processes in the stair-climbing case using the proposed hierarchical model.

A set of snapshots showing the adaptation of the vehicle pattern to environmental changes is shown in Fig. 9. The black cubes are 'unstable' modules. When the stairs in the environment are detected in front of the robot, new target patterns are automatically generated to allow the robots to climb stairs. The v-cells first flow to an intermediate grid (not yet the target grid) using the layer 1 controller, then the modules move to the positions controlled by the layer 2 model. These two processes interleave until the target pattern is realized or the robot successfully climbs over up stairs.

Third case study is to let a Cross-Cube RM robot to traverse a dynamic environment. To optimize the pattern design of the modular robot for a locomotion task, the CMA-ES is applied to evolve the pattern parameters of $f_{ji}(n_j)$, which include the width of the robot (W) and ECM_{up}. Here, for a locomotion pattern, the fitness function of the CMA-ES method is defined as the travel distance within a certain time period. The longer the robot can travel within a certain time period, the better the current pattern is for the locomotion task. We set $\lambda = 20, \mu = 5$ and $\sigma = 50$. The parameters of the best evolved vehicle-like pattern are: $W = 3$ and ECM_{up} =201. Fig. 10 shows a set of snapshots of the self-adaption procedure of a Cross-Cube RM robot to adapt to a changing environment, where the robot dynamically changes its pattern when a narrower passage is detected.

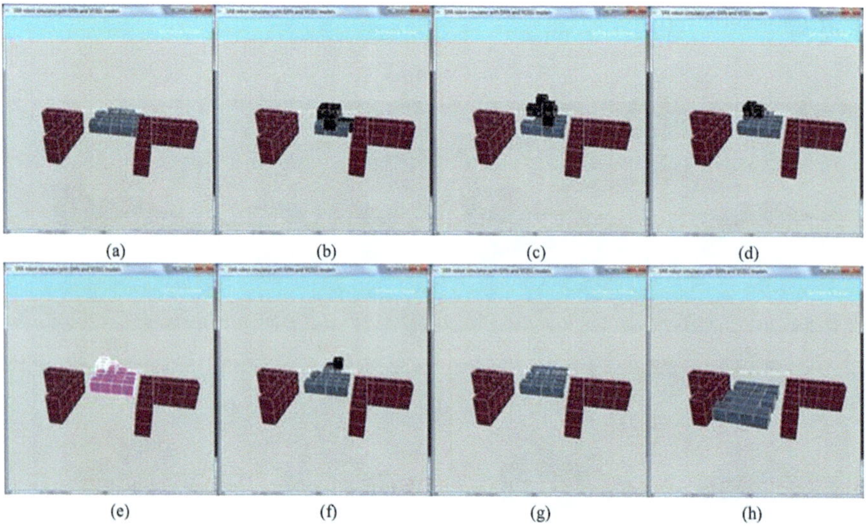

Fig. 10 A set of snapshots demonstrating a series of reconfigurable processes during locomotion in a changing environment where the robot adapted its width to a narrow path.

4.3.4 Intermediate Summary

Cross-Cube can provide very flexible configuration capability. However, the mechanical and electrical design of Cross-Cube is relatively complex to build up the physical prototype. To reduce the design complexity of Cross-Cube while providing good enough flexibility and individual mobility of single module, recently, we improved our mechanical and electrical design of the Cross-Cube and proposed a new RM robot, called Cross-Ball [18] , and also developed the corresponding hierarchical morphogenetic model for the self-reconfiguration of the Cross-Ball RM robots, which will be discussed next.

5 Self-Reconfiguration of Cross-Ball RM Robots

5.1 Hardware Design of Cross-Ball Module

The proposed Cross-Ball module, as shown in Fig. 11, is a sphere with 3-inch diameter. The module is in a ball shape to allow individual mobility and to be more spatial efficient during self-reconfiguration process. It consists of three main components: an arm system in the middle and two halves of a sphere at sides. The two sphere halves can rotate according to the arm system. Six genderless attachment mechanisms are equipped on the six orthogonal directions. Details will be given later. Please refer to [18] for the detailed mechanical design of Cross-Ball and its motion capabilities.

Using the rotary arm and side clasp, a module is able to conduct three types of self-reconfiguration motions: rotating, parallel and diagonal movements. In the following part of the paper, we will call the main rotary arm as "thread". The thread could have three poses: perpendicular to x, y and z axis, respectively.

In the rotating movement, a module can connect its arm to a neighbor's side clasp or arm. Then the module disconnects all its stationary attachments and rotates the main arm clasp. In this way, the whole body will rotate because the other side of the arm has been fixed by the neighboring module. For parallel movement, a module moves to one of its neighboring positions, as shown in Fig. 12(a)-(b). In this movement, two more modules involved besides the moving module, which are called supporting modules. Diagonal movement means a module moves into a neighboring position in the diagonal direction, as shown in Fig. 12(c)-(d).

Based on previous discussion, the Cross-Ball module can rotate, move to parallel positions and diagonal positions along any axis given proper supporting modules. The degree of freedom (DOF) of Cross-Ball is 6. The dimension of a module's connector is 6 due to the stationary attachments. As the module is in a ball shape, the number of cubic space that a module can hold is 1 because it only occupies one cubic space. In general, Cross-Ball can provide competitive reconfiguration capability with Cross-Cube, meanwhile reduces the design complexity significantly.

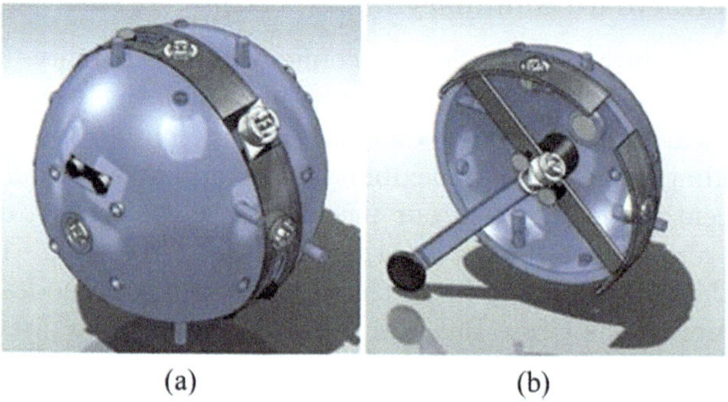

Fig. 11 (a) The proposed Cross-Ball module. Grey part is the rotary arm system with main arm and two clasps. There are also two clasps on the sides of the module. (b) The half model with the wheel extended.

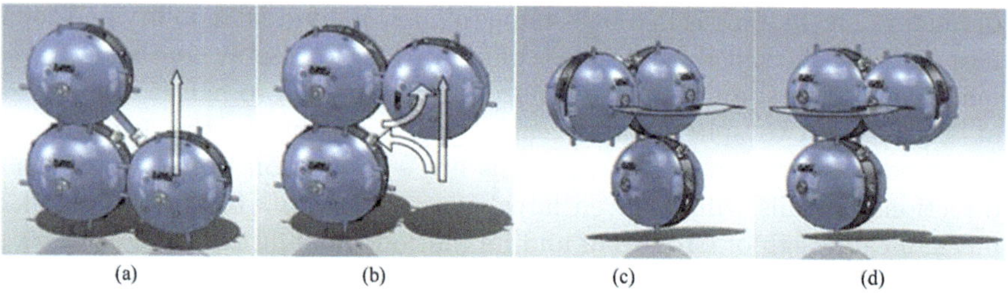

Fig. 12 (a) Before the parallel movement; (b) After the parallel movement. (c) The diagonal movement. The target module is rotated by the arm of the bottom module. Before the diagonal movement; (d) After the diagonal movement.

5.2 The Hierarchical Morphogenetic Model for Self-Reconfiguration

Similar to the Cross-Cube RM robots, a decentralized hierarchical morphogenetic model is proposed for the self-reconfiguration of the Cross-Ball RM robots. However, the hierarchical model for Cross-Ball robots has three layers instead of two layers. The layer 1 controller is a morphogenesis based pattern generator, which is responsible to generate proper target pattern based on the global goal of the current task and the sensory feedback of modules. The layer 2 controller is a GRN-based self-reconfiguration planner, which generates movement decisions for individual module to converge to the target pattern. The layer 3 controller is needed for Cross-Ball robots due to the physical constraints of module movements of Cross-Ball modules, which is responsible for the module motion control.

The layer 1 and layer 2 controllers are genetic and in principle can be applied to any RM robots. Therefore, the mechanochemical model for the layer

1 controller and the GRN-based model for the layer 2 controller proposed for the Cross-Cube RM robots in [17] can be applied directly to the Cross-Ball RM robots. Here, we only focus on the layer 3 controller for the Cross-Ball RM robots.

5.3 Layer 3 Controller: Motion Controller

For Cross-Ball modules, thread configurations are the key to implement the rotating, parallel movement and diagonal movements. During the modules' reconfiguration, their relative positions also change accordingly, which may cause the further movements cannot be implemented due to the lack of proper supporting modules. It is hard to develop comprehensive self-reconfiguration motion controllers for a RM robot due to motion constraints of the modules. Therefore, meta-module based approaches have been introduced in some RM robots [2, 19], where a meta-module is formed up by several SR modules and has less motion constraints than any individual module. Meanwhile, the meta-module enlarges granularity of a RM robot. However, in order to minimize the system granularity and better integrate with the layer 1 and layer 2 controllers, we proposed a skeleton-based algorithm for the layer 3 controller to solve mechanical constraints. First, we define module S as a *skeleton module* if there are 3 adjacent modules that can form up a meta-module with S. Skeleton module S can freely configure its own thread and support the movements of other modules. Fig. 13 shows one example of the meta-module.

The basic idea of the layer 3 controller is to evaluate the self-reconfiguration plan generated from the layer 2 controller and maximize the scale of skeleton modules during self-reconfiguration process. The evaluation criterions include: (1) Skeleton modules will not move if there is non-skeleton "unstable" modules at the current moment; (2) If there is no non-skeleton module, only the skeleton modules with the least number of neighbors will move;(3) For non-skeleton modules and modules who are about to turn to a non-skeleton module by missing neighbors, by priority they should (a) touch the thread to a skeleton module; (b) touch its thread to the thread of a non-skeleton module; or (c) touch its thread to a non-skeleton module. The movements that satisfy all these rules will be executed. Otherwise, they will be ignored. The flow chart of the layer 3 controller is shown in Fig. 14.

By introducing skeleton modules and allowing modules to work in groups (skeleton group and non-skeleton group), a module can easily decide whether to move, and how to choose and move with supporting modules. In other words, the skeleton-based layer 3 controller can significantly reduce the complexity of searching process on the module movement plan. From the system level point of view, by introducing the layer 3 controller, both the layer 1 and layer 2 controllers can be well integrated with the customized hardware design of Cross-Ball and its corresponding locomotion capabilities.

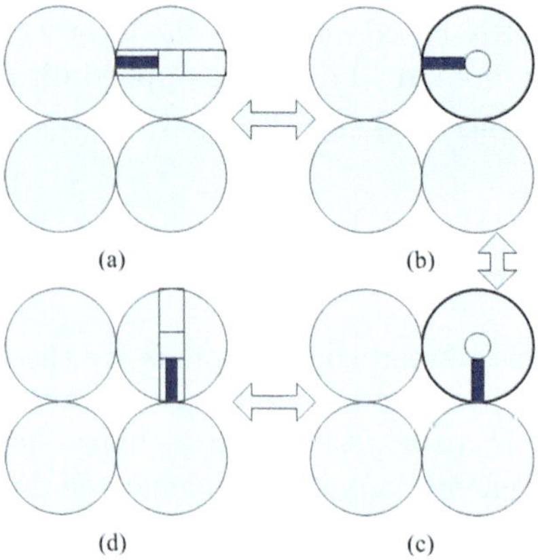

Fig. 13 One meta-module of a Cross-Ball RM robot. (a) The upper right module connects its main arm to the module on the left, disconnects all stationary attachments, and rotates the main arm clasp to change the pose of thread. (b) Reconnect the stationary attachments to finish thread pose adjustment. (c) The module connects its main arm to the module below, disconnect stationary attachments, and rotate the main arm clasp to change the pose of thread. (d) Reconnect stationary attachments to finish thread adjustment. All the process is reversible so the meta-module can freely adjust the pose of thread.

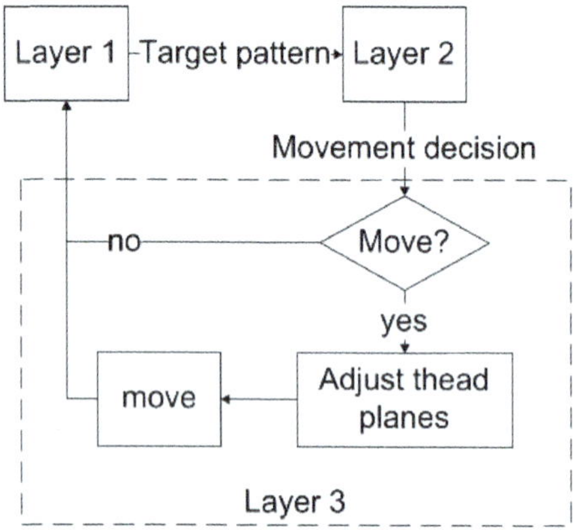

Fig. 14 The flow chart of the layer 3 motion controller.

5.4 Experimental Results

To demonstrate the self-reconfiguration of Cross-Ball RM robots using three-layer morphogenetic controllers, we conduct another experiment where 27 Cross-Ball modules can automatically transform from a snake pattern to a human-like legged pattern. Some snapshots of this self-reconfiguration procedure are shown in Fig. 15(a)-(d). Initially, the space cubes in the human-like legged pattern are set as non-ECM cubes in the layer 1 controller. Then all modules in the non-ECM cubes are initialized with 500 v-cells, as shown in Fig. 15(a). The modules with darker color have more number of v-cells than those with lighter color. When the system starts, the v-cells diffuse to non-ECM cubes so that more space cubes are inserted in the target pattern, as shown in Fig. 15(b). In Fig. 15(b), some modules are relocated into spaces cubes with more v-cells. Therefore, their color becomes darker. Eventually, a human-like legged pattern (see Fig. 15(c)-(d)) is constructed. Then, the human-like legged pattern is reconstructed to a vehicle-like 4 legged pattern with some cargo space on the top. The modules at the end of legs can rotate as wheels for a vehicle-like pattern. Some snapshots of this procedure are shown in Fig. 15(e)-(h). From Fig. 15, we can see that the proposed 3-layer hierarchical morphogenetic framework can efficiently self-reconfigure the Cross-Ball RM robots to various complex patterns.

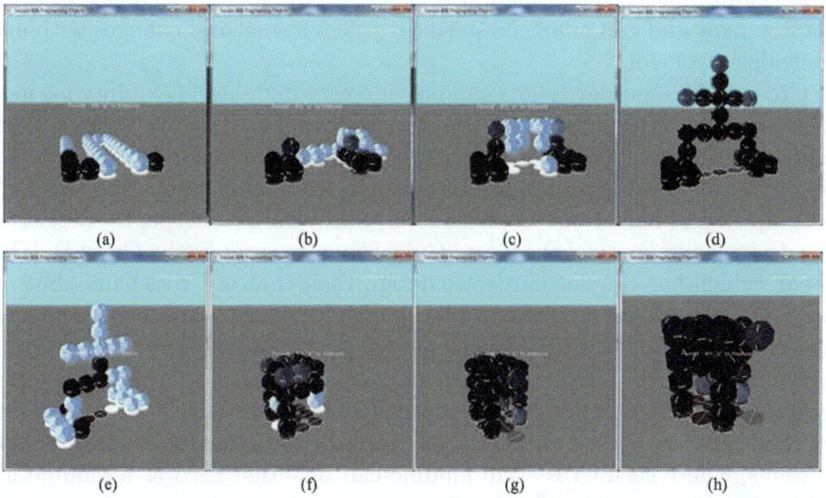

Fig. 15 Snapshots of self-reconfiguration from a snake pattern to a human-like legged pattern (a)-(d), and from a human-like legged pattern to a vehicle-like 4 legged pattern (e)-(f) using the Cross-Ball modules and the proposed 3-layer hierarchical morphogenetic framework.

6 Conclusions

In this chapter, we introduced two new RM robots: Cross-Cube and Cross-Ball. Then we presented a hierarchical morphogenetic approach for the self-reconfiguration of modular robots, which is inspired by multi-cellular morphogenesis. Such a hierarchical structure makes it possible to separate the control mechanisms for defining a target configuration from those for realizing it, similar to biological gene regulatory networks. In response to environmental changes, layer 1 is able to define appropriate target pattern to adapt to new environments, based on which layer 2 generates self-reconfiguration plans for modules to achieve the desired patterns. Depending on the physical constraints of the module movements, layer 3 is dedicated to dealing with physical constraints of module movements so that the target patterns can be eventually achieved. In other words, both layer 1 and layer 2 are generic and in principle can be applied to various RM robots, whereas the controller in layer 3 is hardware specific. Different modular robots require different layer 3 controllers. With more physical constraints of the module movements, the controller in layer 3 becomes more complex.

In the rule-based models for RM robots, heuristic rules have to be predefined to predict all possible situations in the environments, which is impossible in dynamic environments and not flexible enough to adapt to various environmental changes. By contrast, the hierarchical morphogenetic model can self-organize modules autonomously based on the sensory feedback from the environment and can adapt its patterns to environmental changes without a centralized control.

Cross-Cube modular robots can provide very high-level flexibility for module movements without any physical constraints of module movements, therefore, only layer 1 and layer 2 controllers are required for Cross-Cube RM robots for self-reconfigurations. However, the hardware design of Cross-Cube is complex and hard to be implemented in physical prototype. By decreasing the design complexity of the Cross-Cube robot, we proposed a Cross-Ball RM robot, which has simpler hardware design than that of Cross-Cube, but can provide good enough flexibility for module movements. To deal with the physical constraints of module movements of the Cross-Ball modules, a layer 3 controller is added. Furthermore, the independent mobile capability of Cross-Ball module can not only provide more flexibility for the self-reconfiguration, but also provides great potential to be naturally integrated into swarm robotic systems, where each Cross-Ball module can be either used as a module in a RM robot or an individual robot in a swarm robot system.

There are still some unresolved issues in the current hierarchical morphogenetic framework. First, the layer 1 and layer 2 controllers need to be simplified to improve the overall control efficiency of the self-reconfiguration. Second, although layer 3 controller works in a decentralized manner, it still depends on some global information (a module may have to collect position information from all other modules). This may increase the communication and computational costs when the size of modules increases. Third, although the current design of Cross-Ball works in the embodied simulation environment, we still need to build up the physical prototype of the Cross-Ball to further verify the design efficiency and evaluate the proposed hierarchical morphogenetic framework for RM robots. We will investigate all these issues in our future work.

Acknowledgments

This project is partially supported by Honda Research Institute Europe GmbH, 63073 Offenbach/Main, Germany. The authors would also like to thank Yuyang Zhang for the illustrative examples used in this book chapter.

References

1. Alon, U.: Network motifs: theory and experimental approaches. Nature Review Genetics 8, 450–461 (2007)
2. Brandt, D., Christensen, D.J.: A new meta-module for controlling large sheets of atron modules. In: IEEE/RSJ International Conference on Intelligent Robots and Systems, pp. 2375–2380 (2007)
3. Butler, Z., Fitch, R., Rus, D.: Distributed control for unit-compressible robots: goal-recognition, locomotion, and splitting. IEEE/ASME Transactions on Mechatronics 7(4), 418–430 (2002)
4. DeJong, H.: Modeling and simulation of genetic regulatory systems: A literature review. Journal of Computational Biology 9, 67–103 (2002)
5. Endy, D., Brent, R.: Modeling cellular behavior. Nature 409, 391–395 (2001)
6. Fitch, R., Butler, Z.: Million module march: Scalable locomotion for large self-reconfiguring robots. The International Journal of Robotics Research 27(3-4), 331–343 (2008)
7. Fukuda, T., Nakagawa, S., Kawauchi, Y., Buss, M.: Self-organizing robots based on cell structures - cebot. In: IEEE/RSJ Int. Conf.on Intelligent Robots and Systems, pp. 145–150 (1988)
8. Gilpin, K., Kotay, K., Rus, D., Vasilescu, I.: Miche: Modular shape formation by self-disassembly. The International Journal of Robotics Research 27, 345–372 (2008)
9. Guo, H., Meng, Y., Jin, Y.: A uniform framework for self-organized multi-robot pattern formation and boundary coverage inspired from morphogenesis. ACM Trans. on Autonomous and Adaptive Systems (2003) (accepted)
10. Hansen, N., Muller, S.D., Kououtsakos, P.: Reducing the time complexity of the derandomized evolution strategy with covariance matrix adaptation (cma-es). Evolutionary Computation 11, 1–18 (2003)
11. Hiroshi, U., Hiroshi, S., Tetsuji, Y., Yoshiaki, O., Saburo, M., Ryoichi, H., Jun, M.: Ground testbed of reconfigurable brachiating space robot. Advanced robot 14, 355–358 (2000)
12. Hou, F., Shen, W.-M.: Distributed, dynamic, and autonomous reconfiguration planning for chain-type self-reconfigurable robots. In: IEEE International Conference on Robotics and Automation, pp. 3135–3140 (2008)
13. Jorgensen, M.W., Ostergaard, E.H., Lund, H.H.: Modular atron: modules for a self-reconfigurable robot. In: IEEE/RSJ International Conference on Intelligent Robots and Systems, vol. 2, pp. 2068–2073 (2004)
14. Kamimura, A., Kurokawa, H., Yoshida, E., Tomita, K., Kokaji, S., Murata, S.: Distributed adaptive locomotion by a modular robotic system, m-tran ii. In: IEEE/RSJ International Conference on Intelligent Robots and Systems, vol. 3, pp. 2370–2377 (2004)

15. Kurokawa, H., Tomita, K., Kamimura, A., Kokaji, S., Hasuo, T., Murata, S.: Distributed self-reconfiguration of m-tran iii modular. The International Journal of Robotics Research 27, 373–386 (2008)
16. Meng, Y., Zhang, Y., Jin, Y.: A morphogenetic approach to self-reconfigurable modular robots using a hybrid hierarchical gene regulatory network. In: 12th International Conference on the Synthesis and Simulation of Living Systems, ALIFE XII (2010)
17. Meng, Y., Zhang, Y., Jin, Y.: Autonomous self-reconfiguration of modular robots by evolving a hierarchical mechanochemical model. IEEE Computational Intelligence Magazine 6(1), 43–54 (2011)
18. Meng, Y., Zhang, Y., Sampath, A., Jin, Y., Sendhoff, B.: Cross-ball: A new morphogenetic self-reconfigurable modular robot. In: IEEE/RSJ International Conference on Robotics and Automation (2011)
19. Murata, S., Kurokawa, H.: Self-reconfigurable robots. IEEE Robotics and Automation Magazine 14(1), 71–78 (2007)
20. Murata, S., Yoshida, E., Kamimura, A., Kurokawa, H., Tomita, K., Sokaji, S.: M-tran: Self-reconfigurable modular robotic system. IEEE/ASME transactions on mechatronics 7(4) (2002)
21. Murray, J.D.: Modelling biological pattern formation in embryology. ISI Atlas of Science: Animal and Plant Sciences 1, 270–274 (1988)
22. Murray, J.D., Maini, P.K.: Mechanochemical models for generating biological pattern and form in development. Physics Reports 2, 59–84 (1988)
23. Nguyen, A., Guibas, L., Yim, M.: Controlled module density helps reconfiguration planning. In: Workshop on the Algorithmic Foundations of Robotics (2000)
24. Rosa, M., Goldstein, S., Lee, P., Campbell, J., Pillai, P.: Scalable shape sculpting via hole motion: Motion planning in lattice-constrained modular robots. In: Proc. IEEE Int'l Conf. on Robotics and Automation (2006)
25. Salemi, B., Moll, M., Shen, W.-M.: Superbot: A deployable, multi-functional, and modular self-reconfigurable robotic system. In: IEEE/RSJ International Conference on Intelligent Robots and Systems, pp. 3636–3641 (2006)
26. Sastra, J., Chitta, S., Yim, M.: Dynamic rolling for a modular loop robot (2006)
27. Schmickl, T., Hamann, H., Stradner, J., Crailsheim, K.: Hormone-based control for multi-modular robotics. In: Levi, P., Kernbach, S. (eds.) Symbiotic Multi-Robot Organisms: Reliability, Adaptability, Evolution. Springer, Heidelberg (2010)
28. Shen, W.-M., Kirivokon, M., Chiu, H., Everist, J., Rubenstein, M., Venkatesh, J.: Multimode locomotion for reconfigurable robots. Autonomous Robots 20(2), 165–177 (2006)
29. Shen, W.-M., Salemi, B., Will, P.: Hormone-inspired adaptive communication and distributed control for conro self-reconfigurable robots. IEEE Transactions on Robotics and Automation 18(5), 700–712 (2002)
30. Stoy, K., Nagpal, R.: Self-reconfiguration using directed growth. In: Alami, R., Chatila, R., AsamaLevi, H. (eds.) Distributed Autonomous Robotic Systems, vol. 6, pp. 3–12. Springer, Japan (2007)
31. Terada, Y., Murata, S.: Automatic modular assembly system and its distributed control. The International Journal of Robotics Research 27(3-4), 445–462 (2008)
32. Unsal, C., Killiccote, H., Kholsa, P.K.: A modular self-reconfigurable bipartite robotic system: Implementation and motion planning. Autonomous Robots 10, 23–40 (2001)

33. Wolpert, L.: Principles of Development. Oxford Unviersity Press, Oxford (2002)
34. Yim, M., Duff, D.G., Roufas, K.D.: Polybot: a modular reconfigurable robot. In: IEEE International Conference on Robotics and Automation, vol. 1, pp. 514–520 (2000)
35. Yoshida, E., Murata, S., Kamimura, A., Tomita, K., Kurokawa, H., Kokaji, S.: A motion planning method for a self-reconfigurable modular robot. In: Proceedings 2001 IEEE/RSJ International Conference on Intelligent Robots and Systems, pp. 590–597 (2001)
36. Zhang, Y., Fromeherz, M., Crawford, L., Shang, Y.: A general constraint-based control framework with examples in modular self-reconfigurable robots. In: Proc. of IEEE/RSJ Int'l Conf. on Intelligent Robots and Systems (2002)
37. Zykov, V., Mytilinaios, E., Desnoyer, M., Lipson, H.: Evolved and designed self-reproducing modular robotics. IEEE Transactions on Robotics 23(2), 308–319 (2007)

Basic Problems in Self-Assembling Robots and a Case Study of Segregation on Tribolon Platform

Shuhei Miyashita, Aubery Marchel Tientcheu Ngouabeu, Rudolf M. Füchslin, Kohei Nakajima, Christof Audretsch, and Rolf Pfeifer

Abstract. It has been a quite while since people realized that self-assembly technique may be a strong method to manufacture 3D micro products. In this contribution, we investigate some major concerns about realizing such a small sized robot. First we introduce the concept of self-assembly and introduce examples both from nature and artificial products. Followed by the main problems in self-assembly which can be seen in various scales, we classify them into four groups - (A) assembly constraint issues, (B) stochastic motion issues, (C) interactions on physical property issues, and (D) engineering issues. Then we show a segregation effect with our developed platform as an example of self-organizing behavior achieved in a distributed manner.

1 Self-assembly

One of the major features of biological systems is that the activities on the molecular level are realized in a decentralized fashion, namely, without any central control. One aspect of this phenomenon is self-assembly, defined by Whitesides *et al.* as *the autonomous organization of components into patterns or structures without human intervention* [37][1]. Such a new composition method has large potential in manufacturing 3D micro products, where a pick-and-place style fabrication method is still the major approach taken.

Shuhei Miyashita · Aubery Marchel Tientcheu Ngouabeu · Rudolf M. Füchslin · Kohei Nakajima · Christof Audretsch · Rolf Pfeifer
Artificial Intelligence Laboratory Zurich,
University of Zurich, Andreasstrasse 15, 8050 Zurich
e-mail: miya@ifi.uzh.ch

[1] Note the notion self-assembly does not only contain instance of decentralized functionality.

Y. Meng and Y. Jin (Eds.): Bio-Inspired Self-Organizing Robotic Systems, SCI 355, pp. 173–191.
springerlink.com

1.1 Self-assembly in Nature

The instances of self-assembly can be widely found in nature – snow flakes, in which the composing atoms form ordered lattices through the attractive/repulsive interaction forces, employs self-assembly for its spontaneous crystallization. The shapes are commonly hexagonal, but the details of the patterns vary depending on the environmental conditions, such as humidity or temperature. The crystallization begins with a "core" - often a spec of dust in the air - allowing other floating atoms to connect to the seed. Once the atoms connect to the crystal, they change their conformations, exposing the other connection sites, allowing further atoms to attach. In other words, information of the connections is conveyed to external atoms. The system is conservative in terms of energy dispersion, that is, once an atom connects to the cluster and changes the form, it preserves the energy and sustains the formation by means of a hydrogen bond, unless the temperature rises and breaks the bond.

Lately, this kind of automatic assembly has been brought to attention as the difficulty of manufacturing small sized robots starts to be a limiting factor. Now many researchers believe it sounds reasonable to consider self-assembly as a promising tool to be put into practice for the realization of life like machines (e.g. self-repairable machines). However, despite nature's efficiency and precision in assembling supramolecular and mesoscopic structures, most of the attempts on artificial self-assembly components still remain a challenging assignment.

1.2 From Self-assembling Blocks to Self-assembling Robots

Progressive experiments on artificial self-replication were conducted by Lionel and Roger Penrose half a century ago [27], where a provoking mechanical model of natural self-replication in a stochastic environment was presented. Followed by speculations about the clustering patterns of passive elements, focusing on the role of shape on template and components matching [10], and on their time evolution [18]. A series of studies were conducted by the group of Whitesides; for the realization of positional coordinate of molecule-mimetic chemistry [7, 6, 17, 38], circuit functionality [13, 5, 4], reversible aggregation [22], folding structure [3], rotation of magnets [16], rotation of rotors [15]. Similarly, numerous research effort has been devoted to the investigation of morphology [31]. Artificial chemicals that can form in several ways, such as polymers and dimers, depending on the temperature of the system were demonstrated in [8]. Different aggregation patterns with various sizes of components were shown in [39]. An intelligent self-assembling block which can represent multiple states by the units' rotational angle was designed by [34]. The system can physically conduct XOR calculation on a 2D plane.

Currently, there is a growing interest in realizing self-reconfigurable robots relying on stochastic self-assembly. White *et al.* studied two systems in which the modules binding preferences are coded in a program executed by an on-board microcontroller, and thus can easily reconfigure the structure [35]. The modules are initially unpowered and passive, but once they bind to a seed module connected to a power supply, they become active. Griffith *et al.* studied a system of template-replicating modules [14]. They used modules of the same type, which are programmable and can store distinct states. The system demonstrated the self-replication of a five modules polymer. Each module executed a finite-state machine. Klavins *et al.* examined the problems of designing a grammar that causes modules to assemble into desired products, of predicting the time complexity of such processes, and of predicting (and optimizing) the yield of such processes [20]. Emergent self-propulsion mechanisms were investigated by Ishiguro *et al.* [19]. In Ant-inspired robotics, the interest in self-organization has been driven by the observations of the same phenomena in ant colonies, in particular the brood sorting by *Temnothorax* [30]. *Wilson et al.* [21] created an algorithm to realize two colors annular sorting which used differential pull-back distances for different object types. By discriminating between three puck types, the robots could drop the first type of object on colliding with another puck, drop the second object type after pulling back a short distance and drop the third puck type after pulling back a further distance.

2 Major Concerns in Self-assembly

In this section, we outline the problems of self-assembling systems with respect to the scaling behavior of underlying principles.

2.1 The Forward Problem and the Backward Problem

In self-assembly, the problem to derive the final configuration from a given set of components/environments is called the *forward problem* [26][2]. Conversely, the problem of designing components for a targeted configuration is called the *backward problem*. In this "reverse engineering" process, also known as one of the central problems in self-assembly, the designer has to start from the final structure and decompose it.

Several aspects of both, the forward and the *backward problem*, show strong dependence on the length scale of the system. A necessary condition for the prediction of the result of a self-assembly process is a detailed knowledge about the morphology of the components. And this knowledge is easier to get the larger the components are[3]. In contrast to cm-sized components, a molecule usually has many degrees of freedom and therefore the morphology and consequently the interaction between components are not always known with sufficient precision.

Similar considerations hold for the *backward problem*, and, from an engineering perspective, to an even higher degree. Given an object O for which one wants to design a self-assembly process. There are many possibilities to divide a large O into components. The smaller O, the more constraints with respect to the production of

[2] The game *Tetris*® is known as NP-hard problem [12]. Also it may be useful to mention that some situations in self-assembly resemble the Knapsack problem, which is also known as NP-complete problem.

[3] This does not only refer to shape but also to other features of morphology, such as elasticity and degrees of freedom.

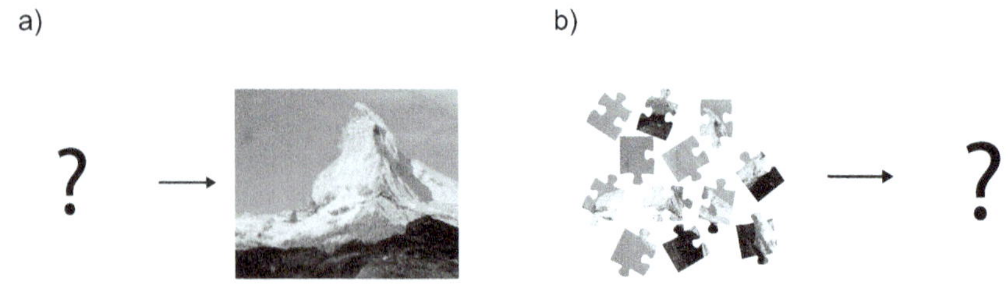

Fig. 1 (a) The backward problem. (b) the forward problem.

the components have to be considered: On the molecular scale, chemical synthesis set narrow limits to what can be accomplished.

We recognize three main problems centering around self-assembly: namely, the issues entailed in (A) assembly, (B) Dynamics, and (C) interactions. The explanations for each problem follow.

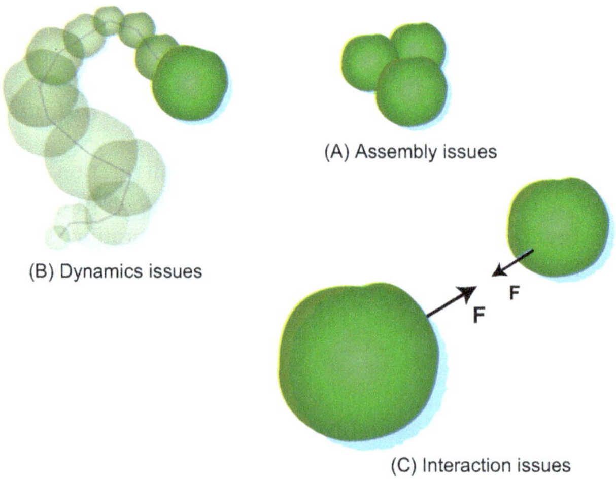

Fig. 2 Three main issues centering around self-assembly.

2.2 *(A) Assembly*

The first issues are properties of assembling processes.

2.2.1 The Mismatch Problem (Error)

In self-assembly, an assembly error (or wrong attachment) is induced when the system converges to a energetically local minimum through interactions between components, mainly due to the low encoding accuracy of bonding sites (Figure 3).

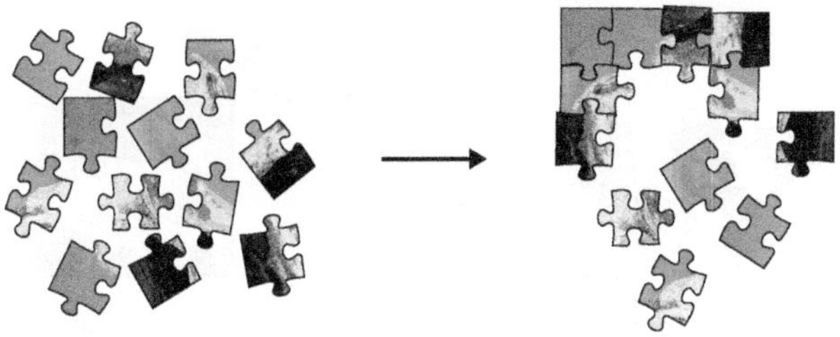

Fig. 3 The *mismatch problem (Error)*.

There are two main strategies for solving the problem; increasing encoding accuracy of bond matching while regulating the agitation level of the system, and implementing internal states to components. Insights from molecular biology remind us of the importance of the fertile encoding capability to attain the adequate bonding affinity level for maintaining connections of molecular bonding; they exploit non-covalent bonds (hydrogen bonds, ionic bonds, and van der Waals attractions) as interaction forces and somehow achieve an amazing specificity in docking with other selected molecules[4]. The agitation level can be regulated by means of temperature or kinetic turbulence magnitudes.

From an engineering perspective, the scaling behavior of the *mismatch problem* exhibits an interesting feature. The relative easiness of the *backward problem* enables one to construct highly specific, literal plug-and-socket connection sites on the *cm* scale.

Molecules, on the other hand, may well be highly flexible (having many degrees of freedom) and, agitated by thermal motion, "sample" their configuration space at a rapid pace. Because of this, e.g. two complementary DNA strands just have be brought in close proximity: If they match, they will eventually bind. Such a "fast configuration sampling" is not possible above the molecular scale. One of the reasons for this is that whereas mechanical structures wear off, molecules don't: A molecular "joint" can be bended infinitely many times (as long as the bond doesn't break, it is in all respects as good as a newly formed one). It is the micro-to (sub-)millimeter scale, at which molecular bonding (and corresponding recognition) is not strong enough anymore, whereas mechanical plug-and-socket connection mechanisms are still hard to produce.

The *mismatch problem* is a fundamental problem in nature. The replication processes of DNA is greatly assisted by self-repair functionalities of enzymes.

2.2.2 The Topological Dead End Problem (Steric Hindrance)

This problem occurs when components assemble in an undesired sequential order. The targeted structure is therefore unreachable, since some earlier assembled components block the way (Figure 4).

[4] The trick of proteins distributing bonding sites around the body, and changes the morphology to pose another bonding site reminds us of the importance of internal states, which enables the component to feature different properties.

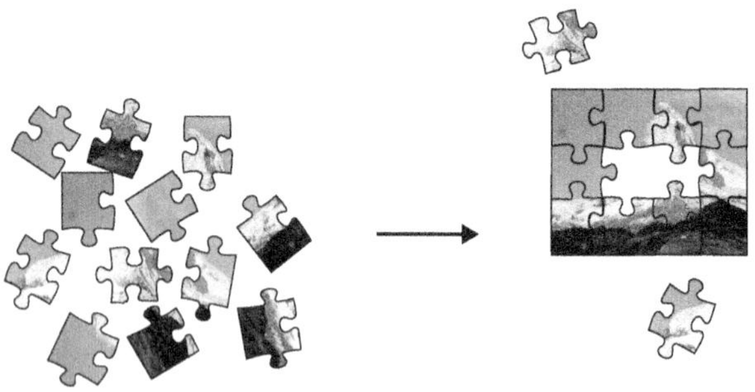

Fig. 4 The *topological dead end problem (Steric hindrance)*.

To solve this problem at scales where the benefits of molecular mechanical flexibility cannot anymore be harvested, the components should reflect the presence of its neighbors e.g. as the internal states. Yet in practice the amount of expressible internal states is limited due to the limited space in a component, leading to a risk of misrecognition by other components.

The unreachable problem can occur irrespective of the heterogeneous/homogeneous level of a system.

2.2.3 The Parallel Yield Problem (Incompletion Problem)

The problem of producing a desired configuration in large quantities (while avoiding incomplete assemblies) by homogeneous system is known as the *parallel yield problem* and has been studied in the context of biological and non-biological self-assembling systems [18][5]. Here, we term the problem that specifically occurs when components assemble in a right manner, however do not complete the targeted final structure, for combinatorial reasons (Figure 5; we assume the circular sector components connect side-by-side). This is because many assembly processes proceed in parallel and components are used in earlier more likely reactions of other assembly processes since reactions leading to the complement of the end product are more unlikely than the preceding reactions. This means that the self-assembly of many products is started but rarely fulfilled. In other words, the likelihood to accomplish the desired end product declines with the rise of the ratio between the likelihood of the earlier and the later reaction.

One approach to improve this problem is, therefore, controlling the system so that the later stages' reactions to be more likely to happen than the early stages. This can be handled by implementing internal states in a component, so that the component behaves differently as a reaction takes place.

Another solution is increasing the heterogeneity of the system, such that in extreme, a product consists of a set of totally different components. Yet a certain

[5] Hosokawa called it "yield problem", while we term it the *parallel yield problem* to avoid confusion.

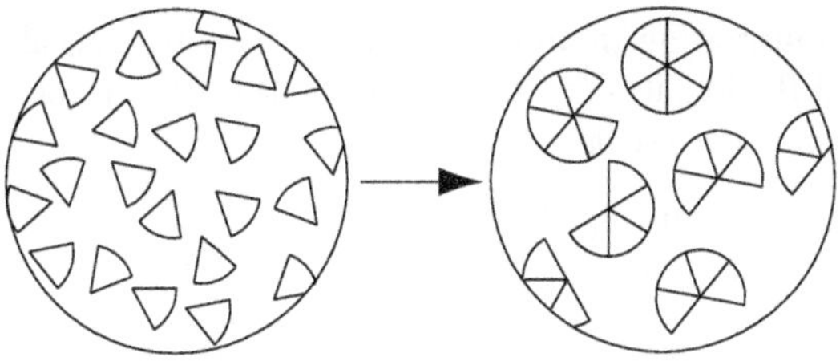

Fig. 5 The *parallel yield problem (Incompletion problem).*

disadvantage of heterogeneity increase is, as described above, that leads to the heightened likelihood of mismatching.

2.3 (B) Dynamics

Self-assembly is commonly believed to range from molecular to cosmological scales. However, it is also agreeable that few examples of self-assembly exist in our human living scales ($cm - m$). The second concern is about stochasticity, which varies to different scales.

2.3.1 Reynolds Number

Biological systems in the $nm - \mu m$ scale often show unique behaviors that cannot be observed in larger scales. This is mostly due to the influence of viscosity, which increasingly becomes dominant with decreasing length scales. The Reynolds number $\Re$ represents a ratio between viscous forces and inertial forces [28];

$$\Re \equiv \frac{\text{inertial forces}}{\text{viscous forces}} \approx \frac{av\rho}{\eta} = \frac{av}{\nu}. \tag{1}$$

where a is radius of a particle, v is its speed, μ is fluid viscosity, and ρ is fluid density. The kinematic viscosity ν is approx. $10^{-2}cm^2/sec$ for water.

The size of 1 cm is a critical size for self-assembling systems. For objects in water at the mm scale, viscosity is as important as inertia (the Reynolds number, that is, the ratio of inertial forces and viscous forces, is ≈ 1). It follows that objects smaller than that size are affected more by viscous forces whereas larger objects are affected more by inertial forces. For objects on the order of 1 μm or less, such as bacteria, exploiting an environmental diffusion is a more effective way of locomotion than active propulsion (e.g., swimming bacteria are slower than diffusing molecules [24]). Good thought-provoking suggestions about the life at low Reynolds number are introduced in [28]. The author states the efficiency of creatures in small scale (μm) such as E. coli to use diffusion through their environment to change their position, rather than self-propelling. Whitesides implies the mechanical system in nanoscale would be different from that in micro scale, and one should learn more from biological systems [36].

The time for transporting anything a distance l by stirring, is about l/v. Whereas, for transport by diffusion, it is l^2/D, where D is the diffusion coefficient in cm^2/sec [28]. Namely in the micro scale,

time for transport by stirring: $\frac{l}{v}$
time for transport by diffusion: $\frac{l^2}{D}$,

and the ratio of these two (termed stirring number; §) is

$$\S \equiv \frac{\text{time for transport by stirring}}{\text{time for transport by diffusion}} = \frac{lv}{D} \approx 10^{-2} \tag{2}$$

which shows the efficiency of diffusion on a small scale.

2.3.2 Navier-Stokes

Incompressible flow of the Navier-Stoks equation is

$$\rho\frac{\partial \mathbf{v}}{\partial t} + \rho(\mathbf{v}\cdot\nabla)\mathbf{v} = -\nabla P + \eta\nabla^2 v \tag{3}$$

where P is the pressure. Ignoring the term of inertia, and considering the large η it can be transformed as

$$\nabla P = \eta\nabla^2 v. \tag{4}$$

It is known that the motions which are that invariant under time reversal does not induce a tow movement.

2.3.3 Diffusion Equation

Consider a particle that exists at $x = 0$ at $t = 0$. The positioning probability ($\rho(x,t)$) of x follows the diffusion equation:

$$\frac{\partial \rho}{\partial t} = D\frac{\partial^2 \rho}{\partial x^2} \tag{5}$$

where D is a diffusion constant.

Considering the initial condition $\rho(x,0) = \delta(x)$, and taking that the ρ satisfies the following normalized condition

$$\int_{-\infty}^{\infty} \rho(x,t)dx = 1, \tag{6}$$

we obtain

$$\rho(x,t) = \frac{1}{\sqrt{4\pi Dt}}\exp\left(-\frac{x^2}{4Dt}\right), \tag{7}$$

which obeys the Gaussian distribution.

The mean-square displacement $\langle x^2 \rangle$ can be derived as

$$\langle x^2 \rangle = \int_{-\infty}^{\infty} x^2 \rho(x,t) dx = 2Dt \propto t, \tag{8}$$

where $D = \frac{k_B T}{\zeta}$, k_B is the Boltzmann constant and ζ is a friction coefficient.

In our scale, where the viscosity is negligible, using agitation for traveling is a good tactic. Whereas in the molecular scale, Brownian motion enables the speedy spacial transitions. It is in the intermediate scale (μm), where those tactics lose validity because of the high viscosity and relatively small momentum.

2.4 *(C) Interaction*

The third concern is about physical interactions among components. The scalability of physical interaction mechanisms – especially electrostatic and magnetic – are well examined in [11, 1]. Here we briefly describe the basics of these two physical quantities.

2.4.1 Electrostatic Interactions

Given that an electric charge q_i exists. The electric field E created by this charge is

$$\mathbf{E} = \frac{q_i}{4\pi\varepsilon_0} \frac{\hat{\mathbf{r}}}{|\mathbf{r}|^2}. \tag{9}$$

where ε_0 is the electric permittivity of free space.

The force F_{ji} that electric charge q_j receives is given by

$$\mathbf{F} = q_2 \mathbf{E} \tag{10}$$

$$= \frac{q_1 q_2}{4\pi\varepsilon_0} \frac{\hat{\mathbf{r}}}{|\mathbf{r}|^2}. \tag{11}$$

Therefore the decay of force over space is identical regardless of the scales.

2.4.2 Magnetism

We consider the magnets as dipoles with a magnetic moment m. The magnetic potential $\phi_j(r)$ at a position r due to the magnetic moment m_j is given by

$$\phi_j(r) = \frac{\mu_0}{4\pi} \frac{m_j \cdot \hat{r}}{r^2} \tag{12}$$

where $\mu_0 = 4\pi \times 10^{-7} Tm/A$ is the permeability of free space, and $\hat{r} \equiv r/|r|$ assuming that $|r| = r$ is much larger than the size of the magnet. The magnetic flux of the dipole is then given by

$$B_j = -\nabla \phi_j \tag{13}$$

and the magnetic potential energy U_{ij} acquired by a second dipole m_i placed in the field of m_j is given by

$$U_{ij} = -m_i \cdot B_j. \tag{14}$$

Then, the force between the two dipoles is found by differentiating (14) with respect to r.

$$F_{ij} = (m_i \cdot \nabla) B_j \tag{15}$$

$$\tau_{ij} = m_i \times B_j \tag{16}$$

We can determine the total potential energy of the system as

$$U_{total} = \frac{1}{2} \sum_{i,j i \neq j} U_{ij}. \tag{17}$$

2.5 ***The Engineering Issues*** - *Actuator Battery Connector Bottleneck*

For modular systems smaller than a few cm, there are three fundamental problems that still await a solution. These problems relate to actuator, battery (or power in general), and connector technology. When designing systems where a high quantity of components of small size is desired, solutions for these problems are of particular relevance. First, actuation endows the parts with the ability to move and re-configure. A common solution is to use electrical servo motors. These actuators, however, are typically big and heavy. Other means of actuation have also been proposed, e.g. pneumatic actuators. Although they are lightweight, they require a source of compressed air (e.g. a compressor). The second problem is concerned with providing power to the actuator(s). A typical solution is to use batteries. Batteries, however, are problematic, because they are only able to provide power for a limited amount of time. Furthermore, their initial charge may vary which leads to heterogeneously actuated components. Another popular solution involves propagating current through the binding locations. Unfortunately, this solution has the drawback that the alignment of the connecting points has to be very precise. In addition, such components can not segregate from the main structure which prohibits this way of powering for mobile type robots. The third problem is the connection mechanism enabling the modular parts to dock to each other. Binding is crucial for reorganization and for a desired structure to hold. The most common ways of binding are magnets and mechanical latches.

A Actuator
B Battery
C Connector

There is a strong interdependency between these issues. The requirements of the connection mechanism as well as the actuator are partly determined by the weight of each component. The heavier the components are, the more force needs to be applied to the binding location. In addition, the actuators have to apply larger torques to displace the components. The use of more powerful components in general leads to even heavier components. Also, the power consumption increases as a result of stronger connection mechanisms and actuators. Surprisingly, small size and weight reduction of modular parts is not a good way to solve this problem, because not only does the power/weight ratio of the most common actuators decrease with a reduction in size, but also so does the strength/weight ratio of common connectors. This implies that the most common ways of actuating, powering and connecting modular robots cannot be applied to small-sized entities. It follows that novel solutions to the *ABC bottleneck* are necessary in order to make progress in small-scale self-assembly robotics.

3 Case Study

In this section, we discuss how to evaluate dynamics of self-assembly based on a case study employed in our group (see [25] for more details). In the experiments, we employed *Tribolon* platform [23] consisting of centimeter-sized modules floating on the water surface. 12 modules were prepared and equipped with permanent magnets (a single cubic permanent magnet each, whose flux density is $1.3T$, and size is $5 \times 5 \times 5mm^3$). They are attached to the bottom of each module orienting the same direction such that the modules repel each other (e.g. north is always pointing up, Fig. 6). Half of the modules were, in addition to the permanent magnet, also equipped with vibration motors (termed active modules, in comparison to passive modules which doesn't feature magnets) such that they can vibrate receiving power from a ceiling via pantograph. The vibrating modules are equipped with a flat coreless vibration motor (T.P.C DC MOTOR FM34F, $12000 \sim 14000 rpm(2.5 - 3.5 Volts)$) on the top of the base plate to induce self-propulsive motion. When an electrical potential was applied to the ceiling plate (Fig. 6 b), current flowed through the pantograph to the vibration motor was applied to the ceiling plate, current returning to ground via electrodes immersed in the conductive water. Due to this setup, all modules receive the same constant power and they are be lightweight ($2.8\,g$ each), which would not be the case if batteries were used.

We conducted 15 trials for the statistical analysis. In Fig. 7, we show a representative result in time sequence of the obtained segregation behavior. The initial starting condition was set as depicted in Fig. 7 (00:00), in which all the modules were symmetrically aligned in a circular form alternately, such that the passive and the vibrating modules have equal chances in the segregation process. The duration time for the experiment was set to 90 seconds.

In order to perform the analysis, the trajectories (positions) of all the modules were tracked using the open source tracking software *Tracker Video Analysis and Modeling Tool* [9]. Our observation is that the red active modules tend to assemble together and go apart from the blue passive modules, such that two different

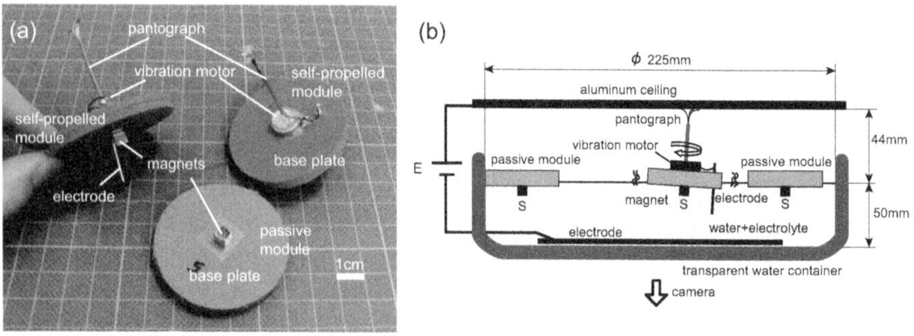

Fig. 6 (a) Self-propelled and passive modules. Each module weighs 2.8 g and has a footprint of 12.25 cm^2.(b) Illustration of the experimental environment with three modules.

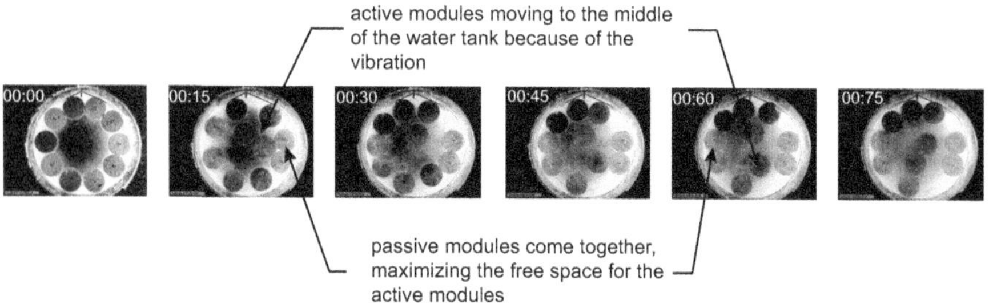

Fig. 7 The experimental results in time sequence. The frames are captured every 15 seconds

modules clusters can be spatially distinguished; the first cluster contains only the active modules and the second cluster the passive modules (Fig. 7 (00:75)).

3.1 Magnetic Potential Energy and Centroid Distance

We defined the magnetic potential energy of the system as $U'_{total} \equiv U_{total}/(\frac{\mu_0}{4\pi}m^2)$ by normalizing the energy defined in Eq. 17, and show the obtained result in Fig. 8 (a) as function of time. The error bars represent the standard deviation of uncertainty within the fifteen experimental trials. Due to the characteristics of the system, namely non-equilibrium system, the magnetic potential energy keeps decreasing. Suppose we have all passive modules, the system is supposed to reach to the state where modules are equally distributed and fixed.

The centroid $(X,Y) = (\frac{1}{N}\sum_{i=1}^{N}(x_i), \frac{1}{N}\sum_{i=1}^{N}(y_i))$ of a group (or cluster) of modules is the center of mass of the modules, where N is the number of modules in the modules group, x_i and y_i are the positions of the i-th component of the considered group, respectively. We calculated the time evolution of the difference between the two modules groups (the passive modules on one side and the active modules on the second side and depicted in Fig. 8 (b). As depicted in Fig. 8 (b), there is an increase in the distance between the centroids of the passive and the vibrating modules. This

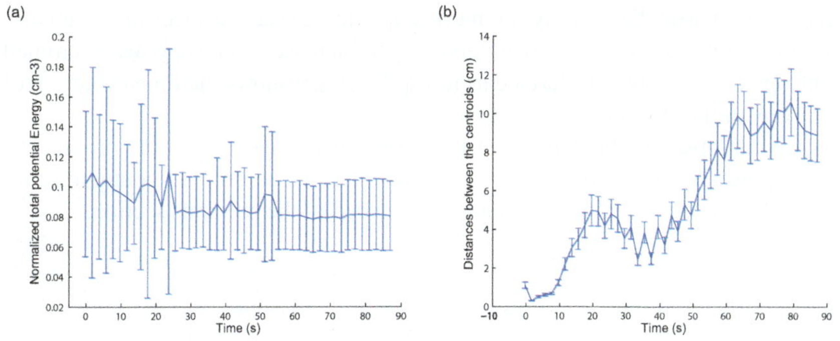

Fig. 8 (a) total energy of the system. (b) time evolution of the distance to between the center of mass of the two clusters ($N = 15$).

corresponds to the formation of two clusters of modules with a final mean distance between the two clusters of approximately 10 centimeters. Given that the diameter of the arena (or tank as you wish) is $22.5\,cm$, this corresponds to the 50% of the whole area.

3.1.1 Entropy for Transient State (Hierarchic Social Entropy)

The definition of entropy differs in scientific fields, depending on to what one applies. Thermodynamics entropy (to heat), statistical mechanics entropy (to object), and information entropy (to event) are probably the three best known entropies in science. In self-assembly, systems that cannot presume some specific physical amounts, such as quantity of heat, employ information entropy for the measurement of their "randomness".

Balch proposed a novel definition of entropy (position order) that can be applied for the measurement of multi-components distributions (or quantitative metric of diversity) [2]. He uses H from Shannon's theory

$$H(h) = -\sum_{i=1}^{N} p_i(h)\log_2(p_i(h)) \tag{18}$$

where p_i is the number of modules in the $i-th$ cluster ($i \in N$) divided by the total number of modules. A component belongs to a cluster if the distance is within the length of h ($||\mathbf{r_i} - \mathbf{r_j}|| < h$; $\mathbf{r_i}$ is the position of the i-th component). He then integrates $H(h)$ over all possible h, and defines it as entropy, namely:

$$S = \int_0^{\infty} H(h)dh. \tag{19}$$

The definition describes the randomness of modules well. Note that in this definition, the entropy may decreases over time. In physics, an entropic force acting in a system is a macroscopic force whose properties are primarily determined not by the character of a particular underlying microscopic force (such as electromagnetism), but by the whole system's statistical tendency to increase its entropy. We examined the entropy of the system as derived as in Eq. 19. Fig. 9 shows the time evolution of the entropy of the system.

As we can observe, the entropy of the system is decreasing as time progresses, which represents the convergence of the system to more ordered configurations. This corresponds to the cluster formation described of the previous section.

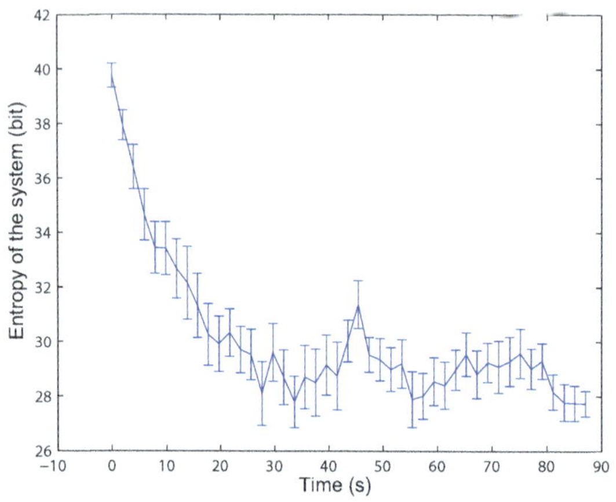

Fig. 9 Transition of entropy.

3.1.2 Transfer Entropy

To see the simple informational structure, we used a measure that aims at extracting directed flow (transfer of information) between time series of both active and pssive modules, called *transfer entropy* [29]. Given two arbitrary time series x_t and y_t, transfer entropy essentially quantifies the deviation from the generalized Markov process: $p(x_{t+\tau}|x_t) \approx p(x_{t+\tau}|x_t, y_t)$, where p denotes the transition probability. If the deviation from a generalized Markov process is small, then the state of Y can be assumed to have little relevance on the transition probabilities of system X. If the deviation is large, however, then the assumption of a Markov process is not valid. The incorrectness of the assumption can be expressed as follows:

$$TE(Y \to X) = \sum_{x_{t+\tau}} \sum_{x_t} \sum_{y_t} p(x_{t+\tau}, x_t, y_t) \log \frac{p(x_{t+\tau}|x_t, y_t)}{p(x_{t+\tau}|x_t)} \qquad (20)$$

where the sums are over all amplitude states, and the index $TE(Y \to X)$ indicates the influence of Y on X. The transfer entropy is explicitly nonsymmetric under the exchange of X and Y, and can thus be used to detect the directed exchange of information between two systems.

Here, Y is a time series of d_t^i of a module i and X is a time series of d_t^j of a module j. The base of log is set to e and the parameter τ is set to 1. By using transfer entropy, we aim to evaluate causal relations between all the pairs of modules.

By using the data, we simply calculated a distance of movement in each time step for each modules (d_t^i) as follows.

$$d_t^i = \sqrt{(x_{t+1}^i - x_t^i)^2 + (y_{t+1}^i - y_t^i)^2} \tag{21}$$

We discretized the value of d_t^i ranging from 0.0 to 12.0 into 10, 20, 50, and 100 bins and all the 15 trials from $0 \leq timesteps \leq 44$ were used and averaged on trials. Since the number of samples is limited, we varied the bin size and saw the relevance.

We show the results in Fig.10 and 11, in which the module number 1 to 6 are passive modules, while 7 to 12 are active modules. As can be seen from the results of Bin=10 (Fig.10(a)), 20 (Fig.10(b)), and 50 (Fig.11(a)), the values of $TE(active \rightarrow active)$ are high. Also, the values of $TE(active \rightarrow passive)$, $TE(passive \rightarrow active)$ are relatively high, and the values of $TE(passive \rightarrow passive)$ are low. These results simply suggest that active elements influence the transitions of active elements and passive elements. On the other hand, passive elements only influence the transitions of active elements. And the values of $TE(Y \rightarrow X)$ are simply the degree of its influence. These results fit well to our natural observations of the system. On the result of Bin=100 (Fig.11(b)), since the number of samples (observed time steps) are relatively small for the size of the state space, it cannot structure the relevant probability density.

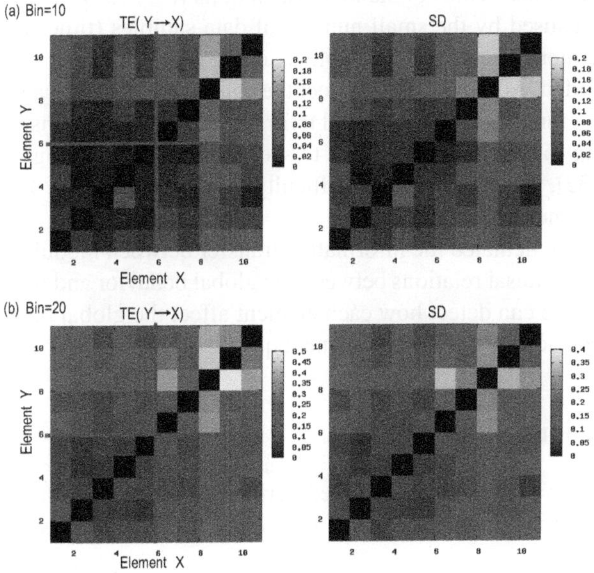

Fig. 10 $TE(Y \rightarrow X)$ and standard deviations (SD). (a) bin=10, (b) bin=20.

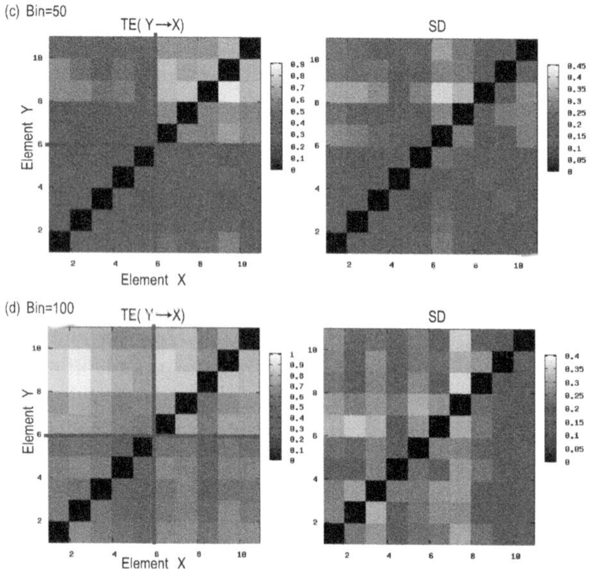

Fig. 11 $TE(Y \rightarrow X)$ and standard deviations (SD). (c) bin=50, (d) bin=100.

In this system, transfer entropy tends to show $TE(active \rightarrow active) > TE(active \rightarrow passive), TE(passive \rightarrow active) > TE(passive \rightarrow passive)$. This result naturally fits to our simple observation of the trajectories.

In our results, the values of standard deviations (SD) for transfer entropy were large. This is caused by the small number of data samples (time steps). Considering the experimental setting, reasonable extensions of the experimental time steps are recommended. Additionally, to detect the causal relation, we can use other measures, such as granger causality, mutual information, symbolic transfer entropy [32], etc., according to what we would like to see. Especially, by using symbolic transfer entropy [32], we can avoid the difficulty to set the bin size. But in this case, extensions of timesteps are inevitable.

Although we calculated the information transfer between modules, it is also possible to measure causal relations between the global behavior and the elements [33]. By doing this, we can detect how each element affect the global behavior, and how the global behavior regulates each element quantitatively.

4 Conclusions

In this paper, we systematically studied various problems on self-assembling systems. Starting from pointing to some fundamental concerns of self-assembly, we categorized them into three basic issues, namely on assembly, dynamics, and interactions. We examined quantification methods utilizing a case study in which modules showed segregation behavior in a distributed way. We further investigated the possible style of description of entropy as well as free energy that can govern macroscopic self-assembly systems. We believe the clarification of basic problems in self-assembly and proper assignment of an approaching method will offer new opportunities to deepen the theoretical understanding of the phenomenon, and will lead to the realization of efficient self-assembly systems.

Acknowledgment

This research was supported by the Swiss National Science Foundation project #200020-118117.

References

1. Abbott, J.J., Nagy, Z., Beyeler, F., Nelson, B.J.: Robotics in the small. IEEE Robotics & Automation Magazine 14, 92–103 (2007)
2. Balch, T.: Hierarchic social entropy: An information theoretic measure of robot group diversity. Autonomous Robots 8, 209–237 (2000)
3. Boncheva, M., Andreev, S.A., Mahadevan, L., Winkleman, A., Reichman, D.R., Prentiss, M.G., Whitesides, S., Whitesides, G.: Magnetic self-assembly of three-dimensional surfaces from planar sheets. PNAS 102, 3924–3929 (2005)
4. Boncheva, M., Ferrigno, R., Bruzewicz, D.A., Whitesides, G.M.: Plasticity in self-assembly: Templating generates functionally different circuits from a single precursor. Angew. Chem. Int. Ed. 42, 3368–3371 (2003)
5. Boncheva, M., Gracias, D.H., Jacobs, H.O., Whitesides, G.M.: Biomimetic self-assembly of a functional asymmetrical electronic device. PNAS 99, 4937–4940 (2002)
6. Bowden, N., Terfort, A., Carbeck, J., Whitesides, G.M.: Self-assembly of mesoscale objects into ordered two-dimensional arrays. Science 276, 233–235 (1997)
7. Bowden, N., Weck, M., Choi, I.S., Whitesides, G.M.: Molecule-mimetic chemistry and mesoscale self-assembly. Acc. Chem. Res. 34, 231–238 (2001)
8. Breivik, J.: Self-oranization of template-replicating plolymers and the spontaneous rise of genetic information. Entropy 3, 273–279 (2001)
9. Brown, D.: Tracker video analysis and modeling tool (2009), `http://www.cabrillo.edu/~dbrown/tracker/`
10. Cohn, M.B., Kim, C.-J.: Self-assembling electrical networks: An application of micromachining technology. In: International Conference on Solid-State Sensors and Actuators, pp. 490–493 (1991)
11. Cugat, O., Delamare, J., Reyne, G.: Magnetic micro-actuators and systems (MAGMAS). IEEE Trans. Magnetics 39(5), 3607–3612 (2003)
12. Demaine, E.D., Hohenberger, S., Liben-Nowell, D.: Tetris is hard, even to approximate. Technical report, Cornell University Library (2002), arXiv.org

13. Gracias, D.H., Tien, J., Breen, T.L., Hsu, C., Whitesides, G.M.: Forming electrical networks in three dimensions by self-assembly. Science 289, 1170–1172 (2000)
14. Griffith, S., Goldwater, D., Jacobson, J.: Robotics: Self-replication from random parts. Nature 437, 636 (2005)
15. Grzybowski, B.A., Radkowski, M., Campbell, C.J., Lee, J.N., Whitesides, G.M.: Self-assembling fluidic machines. App. phys. lett. 84, 1798–1800 (2004)
16. Grzybowski, B.A., Stone, H.A., Whitesides, G.M.: Dynamic self-assembly of magnetized, millimetre-sized objects rotating at a liquid-air interface. Nature 405, 1033 (2000)
17. Grzybowski, B.A., Winkleman, A., Wiles, J.A., Brumer, Y., Whitesides, G.M.: Electrostatic self-assembly of macroscopic crystals using contact electrification. Nature 2, 241–245 (2003)
18. Hosokawa, K., Shimoyama, I., Miura, H.: Dynamics of self-assembling systems: Analogy with chemical kinetics. Artificial Life 1(4), 413–427 (1994)
19. Ishiguro, A., Shimizu, M., Kawakatsu, T.: A modular robot that exhibits amoebic locomotion. Rob. Aut. Sys. 54, 641–650 (2006)
20. Klavins, E.: Programmable self-assembly. IEEE Cont. Sys. Mag. 27, 43–56 (2007)
21. Wilson, M., Melhuish, C., Sendova-Franks, A.: Multi-object segregation: ant-like brood sorting using minimalism robots. In: Proc. Seventh International Conf. on the Simulation of Adaptive Behaviour, Edinburgh, UK, pp. 369–370 (2002)
22. Mao, C., Thalladi, V.R., Wolfe, D.B., Whitesides, S., Whitesides, G.M.: Dissections: Self-assembled aggregates that spontaneously reconfigure their structures when their environment changes. J. Am. Chem. Soc 124(49), 14508–14509 (2002)
23. Miyashita, S., Kessler, M., Lungarella, M.: How morphology affects self-assembly in a stochastic modular robot. In: IEEE International Conference on Robotics and Automation (2008)
24. Motokawa, T.: Time of an elephant, time of a mouse. In: CHUO-KORON-SHINSHA, INC. (1992)
25. Ngouabeu, A.M.T., Miyashita, S., Füchslin, R.M., Nakajima, K., Göldi, M., Pfeifer, R.: Self-organized segregation effect on water based self-assembling robots. In: Artificial Life 12, Odense, Denmark (2010)
26. Pelesko, J.A.: SELF ASSEMBLY. Chapman & Hall/CRC, Boca Raton (2007)
27. Penrose, L.S.: Self-reproducing. Sci. Amer. 200(6), 105–114 (1959)
28. Purcell, E.M.: Life at low reynolds number. Amer. J. Phys. 45, 3–11 (1977)
29. Schreiber, T.: Measuring information transfer. Physical Review Letters 85, 461–464 (2000)
30. Wilson, M., Melhuish, C., Sendova-Franks, A.B., Scholes, S.R., Franks, N.R., Melhuish, C.: Brood sorting by ants: Two phases and differential diffusion. Animal Behaviour 68, 1095–1106 (2004)
31. Stambaugh, J., Lathrop, D.P., Ott, E., Losert, W.: Pattern formation in a monolayer of magnetic spheres. Pysical Leview E. 68, 026207-1–026207-5 (2003)
32. Staniek, M., Lehnertz, K.: Symbolic transfer entropy. Physical Review Letters 100, 158101–158101 (2008)
33. Sumioka, H., Nakajima, K., Lungarella, M., Pfeifer, R.: Complexity detection based on bidirectional information flow (submitted)
34. Tsutsumi, D., Murata, S.: Multistate part for mesoscale self-assembly. In: SICE Annual Conference (2007)

35. White, P., Kopanski, K., Lipson, H.: Stochastic self-reconfigurable cellular robotics. In: Proc. Int. Conf. on Robotics and Automation, vol. 3, pp. 2888–2893 (2004)
36. Whitesides, G.M.: The 'right' size in nanobiotechnology. Nature 21(10), 1161–1165 (2003)
37. Whitesides, G.M., Grzybowski, B.: Self-assembly at all scales. Science 295, 2418–2421 (2002)
38. Wolfe, D.B., Snead, A., Mao, C., Bowden, N.B., Whitesides, G.M.: Mesoscale self-assembly: Capillary interactions when positive and negitive menisci have similar amplitudes. Langmuir 19, 2206–2214 (2003)
39. Yamaki, M., Higo, J., Nagayama, K.: Size-dependent separation of colloidal particles in two-dimensional convective self-assembly. American Chemical Society 11, 2975–2978 (1995)

Part III: Autonomous Mental Development in Robotic Systems

Brain Like Temporal Processing

Juyang Weng

Abstract. This chapter presents a general purpose model of the brain, called Developmental Networks (DN). Rooted in the biological genomic equivalence principle, our model proposes a general-purpose cell-centered in-place learning scheme to handle all levels of brain development and operation, from the cell level all the way to the brain level. It clarifies five necessary "chunks" of the brain "puzzle": development, architecture, area, space and time. Then, this chapter analyzes how such a model enables a developmental robot to deal with temporal contexts. It deals with temporal context of any length without a dedicated temporal component.

1 Introduction

It is known that a fully programmed robot has only very limited capabilities for dealing with the real world environments. For a task that is difficult to program well, machine learning techniques can be used. However, traditional machine learning techniques are task-specific, meaning that the human programmer requires that the task that the robot executes be given to him before he finishes the programming. It is the human programmer who understands the task and handcrafts a task specific representation into the robot's control program. However, such robots tend to be brittle in real-world environments, since it is difficult for the human programmer to sufficient predict all the task settings and all the environmental situations.

Demonstrated by human cognitive and behavioral development from infancy to adulthood, autonomous development is nature's approach to human intelligence (Piaget 1954 [13], Elman et al. 1997 [4], Weng et al. 2001 [22]).

There have been several impressive attempts to model the brain as a symbolic information processor (Albus 1991[1], Hecht-Nielsen 2007 [7], Albus 2010 [2]). However, they are without sufficient, biologically plausible learning to account for

Juyang Weng
Michigan State University, East Lansing, MI, USA
e-mail: weng@cse.msu.edu

Y. Meng and Y. Jin (Eds.): Bio-Inspired Self-Organizing Robotic Systems, SCI 355, pp. 195–212.
springerlink.com

the overwhelming brain complexity. As symbol-only modeling is insufficient to deal with uncertainty, the Bayesian probability framework was added to such symbolic models, using either probability models for spatial aspects or Markov chains for temporal aspects (Jelinek 1990 [8], Lee & Mumford 2003 [9], Emami & Jelinek 2005 [5], Tenenbaum et al. 2006 [16], George & Hawkins 2009 [6]). A major advantage of such Bayesian models is that they are intuitive for human to understand, but they face a fundamental problem: They are not *developmental* —- the symbolic boundaries ("walls") between different internal units (nodes or Markov chains) were handcrafted (defined) by human programmers.

In fact, they all correspond to an "skull-open approach" to the brain — it is the human teacher who understands a given task and the concepts it needs. Then he directly manipulates (defines) the "brain's" internal representation through its open "skull". Among many limitations of such "skull-open" approaches are:

1. It is a static handcrafted information processor that cannot explain the miracle of brain development. It cannot explain the following process: With his "skull closed," a child autonomously interacts with the environment and he autonomously learns and discovers concepts that his parents do not even know about. Autonomous discovery is a miracle of such a developmental brain.
2. It is labor-intensive to build and brittle for the real world environments. Given a task, the process of handcrafting a brain-like information processor requires a large amount of man-hours for manual instantiation of an internal representation for each task. The resulting system is known to be brittle due to the inability of a human to sufficiently predict the dynamic real world.

Many computer vision researchers thought that the human vision system can be sufficiently modeled by a static object recognizer. It cannot. The brain learns new objects and new object variations all the time.

Inspired by human mental development from conception, a developmental robot is one that autonomously develops its mental skills and task performance capabilities from interactions with the environments. Fig. 1 illustrates the paradigm of autonomous mental development (Weng et al. 2001 [22]).

As far as we know, Cresceptron 1993 [19, 20] was the first developmental model for visual learning from complex natural backgrounds. By developmental, we mean that the internal representation is fully emergent from interactions with environment, without allowing a human to manually instantiate a task-specific representation.

In this chapter, I our latest general purpose model of the brain, called Developmental Network (DN), concentrating on a basic unit: area. Then, I will concentrate on the temporal properties of the DN model.

2 Five Chunks of a Brain Model

Rooted in the biological genomic equivalence principle, our model proposes a general-purpose cell-centered in-place learning scheme to handle all levels of brain development and operation, from the cell level all the way to the brain level. It clarifies five necessary "chunks" of the brain "puzzle": development, architecture, area,

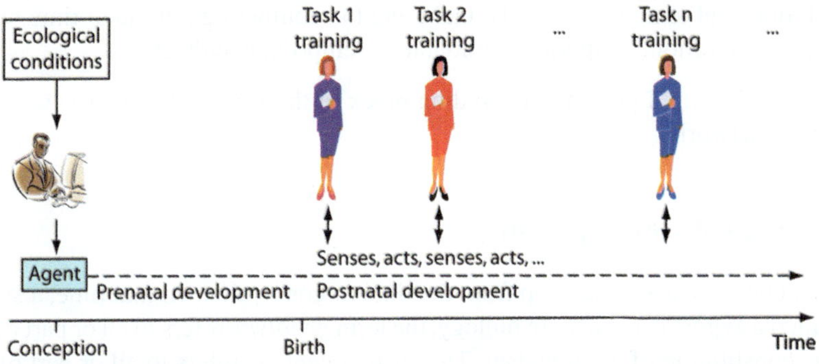

Fig. 1 The paradigm of autonomous mental development by machines, inspired by human mental development. No task is given during the programming (i.e., conception) time. A general-purpose task-nonspecific developmental program must be ready and be loaded onto the agent's "brain". Prenatal development may preliminarily wire the "brain" before "birth" using "spontaneous" (internally generated) signals, e.g., from sensors and motors. After the "birth," the agent starts to learn an open variety of skills and tasks through interactions with the physical world. During the development, the "brain" is "skull-closed" meaning that there is no need for the programmer to direct intervene in the brain's internal representation after the conception. The tasks that the agent learns during lifetime are determined after the birth by other users and, therefore, the brain's internal self-organization is totally autonomous (i.e., emergent representation).

space and time. Although there are many characteristics of the biological brain, these five "chunks" are probably the most fundamental in the brain "puzzle":

1. **The "development" chunk** means that any practical brain, natural or artificial, needs to autonomously develop through interactions with the natural environments without any previously given set of tasks.
2. **The "architecture" chunk** handles (1) complex backgrounds where the signal-to-noise ratio is at least smaller than 1 (< 0 db), or equivalently, more input components are irrelevant to immediate actions than those that are relevant; (2) abstraction, reasoning and generalization with any abstract and concrete contexts; (3) multiple sensory modalities and multiple motor modalities and their integration.
3. **The "space" chunk** deals with any practical foreground objects with any practical complex backgrounds, and includes conflicting invariance and specificity criteria for type, location, size, orientation, expression, etc. Learned context-dependent spatial attention is a key capability for dealing with all these conflicting spatial criteria.
4. **The "area" chunk** addresses the issue of feature development and area representation, without rigidly specifying what each neuron does.
5. **The "time" chunk** indicates that the brain uses its intrinsic spatial mechanisms to deals with time, without dedicated temporal components. The model copes

with practical temporal contexts, including the conflicting criteria of time warping, time duration, temporal attention, long temporal length, etc.

In the following, I present the building block of the DN model: a basic unit as a three-area network.

3 Biological Development

A human being starts to develop from the time of conception. At that time, a single cell called a zygote is formed. In biology, the term *genotype* refers to all or part of the genetic constitution of an organism. The term *phenotype* refers to all or part of the visible properties of an organism that are produced through the interaction between the genotype and the environment. In the zygote, all the genetic constitution is called genome, which mostly resides in the nucleus of a cell. At the conception of a new human life, a biological program called the *developmental program* starts to run. The code of this program is the genome, but this program needs the entire cell as well as the cell's environment to run properly.

The biological developmental program handles two types of development, *body development* and *mental development*. The former is the development of everything in the body excluding the brain. The latter is the development of the brain (or the central nervous system CNS). Through the body development, a normal child grows in size and weight, along with many other physical changes. Through the mental development, a normal child develops a series of mental capabilities through interactions with the environment. Mental capabilities refer to all known brain capabilities, which include, but not limited to, perceptual, cognitive, behavioral and motivational

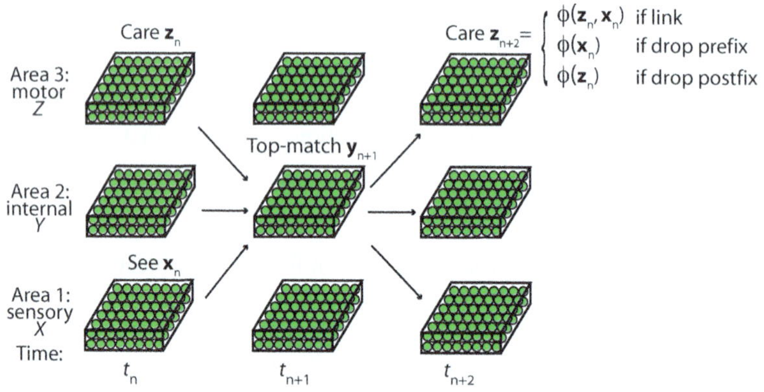

Fig. 2 The spatial DN for both spatial processing and temporal processing without dedicated temporal components. At each temporal unit shown above (two time frames), three basic operations are possible: link, drop prefix, and drop postfix. After proper training, the DN is able to attend any possible temporal context up to the temporal sampling resolution.

capabilities. In this chapter, the term *development* refers to mental development unless stated otherwise. The biological mental development takes place in concurrence with the body development and they are closely related. For example, if the eyes are not normally developed, the development of the visual capabilities is greatly affected. In the development of an artificial agent, the body can be designed and fixed (not autonomously developed), which helps to reduce the complexity of the autonomous mental development.

The *genomic equivalence* principle [14] is a very important biological concept for us to understand how biological development is regulated. This principle states that the set of genes in the nucleus of every cell (not only that in the zygote!) is functionally complete — sufficient to regulate the development from a single cell into an entire adult life. This principle is dramatically demonstrated by cloning. This means that there are no genes that are devoted to more than one cell as a whole. Therefore, development guided by the genome is *cell-centered*. Carrying a complete set of genes and acting as an autonomous machine, every cell must handle its own learning while interacting with its external environment (e.g., other cells). Inside the brain, every neuron develops and learns in place. It does not need any dedicated learner outside the neuron. For example, it does not need an extra-cellular learner to compute the covariance matrix (or any other moment matrix or partial derivatives) of its input lines and store extra-celullarly. If an artificial developmental program develops every artificial neuron based on only information that is available to the neuron itself (e.g., the cellular environment such as pre-synaptic activities, the developmental program inside the cell, and other information that can be biologically stored intra-cellularly), we call this type of learning *in-place learning*.

This in-place concept is more restrictive than a common concept called "local learning." For example, a local learning algorithm may require the computation of the covariance matrix of the pre-synaptic vector which must store extra-cellularly. In electronics, the in-place learning principle can greatly reduce the required electronics and storage space, in addition to the biological plausibility. For example, suppose that every biological neurons requires the partial derivative matrix of its pre-synaptic vector. As the average number of synapses of a neuron in the brain is on the order of $n = 1000$. Each neuron requires about $n^2 = 1,000,000$ storage units outside every neuron. This corresponds to about 1,000,000 of the total number of synapses (10^{14}) in the brain!

Conceptually, the fate and function of a neuron is not determined by a "hand-designed" (i.e., genome specified) meaning of the external environment. This is another consequence of the genomic equivalence principle. The genome in each cell regulates the cells mitosis, differentiation, migration, branching, and connections, but it does not regulate the meaning of what the cell does when it receives signals from other connected cells. For example, we can find a V1 cell (neuron) that responds to an edge of a particular orientation. This is just a facet of many emergent properties of the cell that are consequences of the cells own biological properties and the activities of its environment. A developmental program does not need to, and should not, specify which neuron detects a pre-specified feature type (such as an edge or motion).

4 Why Autonomous Mental Development?

One can see that biological development is very "low level", regulating only individual neurons. Then, why is it necessary to enable our complex electronic machines to develop autonomously? Why do we not design high-level concepts into the machines and enable them to carry out our high-level directives? In fact, this is exactly many symbolic methods have been doing for many years. Unfortunately, the resulting machines are brittle — they fail miserably in real world when the environment fall out of the domains that have been modeled by the programmer.

To appreciate what are faced by a machine to carry out a complex task, Weng [17] introduced a concept called *task muddiness*. The composite muddiness of a task is a multiplicative product of many individual muddiness measures. There are many possible individual muddiness measures. Those individual muddiness measures are not necessarily mutually independent or at the same level of abstraction, since such a requirement is not practical nor necessary for describing the muddiness of a task. They fall into five categories: (1) external environment, (2) input, (3) internal environment, (4) output and (5) goal, as shown in Table 1. The term "external" means external with respect to the brain and "internal" means internal to the brain.

The composite muddiness of a task can be considered as a product of all individual muddiness measures. In other words, a task is extremely muddy when all the five categories have a high measure. A chess playing task with symbolic input and output is a clean problem because it is low in categories (1) through (5). A symbolic language translation problem is low in (1), (2) and (4), moderate in (3) but high in (5). A vision-guided navigation task for natural human environment is high in (1), (2), (3) and (5), but moderate in (4). A human adult handles extremely muddy tasks that are high in all the five categories.

From the muddiness table Table 1 we have a more detailed appreciation what a human adult deals with even in a daily task, e.g., navigating or driving in a city environment. The composite muddiness of many tasks that a human or a machine can execute is proposed by Weng [17] as a metric for measuring required intelligence.

A human infant is not able to perform those muddy tasks that a human adult performs everyday. The process of mental development is necessary to develop such a wide array of mental skills. Much evidence in developmental psychology has demonstrated that not only a process of development is necessary for human intelligence, the environment of the development is also critical for normal development.

Likewise, it is not practical for a human programmer to program a machine to successfully execute a muddy task. Computers have done very well for clean tasks, such as playing chess games. But they have done poorly in performing muddy tasks, such as visual and language understanding. Enabling a machine to autonomously develop task skills in its real task environments is the only approach that has been proved successful for muddy tasks — no existing higher intelligence for muddy tasks is not developed autonomously.

Table 1 A list of muddiness factors for a task

Category	Factor	Clean	⟷ Muddy
External Env.	Awareness	Known	Unknown
	Complexity	Simple	Complex
	Controlledness	Controlled	Uncontrolled
	Variation	Fixed	Changing
	Foreseeability	Foreseeable	Nonforeseeable
Input	Rawness	Symbolic	Real sensor
	Size	Small	Large
	Background	None	Complex
	Variation	Simple	Complex
	Occlusion	None	Severe
	Activeness	Passive	Active
	Modality	Simple	Complex
	Multi-modality	Single	Multiple
Internal Env.	Size	Small	Large
	Representation	Given	Not given
	Observability	Observable	Unobservable
	Imposability	Imposable	Nonimposable
	Time coverage	Simple	Complex
Output	Terminalness	Low	High
	Size	Small	Large
	Modality	Simple	Complex
	Multimodality	Single	Multiple
Goal	Richness	Low	High
	Variability	Fixed	Variable
	Availability	Given	Unknown
	Telling-mode	Text	Multimodal
	Conveying-mode	Simple	Complex

5 Building Blocks

The biological genomic equivalence principle implies that a cell is a general-purpose machine during its development and operation as far as its genome is concerned. All generated from the single cell zygote through many rounds of mitosis, many cells in the brain become increasingly *differentiated* [14], meaning that they become more specialized while migrating after being generated from the progenitor cells in the ventricular zone. However, mitosis continuously goes on in a developing brain. Where each cell goes (cell migration), how it grows (cell expansion), how it extends (axon and dendrite pathfinding), whether it survives (neurotrophic factors), synapse formation and synapse elimination (synaptogenic factors) are all activity dependent. This cell-centered autonomy, while interacting with nearby environment, gives the basis for our DN model to treat any set of cells (neurons) as a unit. This scheme facilitates understanding without being overwhelmed by the apparent high complexity of a developed brain.

We consider such a generic area Y which has its sensory area X and its motor area Z, as illustrated in Fig. 2. For every unit, its sensory area is also an output port for its top-down attention (self-effecting), and its motor area is also an input port for its top-down sensing (self-aware). A lower brain(e.g., mid-brain) is developed earlier, so that the higher brain (e.g., forebrain) as basic units can innervate into lower ones later.

Because of the need to address the complex background problem, the DN model provides a deeper need: to provide receptive field and effective field that are smaller than X and Z. If the receptive field of a neuron matches the foreground object well, the response of the neuron is not very sensitive to the background.

The general-purpose, spatial object recognition from complex backgrounds using this basic unit has been tested and the experimental results will appear elsewhere. In this chapter, we will focus on temporal aspect of the DN.

DN learns while performing. Suppose that DN has l layers, 1 to l. At time $t = 0$, every layer has the weights of all its neuronal initialized with Gaussian weights at different positions. Then, DN computes as below.

Algorithm 1 (DN). *Input areas: X and Z. Output areas: X and Z. The dimension and representation of X and Y areas are hand designed based on the sensors and effectors of the species (or from evolution in biology). Y is the skull-closed, internal brain, not directly accessible by the world outside the skull.*

1. *At time $t = 0$, for each area A in $\{X,Y,Z\}$, initialize its adaptive part $L = (V,A,r)$, where V contains all the synaptic weight vectors, A stores all the neuronal ages, and r is the radius of the excitation sphere of each neuron. L is initialized according to an initialization method.*
2. *At time $t = 1,2,...,$ for each area A in $\{X,Y,Z\}$, do the following steps repeatedly forever:*
   a. *Take input from bottom-up input* **b** *and top-down input* **t**, *if it exists.*
   b. *Every area A performs mitosis if it is needed, using its bottom-up and top-down inputs* **b** *and* **t**, *respectively.*
   c. *Every area A computes its area function f, described in the next subsection,*
   $$(\mathbf{r}',L') = f_{\mathrm{LCA}}(\mathbf{b},\mathbf{r},\mathbf{t},L).$$
   where **r** *and* **r** *are the input and output responses, respectively.*
   d. *For every area A replaces: $L \leftarrow L'$ and $\mathbf{r} \leftarrow \mathbf{r}'$.*

In the algorithm, LCA denotes the Lobe Component Analysis [21], a dually optimal model for updating a neuronal layer:

1. Spatially optimality: the target neuronal weights are best in minimizing the representation error of the layer input space $P = X \times Z$ (parallel input space of the bottom-up input space X and the top-down input space Z), using a limited number of neurons.
2. Temporal optimality: the learning directions and step sizes at different steps of all neurons are best to minimize the distance between the estimated weights and their target weights.

The spatial optimality implies that the network is smallest in each layer. The temporal optimality means that the network learns fastest using a limited amount of learning experience.

6 Lobe Component Analysis

A general neural area may contain several levels, represented by the cortical laminar structure, as discussed in Luciw & Weng 2010 [10] in order to avoid top-down hallucination. Here, for simplicity we consider a simpler case where each area has only one layer.

Each area performs the Candid Covariance-free Incremental (CCI) LCA algorithm f_{LCA}. It incrementally updates the adaptive part of its neuronal level $L = (V, A, r)$, where $V = (\mathbf{v}_1, \mathbf{v}_2, ..., \mathbf{v}_c)$ contains c synaptic vectors, $A = (n_1, n_2, ..., n_c)$ consists of the corresponding firing ages, and r is the radius of the excitation sphere of each neuron.

Each input sample is in the form $(\mathbf{x}, \mathbf{z}) = X \times Z$, where X is the bottom-up space and Z is the top-down input space. The vectors $\mathbf{x}$ and $\mathbf{z}$ are normalized before feeding into f_{LCA}, to be discussed later. In a network, X or Z may have multiple parallel input subspaces. Algorithmically, f_{LCA} takes L as the current level and $(\mathbf{x}, \mathbf{z})$ as the bottom-up and top-down input, respectively, to generate the response vector $\mathbf{y} = (y_1, y_2, ..., y_c)$ and update its representation L:

$$(\mathbf{y}, L) \leftarrow f_{\mathrm{LCA}}(\mathbf{x}, \mathbf{y}, \mathbf{z}, L).$$

The CCI LCA algorithm is as follows.

Algorithm 2 (LCA initialization). *Initialize synaptic vectors $\mathbf{v}_i$, using c Gaussian functions, each Gaussian having a different mean vector, response y_i with a small random number in $[0, 0.1]$, and firing age $n_i = 0$, $i = 1, 2, ..., c$.*

The CCI LCA update is cell-autonomous, meaning that when each neuron updates, it simply uses the currently available response values of other neurons and does not wait using any global clock. This is critical for the temporal processing.

Algorithm 3 (CCI LCA update). *Input:* $\mathbf{x}, \mathbf{y}, \mathbf{z}, L$. *Output:* $\mathbf{y}, L$.

1. **Compute pre-responses**. *Depending on the level's location in the network, take external or internal input*[1] $\mathbf{p} = ((1-\alpha)\mathbf{x}, \alpha\mathbf{z})$, *where $\mathbf{z}$ may have been just overridden by a teacher if $\mathbf{z}$ is an external port and α, $0 \leq \alpha \leq 1$ is the relative weight of the top-down input. Compute the pre-competition response*[2]*:*

---

[1] A contrast normalization for bottom-up input $\mathbf{x}$ often improves the performance. For example, $\mathbf{x}$ is normalized so that the minimum component is zero and the maximum component is 1.

[2] Sigmoidal function is approximated by the cell nonlinearity in the next step. The model here can be extended to a spiking neuronal model which is useful for a higher temporal resolution.

$$y_i = \frac{\mathbf{v}_i \cdot \mathbf{p}}{\|\mathbf{v}_i\|}, i = 1, 2, ..., c. \quad (1)$$

2. **Neurons mutually inhibit for dynamic sparse coding**. *For simulating lateral inhibition with a relatively lower update frequency, use this non-iterative ranking-and-scaling mechanism.*[3] *Rank only $k+1$ top winners so that after ranking, $y_1 \geq y_2 ... \geq y_c$, as ranked responses. Use a piecewise linear but globally nonlinear function to scale the responses:*

$$y_i \leftarrow \frac{y_i - y_{k+1}}{y_1 - y_{k+1}}, i = 1, 2, ..., k. \quad (2)$$

All other neurons do not fire $y_i = 0$ for $i = k+1, k+2, ..., c$. This ranking-and-scaling mechanism replaces repeated iterations that take place among two-way connected neurons in the same level. This simulates L5 assisting L2/3 for lateral inhibition and L6 assisting L4 for lateral inhibition[3].

3. **Optimal Hebbian learning**. *Update only the top k winner neurons $\mathbf{v}_j$, $j = 1, 2, ..., k$, using the pre-synaptic activity* $\mathbf{p}$, *the post-synaptic activity y_j, and its firing-age dependent plasticity:*

$$\mathbf{v}_j \leftarrow w_1(n_j)\mathbf{v}_j + w_2(n_j) y_j \mathbf{p}, \quad (3)$$

where the learning rate w_2 and the retention rate w_1, respectively, are determined by:

$$w_2(n_j) = \frac{1 + \mu(n_j)}{n_j}, w_1(n_j) = 1 - w_2(n_j) \quad (4)$$

where $\mu(t)$ is a non-negative amnesic function.[4] *Note $y_1 = 1$ for the top winner. Update the real-valued neuron "firing age" n_j only for the winners: $n_j \leftarrow n_j + y_j$, $j = 1, 2, ..., k$.*

4. **Lateral excitation for cortical smoothness**. *Mutual excitatory connections [3] among neurons in the same level are useful for the nearby neurons to detect similar features. In the computer simulation of lateral excitation, there is a sphere of excitation with radius r from each neuron. Not only the top-k winners update, but also the neurons within the sphere of excitation. The scope radius r starts from the half size of the neuronal layer during initialization. It slowly decreases to $r = 0$ when the network matures. The input* $\mathbf{y}$ *is used for lateral excitation.*

[3] This mechanism of ranking-and-scaling is an approximation of biological in-place inhibition. It is not in-place, as it requires extra-cellular sorting. But it is very effective computationally by eliminating iterations within an LCA level.

[4] The amnesic function $\mu(t)$ is for a general nonstationary process $\mathbf{u} = y\mathbf{p}$ where the probability distribution for $\mathbf{u}$ slowly changes. $\mu(t) = 0$ when $t \leq t_1$, so that $w_2(t) = 1/t$; $\mu(t) = 2(t - t_1)/(t_2 - t_1)$ when $t_1 < t \leq t_2$ so that $w_2(t)$ linearly changes from 0 to 2; $\mu(t) = 2 + (t - t_2)/r$ when $t_2 < t$ so that $w_2(t)$ approaches $1/r$ when t grows without bound. We chose $t_1 = 20, t_2 = 200$ and $r = 2000$ in our experiments. These numbers were selected based on the need for stability and fast adaptation.

The lateral inhibition is modeled by top-k competition in terms of pre-response values to avoid undesirable oscillations.

5. **Long-term memory**. *All other neurons that do not update keep their firing age and synapses unchanged. They are long term memory for this context of* **p**. *Other updated neurons are working memory for this context.*

The algorithm is called the Candid Covariance-free Incremental (CCI) LCA algorithm. The term "Candid" means that the algorithm uses candid information in the inputs for the dual optimality. LCA was compared with some well known methods in [21].

7 Representation Emergence

Unlike FSM or its probabilistic versions HMM and POMDP, DN uses emergent representation as shown in Fig. 2 (for the three layer DN, the lowest area is considered as image input layer which does not conduct computation). DN has multilevel distributed state representation $(\mathbf{y}_t^{(2)}, \mathbf{y}_t^{(3)})$, where $\mathbf{y}_t^{(l)}$ represents the response vector of layer l. Each layer l has its own state presentation as a distributed representation. It takes distributed input from space $P = X \times Z$, and generate distributed response $\mathbf{y} \in Y$.

7.1 Soft-Logic AND in Layer 2

Given a limited number c of neurons in layer l, the bottom-up input space of the layer is X and the top-down input space of the layer is Z. Thus, the input space of the layer is $P = X \times Z$. The theory of LCA indicates that the weight vectors of the c neurons $\mathbf{p}_1, \mathbf{p}_2, ..., \mathbf{p}_c$ in the layer are optimally distributed in the observed signal manifold of P though developmental learning so that a region in P with a high probability density recruits more neurons than a region with a lower probability density. Given any input pair $\mathbf{p} = (\mathbf{x}, \mathbf{z})$ to the layer, LCA finds the top neuron(s) who gives the highest response $r(\mathbf{p}, \mathbf{p}_i)$ (i.e., best matching):

$$j = \arg \max_{1 \le i \le c} r(\mathbf{p}, \mathbf{p}_i). \tag{5}$$

Thus, the best matched neuron j serves as the representative of unknown input $\mathbf{p}$. (When the number of neurons c is large so that the top matched neurons are near enough to be "experts" of the current input case $\mathbf{p}$, more than 1 neurons can be allow to fire, so that the firing pattern $\mathbf{y}$ from the layer allows sub-neuronal precision estimation of the input $\mathbf{p}$.) Thus, Layer 2 serves as a soft-logic AND: all the corresponding components in $\mathbf{p}$ and $\mathbf{p}_j$ must match well in terms of normalized inner product which gives the response $r(\mathbf{p}, \mathbf{p}_i)$.

7.2 *Soft-Logic OR in Layer 3*

During supervised learning for Layer 3, suppose that the neuron k in Layer 2 is set to the highest value 1 at time t. Further suppose that the neuron j in Layer 2 fires at time $t-1$. In other words, the j's component of the bottom-up input $\mathbf{x}_{t-1}$ of the neuron k is high. Consequently, with regard to the synapse that links neuron j with neuron k, the pre-synaptic activity r_j and post-synaptic action r_k are both high. Then the weight that links these two neurons is strengthened. Mathematically, according to the theory of LCA, the weight that links neuron j in the lower layer to neuron k in the higher layer is

$$w_{j,k} \approx E\{r_j r_k \mid \text{neuron } k \text{ fires}\}. \tag{6}$$

Therefore, the more often neuron j fires when neuron k fires, the higher the weight from j to k. Therefore, all the neurons in Layer 2 that fire with neuron k in Layer 3 have non-zero weights. As only one (or few) neuron fires in Layer 2 at any time, this is a soft-logic OR for Layer 3: all the cases in Layer 2 that correspond to k firing neuron are connected to the target neuron in Layer 3.

7.3 *No Local Extrema*

Intuitively, as long as there are a sufficient number of neurons in Layer 2 and there is a sufficient amount of training experience, the trained DN can approximate any smooth high dimensional input-output mapping to a desired precision rate, as long as the bottom-up and top-down pair $\mathbf{p}_t = (\mathbf{x}_t, \mathbf{z}_{t-1})$ uniquely determines the desired action output from the layer 3. This distribution scheme largely avoids the problem of local extrema with the error back-propagation methods.

7.4 *Discriminant Features*

Furthermore, the dually optimal distribution of c neurons in the space $P = X \times Z$ enables development of discriminant feature neurons: Which neuron wins in Eq. (5) is less sensitive to distractor components in $\mathbf{x}$ that do not affect output $\mathbf{z}$ but are more sensitive to features in $\mathbf{x}$ that significantly affect output $\mathbf{z}$. Consequently, the dually optimal self-organization of c neuronal weights $\mathbf{p}_1, \mathbf{p}_2, ..., \mathbf{p}_c$ lead to nearly smallest error in the estimated output space of DN.

8 Properties

Based on the above discussion, we are ready to present major properties that are of paramount importance to temporal processing in DN.

8.1 Context Dependent Attention

Corollary 1 (Context-dependence). *Given external bottom-up input* $\mathbf{x}_t$ *and top-down context* $\mathbf{z}_{t-1}$*, the three-layer DN network has three classes of internal behaviors, external (E), internal (I) and mix (M), which means that the motor output* $\mathbf{z}_t$ *is dependent on external input* $\mathbf{x}_t$ *only, dependent on internal top-down context* $\mathbf{z}_{t-1}$ *only, and dependent on both, respectively, subject to learning.*

Proof. As discussed above, the best matched feature $\mathbf{p}_j$ in Layer 2 fits both $\mathbf{x}_t$ and $\mathbf{z}_{t-1}$. There are three cases for the Layer 3 to learn the next action $\mathbf{z}_t$, (1) $\mathbf{z}_t$ depends on $\mathbf{x}_t$ only, (2) $\mathbf{z}_t$ depends on $\mathbf{z}_{t-1}$ only, and (3) $\mathbf{z}_t$ depends on both. They corresponding to E, I, and M, respectively. The quality and amount of learning experience affect the ideal independence of the output $\mathbf{z}_t$ on $\mathbf{z}_{t-1}$ (for the E case) or $\mathbf{x}_t$ (for the I case).

8.2 Active Time Warping

Suppose that the network is updated at discrete times, $t = 1, 2, \ldots$. This series of discrete times can represent any network update frequency. For example, if the time interval between every two consecutive time instances is 1ms, the network is updated at 1000Hz. Denote the sensory input at time t to be $\mathbf{x}_t$, $t = 1, 2, \ldots$. The series of discrete times can be considered the sampling times on the continuous physical world. The phenomenon that a dynamic event can proceed at different speed at different stages is called *time warping*. It is desirable that physical events with different time warpings are recognized at the motor layer as the same type of event (e.g., in speech recognition and dynamic visual recognition). Of course, at Layer 2, the representations with different time warpings are different, reflecting an advantage of the multi-area representation, in contrast with the single state representation of HMM and POMDP.

For example, the following two sequences should be recognized as the same sequence at the motor layer:

```
|w|w|w|w|w|w|_|u|u|u|_|u|u|_|_|z|z|z|z|_|
|w|w|_|u|u|u|u|_|u|u|u|u|_|z|z|_|
```

where w, u, z are words and | is a delimiter of time frames, each of which corresponds to a different t.

Fort this property, we have the following theorem:

Theorem 1 (Active Time Warping). *The motor layer of DN can be taught to carry out active time warping, subject to learning.*

Proof. Use the Corollary 1. For duplicate word, learn the "drop postfix" case illustrated in Fig. 2. Dropping duplicate spaces _ is similar. For detecting two w's separated by a space _, $\mathbf{z}_t$ should correspond to a context that requires a trailing space _ before receiving the second w.

8.2.1 Active Context

The major goal is to interactively train DN so that it make sequential decisions that require spatiotemporal, attended context in a dynamic range of the past.

Theorem 2 (Context of any temporal length). *The temporal length of top-down context Z of the DN is arbitrary, depending on learned attention.*

Proof. According to the Corollary 1, any length of history can be generated at Layer 3 by recursively apply the "link" shown in Fig. 2. That is, the context $\mathbf{z}_t$ indicates a label that requires appending the last bottom-up input $\mathbf{x}_t$.

Theorem 3 (Context of any temporal subset). *The temporal context of top-down context $\mathbf{z}_t$ of DN can represent any subset of the bottom-up stimuli, depending on learned attention.*

Proof. Simply drop the parts that do not belong to the subset, using the "drop postfix" or "drop prefix" function shown in Fig. 2.

Theorem 4 (Flush). *The temporal context of top-down context $\mathbf{z}_t$ of DN can be reset to represent only the last bottom-up input, depending on learned attention.*

Proof. According to Corollary 1, apply the internal (I) behavior.

Combining above two properties gives the following theorem.

Theorem 5 (Any context). *The temporal context of top-down context $\mathbf{z}_t$ of DN can represent any subset of the bottom-up stimuli of any length of the history, depending on learned attention.*

Proof. Combining the above three theorems, use Theorem 4 to start at a desired frame of the history, use Theorem 3 to keep the desired subset, and use Theorem 2 to keep it of any desired length.

Theorem 5 implies that the DN agent can learn to attend to any part of spatiotemporal context, a necessity of complex language understanding.

8.2.2 Time Duration

The time duration task is an opposite problem of time warping. The goal of the task is to count the length of time between two events `a` and `b`.

Different from the above models, DN can learn to count, using its intrinsic property of response time (e.g., sampling time) as the basic time unit.

Theorem 6 (Time duration). *The action $\mathbf{z}_t$ of DN can represent any finite length of time between two specified events.*

Proof. Suppose $\mathbf{z}_t = i$ means time i, $i = 1, 2, ..., k$ from `a`. The DN starts to count using is action $\mathbf{z}_t$ as soon as it senses `a`, and terminates counting when it senses `b`. Suppose $*$ is a distractor, other than `a` and `b`. DN needs to continue to count when it sees a distractor. The following shows how DN responds.

```
x:|*|*|a|*|*|*|*|*|*|*|b|*|*|...
z:|_|_|_|1|2|3|4|5|6|7|8|9|_|...
```

8.2.3 New Sentences

The above results can be used to understand how to use abstract motor states to generalize to new sentences that the system has never learned.

Problem 1 (New Sentences). Suppose that there are four word meanings, W_1, W_2, W_3, W_4. Each word meaning W_i has ten synonyms $\{w_{ij} \mid j = 1, 2, ..., 10\}$, $i = 1, 2, 3, 4$. Then, there are 10000 equivalent 4-word sentences in the form of $(w_{1h}, w_{2i}, w_{3j}, w_{4k})$, $h, i, j, k = 1, 2, ..., 10$. Interactively learn the synonyms in the corresponding sentence contexts, so that the developmental agent recognizes all new sentences of length 1 to 4 of the form of $W_1, W_2, ...W_j$, with $j = 1, 2, 3, 4$.

This problem is addressed in the following way. In Lesson 1, learn individual words. The x-row below denotes sensory input at each time frame, while The z-row below denotes motor output (label) at the corresponding time frame.

```
x:|a1|a1|_|_|a2|a2|a2|_|a3|a3|_|a4|a4|...
z:|_ |A |A|_|_ |A |A |A|_ |A |A|_ |A |...
```

The delay in the corresponding motor output is due to the fact that it takes two updates for the signal of sensory input to pass the two levels to reach the motor. Do the same for B, C and D. In Lesson 2, learn two-word sentences:

```
x:|a1|b1|_ |a1|b2|_ |a1|b3|_ |a1|b4|...
z:|_ |A |AB|_ |A |AB|_ |A |AB|_ |A |AB|...
```

In Lesson 3, learn 3-word sentences in a similar way. In Lesson 4, learn 4-word sentences:

```
x:|a1|b1|c1 |d1  |_  |a1|b1|c1 |d2  |_  |...
z:|_ |A |AB |ABC |ABCD|_ |A |AB|ABC |ABCD|...
```

The number of sentences learned in these lessons are 40, 10, 10, 10, respectively. The number of new sentences to be recognized are 0, $100-10$, $1000-10$, $10000-10$, respectively. Totally, the DN learns 70 sentences, but recognizes $90+990+9990 = 11070$ new sentences. Of course, no two English words are exactly synonyms. The subtle difference is represented in Layer 2 response in DN, but the motor outputs are the same.

In our experiment to appear elsewhere, Dr. Qi Zhang at Fudan University used 64×64 neurons in Layer 2. He tested the updated DN after every epoch through the training set. The DN perfectly (100%) recognized all the 70 trained senses and all the 11070 new sentences from epoch 23.

8.2.4 Complexity

Theorem 7 (Exponential Capacity). *The number of distinguishable patterns by a cortical level with n neurons is exponential* $\mathcal{O}(2^n)$.

Proof. Each neuron has least two status, firing and nor firing. The number of firing patterns of n neurons is at least $\mathcal{O}(2^n)$.

This is a great advantage of distributed representation compared to a symbolic representation. While a symbolic representation potentially requires an exponential number of symbols, no symbol is assigned to these 2^n patterns using the emergent, distributed representations.

Theorem 8 (Linear complexity). *The amount of computations required by a cortical layer with n neurons is linear $\mathcal{O}(cn)$, assuming a constant number c of average synapses per neuron.*

Proof. The proof is obvious, as only n neurons need to be computed and learned.

With learning time t, the time complexity is $\mathcal{O}(cnt)$.

The above two properties lead to the conclusion that the DN uses a linear space complexity and linear time complexity to solve a general-purpose exponential size problem.

9 Experimental Results

The DN model has been tested for recognizing temporal visual events (Luciw et al. 2008 [11]), spatiotemporal disparity from stereo without explicit stereo images matching Solgi & Weng 2009 [15], text processing as temporal sequences for generalization to new sentences based on synonyms, the part-of-speech tagging problem and the chunking problem using natural languages from the Wall Street Journal in Weng et al. 2009 [23], and modeling early language acquisition and generalization in Miyan & Weng 2010 [12].

10 Conclusions

The motor area of DN is not only the hub of actions, but also for *active* abstraction — converting different temporal contexts into the equivalent state represented by the firing pattern of the motor area. Temporal context of any length and of any subset can be attended and abstracted by the spatial DN without a need for any dedicated temporal component. Recently, Weng 2010 [18] stated that a DN can simulate any Finite Automaton.

According to task context, DN can deal with time warping or determine time duration. Unlike the symbolic HMM, POMDP, and Bayesian nets which require a handcrafted network structure, DN self-generates its network structure. DN performs while learning, regardless supervisory signals are available at the motor end or not. The exponential complexity, in terms of the scan window length in online sensory processing, is converted by DN into a linear time complexity and linear space complexity. The experimental results cited here but presented elsewhere have shown that this general purpose model can deal with a variety of temporal context problems, including vision and natural language processing. The future work includes testing such temporal capabilities on real developmental robots.

References

1. Albus, J.S.: Outline for a theory of intelligence. IEEE Trans. Systems, Man and Cybernetics 21(3), 473–509 (1991)
2. Albus, J.S.: A model of computation and representation in the brain. Information Science 180(9), 1519–1554 (2010)
3. Callaway, E.M.: Feedforward, feedback and inhibitory connections in primate visual cortex. Neural Networks 17, 625–632 (2004)
4. Elman, J.L., Bates, E.A., Johnson, M.H., Karmiloff-Smith, A., Parisi, D., Plunkett, K.: Rethinking Innateness: A connectionist perspective on development. MIT Press, Cambridge (1997)
5. Emami, A., Jelinek, F.: A neural syntactic language model. Machine Learning 60, 195–227 (2005)
6. George, D., Hawkins, J.: Towards a mathematical theory of cortical micro-circuits. PLoS Computational Biology 5(10), 1–26 (2009)
7. Hecht-Nielsen, R.: Confabulation Theory. Springer, Berlin (2007)
8. Jelinek, F.: Self-organized language modeling for speech recognition. In: Waibel, A., Lee, K. (eds.) Readings in Speech Recognition, pp. 450–506. Morgan Kaufmann, San Mateo (1990)
9. Lee, T.S., Mumford, D.: Hierarchical bayesian inference in the visual cortex. J. Opt. Soc. Am. A 20(7), 1434–1448 (2003)
10. Luciw, M., Weng, J.: Where What Network 3: Developmental top-down attention with multiple meaningful foregrounds. In: Proc. IEEE International Joint Conference on Neural Networks, Barcelona, Spain, July 18-23, pp. 4233–4240 (2010)
11. Luciw, M., Weng, J., Zeng, S.: Motor initiated expectation through top-down connections as abstract context in a physical world. In: IEEE International Conference on Development and Learning, August 9-12, Monterey, CA, pp. +1–6. (2008)
12. Miyan, K., Weng, J.: WWN-Text: Cortex-like language acquisition with What and Where. In: Proc. IEEE 9th International Conference on Development and Learning, Ann Arbor, August 18-21, pp. 280–285 (2010)
13. Piaget, J.: The Construction of Reality in the Child. Basic Books, New York (1954)
14. Purves, W.K., Sadava, D., Orians, G.H., Heller, H.C.: Life: The Science of Biology, 7th edn. Sinauer, Sunderland, MA (2004)
15. Solgi, M., Weng, J.: Developmental stereo: Emergence of disparity preference in models of visual cortex. IEEE Trans. Autonomous Mental Development 1(4), 238–252 (2009)
16. Tenenbaum, J.B., Griffithsb, T.L., Kemp, C.: Theory-based bayesian models of inductive learning and reasoning. Trends in Cognitive Sciences 10(7), 309–318 (2006)
17. Weng, J.: Task muddiness, intelligence metrics, and the necessity of autonomous mental development. Minds and Machines 19(1), 93–115 (2009)
18. Weng, J.: A 5-chunk developmental brain-mind network model for multiple events in complex backgrounds. In: Proc. Int'l Joint Conf. Neural Networks, Barcelona, Spain, July 18-23, pp. 1–8 (2010)
19. Weng, J., Ahuja, N., Huang, T.S.: Learning recognition and segmentation of 3-D objects from 2-D images. In: Proc. IEEE 4th Int'l Conf. Computer Vision, pp. 121–128 (May 1993)

20. Weng, J., Ahuja, N., Huang, T.S.: Learning recognition and segmentation using the Cresceptron. International Journal of Computer Vision 25(2), 109–143 (1997)
21. Weng, J., Luciw, M.: Dually optimal neuronal layers: Lobe component analysis. IEEE Trans. Autonomous Mental Development 1(1), 68–85 (2009)
22. Weng, J., McClelland, J., Pentland, A., Sporns, O., Stockman, I., Sur, M., Thelen, E.: Autonomous mental development by robots and animals. Science 291(5504), 599–600 (2001)
23. Weng, J., Zhang, Q., Chi, M., Xue, X.: Complex text processing by the temporal context machines. In: Proc. IEEE 8th International Conference on Development and Learning, Shanghai, China, June 4-7, pp. +1–8 (2009)

Part IV: Special Applications

Towards *Physarum* Robots

Jeff Jones, Soichiro Tsuda, and Andrew Adamatzky

Abstract. The true slime mould *Physarum polycephalum* is a suitable candidate organism for small scale robotics because it spontaneously generates transport, movement and navigation, exhibiting complex behaviour from very simple component interactions. *Physarum* may be considered as a smart computing material as its motor and control systems are distributed within a simple tissue type and can survive trauma such as excision, fission and fusion of plasmodia. We demonstrate experimentally how the plasmodium of *Physarum* may be configured to generate complex and controllable oscillatory transport behaviour which may prove useful in small robotic devices. We measure the lifting force of the plasmodium and demonstrate how protoplasmic transport can be influenced by externally applied illumination stimuli. We provide an exemplar vehicle mechanism by coupling the oscillations of the plasmodium to drive the wheels of a Braitenberg vehicle and use light stimuli to effect a steering mechanism. To explore the generation of complex behaviour from such simple component parts we present a particle based model of *Physarum* which spontaneously generates complex oscillatory patterns from simple local interactions, is distributed in terms of the origin and control of motor behaviour, is morphologically adaptive, is amenable to external influence, and is robust to environmental insult and thus can itself be considered as a virtual smart material. We demonstrate different forms of controllable motion, including linear, reciprocal, rotational, helical, and amoeboid movement. We enable external control of the robotic movement by simulated chemo-attraction ('pulling') and simulated light hazards ('pushing'). The amorphous and distributed properties of the collective are demonstrated by cleaving it into two independent entities and fusing two separate entities to form a single device, thus enabling it to traverse difficult or separate paths. We conclude by examining ways in which future robotic devices may be developed using physical instances of smart materials.

Jeff Jones · Soichiro Tsuda · Andrew Adamatzky
Unconventional Computing Group, University of the West of England,
Bristol, BS16 1QY, UK
e-mail: {jeff.jones, so.tsuda, andrew.adamatzky}@uwe.ac.uk

Y. Meng and Y. Jin (Eds.): Bio-Inspired Self-Organizing Robotic Systems, SCI 355, pp. 215–251.
springerlink.com

1 Introduction

Very small scale mechanisms made possible by recent advances in micro-fabrication still require a means of energy conversion to convert a fuel supply into useful motive force. Although it is possible to reduce conventional engines and transmission designs in scale, doing so introduces challenges with regard to friction [1], heat [2] and more general miniaturisation problems [3] that may differ from or exacerbate those observed at the larger scale. Meanwhile, small scale engines are still subject to the necessities of complex engine design, control and assembly, including maintenance schedules and repair (however, one goal of micro-fabrication is in the forecast minimal production costs which may render repair economically unnecessary).

Recent approaches have sought inspiration from biological sources of to provide the transformation of energy conversion to motive force, examples from flagellated movement of bacteria [4] and eukaryotes [5], ciliated transport for microscopic mixing [6] and transport [7], peristaltic propulsion [8], [9], amoeboid movement [10] and collective mass transport [11] have all been suggested. Biologically inspired mechanisms are attractive since they are composed (at their lowest level) of simple and plentiful materials and show impressive feats of redundancy, fault tolerance and self repair. However, even though the above mechanisms are technologically elegant and the product of impressive biological self-assembly, the internal arrangement of components underlying some mechanisms - for example the bacterial flagellar motor system - appears to be almost as complex as those in non-biological machines. Progress in the development of biologically inspired machines has also been made by considering the use of even more simple biological structures which straddle the boundary of non-living physical materials and living organisms. Such structures include biological fibres and membranes [12], lipid self assembly in terms of networks [13], pseudopodium-like membrane extension [14] and even basic chemotaxis response [15]. Some engineering and biological insights have already been useful by studying the structure and function of what might be termed 'semi-biological' materials and the complex behaviour seen in such minimal examples raises questions about the lower bounds necessary for the emergence of apparently intelligent behaviour.

Of particular interest are materials which embody oscillatory states, since the periodic transitions between two or more states can provide differential behaviour and impetus, if suitably coupled to a propulsion mechanism. The resulting transport may be relatively simple, such as flagellar or ciliary motion [16], or as complex as the patterns produced by neural assemblies of central pattern generators [17]. When oscillatory states are combined with local communication waves of excitation may propagate which can also result in motion if the excitation is coupled to propulsion mechanism. Reaction-diffusion phenomena have proven to be useful in this approach because the emergence of oscillatory states can be a self organised and robust process and also feature travelling wave phenomena. Reaction-diffusion approaches to robotics include the use of propagating waves for robotic control and navigation [18], for the manipulation and transport of objects [19], and oscillatory wave transport within a gel-like material for forward movement [20].

Biological (complete living systems) and semi-biological (materials produced by living systems) approaches can each provide valuable insights and techniques in the pursuit of small scale machines. The ideal hypothetical candidate for a biological machine would be an organism which is capable of the complex sensory integration, movement and adaptation of a living organism, yet which is also composed of a relatively simple material that is amenable to simple understanding and control of its properties.

We suggest that the myxomycete organism, the true slime mould *Physarum polycephalum*, is a suitable candidate organism which meets both criteria; i.e. it is a complex organism, but which is composed of relatively simple materials. A giant single-celled organism, *Physarum* is an attractive biological candidate medium for emergent motive force because the basic physical mechanism during the plasmodium stage of its life cycle is a self-organised system of oscillatory contractile activity which is used in the pumping and distribution of nutrients within its internal transport network. The organism is remarkable in that the control of the oscillatory behaviour is distributed throughout the almost homogeneous medium and is highly redundant, having no critical or unique components.

The plasmodium of *Physarum* is amorphous, typically jellied and flat in appearance and ranges from the microscopic scale to up to many square metres in size. The plasmodium is a single cell syncytium formed by repeated nuclear division and is comprised of a sponge-like actin-myosin meshwork co-occurring in two physical phases. The gel phase is a dense matrix subject to spontaneous contraction and relaxation under the influence of changing concentrations of intracellular chemicals. The sol phase is a liquid protoplasm transported through the plasmodium by the force generated by the oscillatory contractions within the gel matrix. There is a complex interplay between the gel and sol phases and both phases can change between each form when subject to changes in pressure, temperature, humidity and local transport. The internal structure of *Physarum* can thus be regarded as a complex functional material capable of both sensory and motor behaviour. Indeed *Physarum* has been described as a membrane bound reaction-diffusion system in reference to both the complex interactions within the plasmodium and the rich computational properties afforded by its 'material' properties [21].

In its natural habitat (shaded and damp forest environments) *Physarum* utilises the properties of this computational material to extend a growth front towards local nutrient sources. It feeds on bacteria found on other living matter, extending its amorphous body to engulf and digest its food before adapting its body plan behind the active growth front to form protoplasmic vein structures which transport nutrients to other parts of the plasmodium. A plasmodium may be cut into two separate independent and functioning plasmodia or, alternately, two separate plasmodia may fuse and become a single entity. This ability alone suggests a complex, distributed and redundant mechanism of sensory and motor control and led to intense recent research into its computational ability, initiated by Nakagaki et al., who found that independent plasmodial fragments, when fused, could solve maze problems by finding the shortest connecting path between two sources of food [22]. Other examples of the computational ability of the plasmodium include its use as a spatially represented unconventional computer suitable for approximation of proximity graphs [23], shortest path and collision free path problems [24], division of the plane [25], vehicle routing [26], approximation of urban path planning [27], [28] and as a spatially represented general purpose computing machine [29].

From a robotics perspective it has previously been shown that *Physarum* plasmodium, by its adaptation to changing conditions within its environment, may be considered as a prototype micro-mechanical manipulation system, capable of simple and programmable robotic actions including the manipulation (pushing and pulling) of small scale objects [30] and as a guidance mechanism in a biological/mechanical hybrid approach where the response of the plasmodium to light irradiation was used to provide feedback control to a robotic system [31]. A *Physarum* inspired approach to robotics using amoeboid movement generated by a series of fluid coupled oscillators has been also been demonstrated [10].

In classical engineering the role of the source of motive force and its distribution and control are usually separated, for example the generation of power by an engine and its useful distribution by the drive system are controlled by separate (although linked) control systems. Furthermore these systems are carefully designed to ensure that forces are applied at the correct times and within specific ranges. In the *Physarum* plasmodium the forces (i.e. the emergence of oscillatory behaviour) emerge spontaneously and there is no distinction between the generation of force and the control of its distribution. It has been shown that the patterns of force generation (emergence of oscillatory behaviour) are influenced by the shape of the plasmodium and the shape is, in turn, affected by the oscillatory behaviour [32]. Thus it may be possible to govern - to some degree - the pattern of oscillations by restricting the plasmodium to a chambered environment in order to utilise particular aspects of its oscillatory behaviour.

In this report we discuss the use of *Physarum* as a candidate organism and material for the spontaneous generation and distribution of physical forces for engine-like and transport mechanisms. The layout of the paper is as follows: In Section 2 we show experimental results demonstrating that *Physarum* plasmodium is capable of generating motive lifting force which is both measurable and externally ontrollable. Using a dumbbell patterned design we demonstrate complex phase relationships from the oscillatory dynamics and we design a prototype example mechanism where the motive output of *Physarum* plasmodium oscillations is used to drive the wheels of a simulated Braitenberg vehicle, using external light stimuli to steer the vehicle. Building upon the experimental results we investigate in section 3 the emergence of oscillatory transport in the computational modelling of a virtual material composed of a particle collective where - as with the plasmodium - oscillatory phenomena emerge from simple and local interactions. Taking this emergent behaviour as our base point we explore the effect of patterning the plasmodium in constrained and unconstrained environments on movement patterns composed of oscillatory waves. In section 4 we demonstrate a range of transport types (linear, rotary, reciprocal and helical) and coupling methods. We also consider smaller isolated 'blobs' of the material which exhibit spontaneous amoeboid movement, investigating mechanisms whereby the virtual 'blobs' may be guided externally by repulsion (simulated light irradiation) and attraction (by the deposition of attractants). We summarise the experimental and theoretical findings in section 5 and suggest future research by which small scale transport devices may be constructed from materials which exhibit emergent properties from simple component interactions.

2 Experimental

This section focus on experimental investigation into the possibility of *Physarum* engine implementation. The *Physarum* plasmodium is famously known for fast protoplasmic streaming within the cell body and the motive force of the cell has been studied extensively in the field of cell biology for more than 50 years (cf. [33, 34, 35]). The mechanism of streaming generation is based on the actin polymerization/depolymerization cycle, which switches every 30 s regulated by calcium ions [36]. This periodic cycle controls the direction of the protoplasmic streaming, and as a result, the whole cell shows rhythmic cell thickness changes [34]. The flow speed of protoplasmic streaming of the *Physarum* plasmodium becomes up to 1 mm/s, whereas that of other organisms, such as plant and amoeba cells, are about tens of μm/s [37]. This is one of the fastest protoplasmic streamings known so far, and together with the periodicity of the protoplasmic streaming, the plasmodium of *Physarum polycephalum* could be ideal parts of small-scale biological devices, such as actuator and transporter.

In order to explore the possibility of using the *Physarum* cell as source of power, it is important to know (1) how much force a *Physarum* plasmodium can generate and (2) how the force can be steered externally. Before going to these two points, we first discuss characteristics of the cell shape and the contraction oscillation rhythm in the *Physarum* plasmodium.

2.1 Cell Shape and Oscillation Pattern

The authors have previously investigated the generation of periodic rhythm of the cell confined in a small circular well on the agar gel [38]. When observed under a microscope, a piece of *Physarum* cell just placed in the well consists of smaller pieces of granular protoplasm, each of which can potentially become an individual slime mold cell. These granules start contracting about 10 minutes after transplanted and neighboring granules gradually merge together to synchronize the contraction rhythm. They eventually fuse into one single *Physarum* plasmodium and show several types of synchronized oscillation rhythms, such as bilateral shuttle streaming of protoplasm (Fig. 1a) and clockwise/anti-clockwise rotation streaming (Fig. 1b). What is remarkable about the oscillation regeneration in the *Physarum* plasmodium is "size-invariant" behavior: Even if the cell size becomes larger (the size of well becomes bigger), a whole cell fully synchronizes within a certain amount of time. We have tested with 1.5, 3.0, 4.5, 6.0, and 7.5mm diameter wells, and in all cases the *Physarum* cell fully synchronized approximately in 70 minutes [38]. In other words, granular "swarms" can self-organize a synchronized oscillation rhythm in a fixed time, no matter how large the cell becomes. This result suggests that the organism is good at coordinating the internal structure to maintain itself as a single cell. Even when the body size changes, it is able to recover the synchronization using a distributed-computing type control.

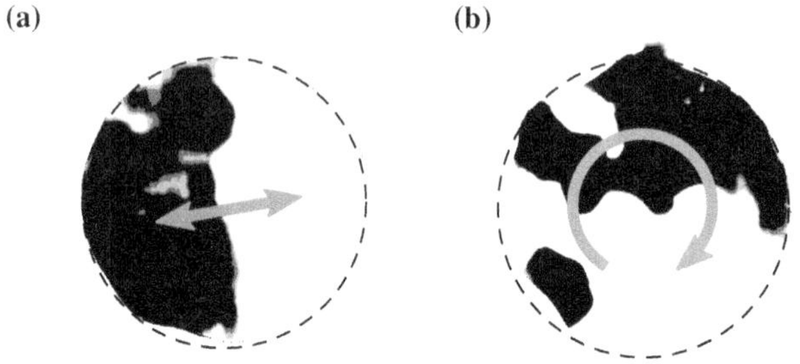

Fig. 1 Oscillation patterns of the *Physarum* plasmodium in a single well. (a) bilateral shuttle oscillation (b) clockwise rotation wave. Black and white regions indicate areas where thickness is increasing and decreasing, respectively.

Although a cell in the well does show oscillations that potentially can be exploited as source of power, the single well design would not be appropriate for *Physarum* engine. As we will show in the next section, given that a whole plasmodium in a well is used to apply force to another object, a total displacement (i.e. thickness change) of the cell will be crucial. However, in the case of single well, the total displacement will be balanced out: For example, if a cell is showing bilateral oscillation (Fig. 1a), the thickness of a half of the cell is increasing while that of the other half is decreasing. As a result, the total displacement becomes almost zero. This fact led us to the design of a dumb-bell shape, in which two circular wells are connected by a narrow channel (cf. Fig. 4a). This design is originally made by Takamatsu et al [39]. When the width of the connecting channel is 0.4 mm, anti-phase thickness oscillation between two wells can be observed [40]. This means protoplasm of the cell flows to and from two wells, and thus the thickness displacement in one well will not be canceled out.

In the following section, we adopt the dumb-bell shape design and keep the cell to take the shape by hydrophobic structures. Because it prefers wet regions to dry ones, the plasmodium stays inside the structures and keeps the shape for a long time (provided that the humidity for the cell is in an appropriate range). *Physarum* cells used for experiments below were cultured on 1.5 % agar gel in the thermostatic chamber (Lucky Reptile Herp Nursery II Incubator, Net Pet Shop, UK) at 26°C and fed with oat flakes once or twice a day. They were starved at least for 12 hours prior to experiments.

2.2 Force Generated by the Physarum Plasmodium

This section investigates how much force can be produced by *Physarum*'s contractile cell oscillation. We estimate the force by loading some weight on a plasmodium. In particular, we placed water on one of two wells of a dumb-bell shaped plasmodium (Fig. 2a). By changing the height of water, the load on the cell cen be controlled.

A dumb-bell shaped plasmodium was prepared as follows: First, small pieces of *Physarum* are taken from a larger culture and set in dumb-bell shaped wells of a thin printed circuit board (all copper coating was removed). They are then placed on a 1.5 % agar gel and kept in a dark place at least for two hours so that the *Physarum* cells fuse into one single cell. After fusion, the cell and the PCB are clamped together with plexiglasses and a PDMS block, as illustrated in Fig. 2a. A glass tube, which inner diameter is same as the diameter of a plasmodium well, is glued to the top plexi glass and filled with distilled water to a certain height. The whole setup is covered with a wet glass beaker to avoid evaporation of water in the glass tube and placed on the three-dimensional micro-stage (Fig. 2b). The displacement of water surface is recoded for 20 minutes by a microscope from the size.

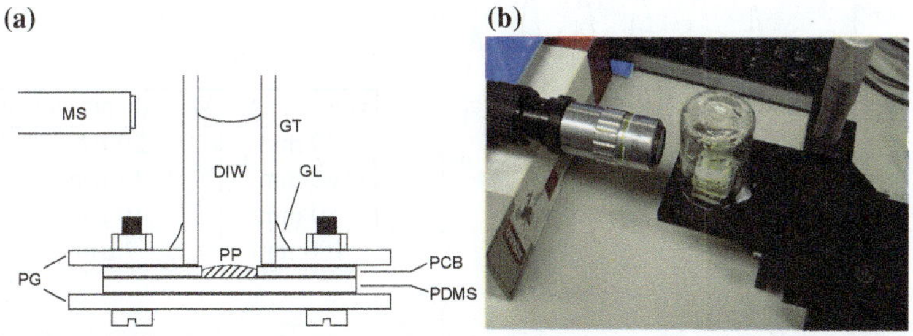

Fig. 2 (a) A schematic illustration of plasmodial force measurement setup. MS: Microscope, GT: Glass tube, DIW: Distilled water, GL: Glue, PP: *Physarum* plasmodium, PG: Plexiglass, PCB: Printed circuit board, PDMS: Polydimethylsiloxane silicon rubber. (b) A photo of the setup.

Figure 3a shows a result of surface displacement experiment in the case of 100 mm water height. We have tested with 40, 60, 80, and 100 mm water height (71, 106, 141, and 177 mg in weight, respectively). In all the cases the height of water surface periodically oscillates approximately every 100 s due to the thickness oscillation of the plasmodium at the bottom. In many cases, water height slightly decreased (approximately 10–40 μm) during a 20 minute experiment. This is possibly because the *Physarum* cell slightly escaped to the side of the well not connected to the water tube. The maximum displacement of water surface was estimated by detecting peaks and bottoms of the oscillation (marked as "$*$" in Figure 3a) and calculated differences from a bottom to the next peak. Figure 3b summarizes the maximum displacement of various water heights. As the water height become higher, the displacement decreased monotonically. By extrapolating this result, the maximum height the dumb-bell shaped *Physarum* plasmodium could bear with would be 140 mm, and it could lift up to approximately 250mg load.

The work by the plasmodium in the case of 100 mm water height can be calculated as $W = (0.75)^2 \times \pi \times 100 \times \rho \times g \times 0.007 \approx 12.1$ (mJ), where ρ is the density of water (1mg/mm^3) and g is the gravity acceleration (9.8 m/s^2). Assuming that the speed of thickness change is constant, the power of the cell can be estimated as $P = W/T = 12.1/50 \approx 0.24$ (mW).

At the micron scale, however, one has to bear in mind that some factors, which do not need to be considered at a macro scale, become crucial. For example, friction becomes relatively large compared to the force generated by the cell. Thus, for the effective use of the force, a careful attention to things like a choice of lubricant and the design of stop valve have to be paid. Nevertheless, it is noteworthy that a tiny cell, which weigh approximately 5 mg, can lift up a load over 36 times heavier than its own weight.

(a)

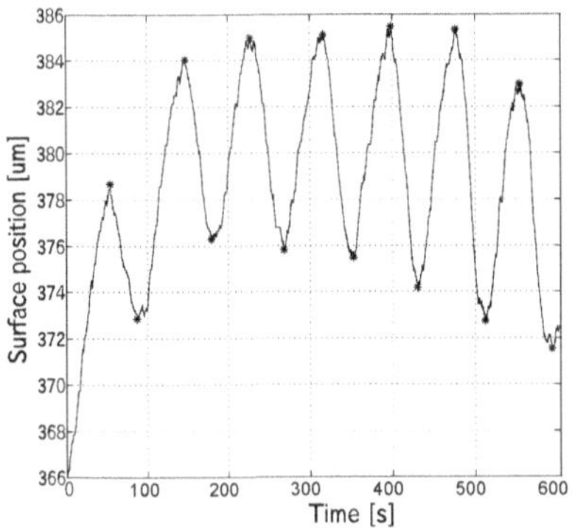

(b)

Water height	Max. displacement
40 mm	20 μm
60 mm	14 μm
80 mm	10 μm
100 mm	7 μm

Fig. 3 (a) A time course of the surface displacement in the case of 100 mm water height. First 10 min is shown. (b) A table of the maximum water displacement with different water height. It decreases monotonically as the water height increases.

2.3 Steering Control of Physarum Engine

There are several ways to externally stimulate the *Physarum*'s oscillation, such as light [41], temperature [42], and electric stimuli [43]. Among others, light stimulus is the most commonly used one because it can be easily introduced or removed simply by switching light on and off. It is known that, when a local part of a plasmodium is exposed to white light, the frequency of thickness oscillation at the local part decreases [44] . Hence we adopt white light as external controller for steering of *Physarum* engine.

To confirm if white light affects the oscillation rhythm of the dumb-bell shaped cell, we conducted a following experiment: The dumb-bell shaped *Physarum* plasmodium is constructed on on a 1.5 % agar gel using a transparency sheet mask (Fig. 4a). The dumb-bell shaped hole in the mask was cut out with a 1/16" hand punch, which makes approximately 1.5 mm diameter hole, and the connecting channel in between was cut with a scalpel under a stereo-microscope. By controlling the humidity, the plasmodium stays inside the mask (where agar gel is exposed) as it prefers wet regions to dry ones. The cell is placed under a microscope (Leica Zoom 2000, Germany) and illuminated from the bottom with bandpass-filtered light near the 600 nm wavelength (Fig. 4b). This wavelength of light does not affect the *Physarum*'s oscillation activity [45]. The thickness of the cell is measured by the light transmission through the cell, which is inversely proportional to the thickness. The controller, an ultra bright white LED coated with a heath shrink tubing, illuminates only one of two wells of the plasmodium (a dotted circle in Fig. 4a). A snapshot image of the cell is saved every 3 s in a PC for subsequent image analysis.

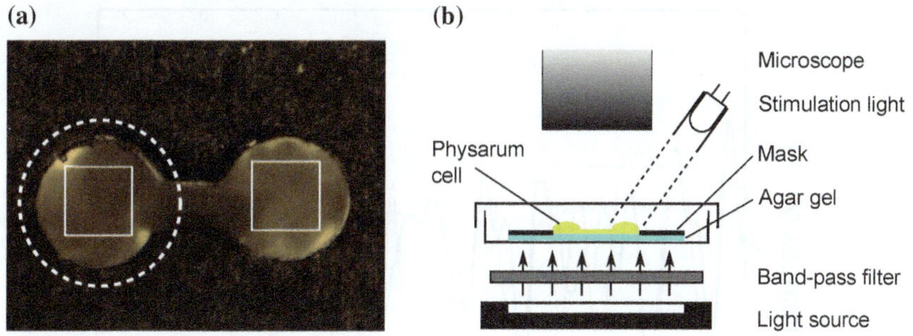

Fig. 4 (a) A dumb-bell shaped plasmodium on a 1.5% agar gel. The black mask surrounding the cell is made of transparency sheet. A dotted circle indicates the area exposed to white light. The areas indicated by two solid squares are used to calculate thickness oscillations of the cell. (b) A schematic illustration of light illumination experiment. The cell is exposed to bandpass-filtered light from the bottom for monitoring and one of two wells is exposed to white LED light from the top for stimulation. The cell is kept in a Petri dish to maintain constant humidity condition.

The thickness of the cell is calculated as follows: (1) Each recorded snapshot is converted from RGB to gray scale image. (2) Calculate a difference of images taken at t and $t-\Delta t_1$, which gives thickness change of the *Physarum* plasmodium in Δt. We used $\Delta t_1 = 7$ in this analysis. (3) Apply a moving average filter spatially over 91×91 pixels (indicated by a solid square in Fig. 4a) and temporally over 15 images on each well. This works as an image smoothing filter to reduce camera noise.

Figure 5 shows a typical reaction of the cell to white light. Before the point the light becomes on (the first half of Fig. 5), oscillations of two wells are sychronized in terms of amplitude and phase most of the time. However, as soon as one of the wells is exposed to the light (blue curve), they becomes out of phase and the oscillation. Light works as a negative stimulus to the cell and slows down the contractile oscillation at a local part of a plasmodium. In fact, periods of oscillation cycle becomes longer in the latter half of the plot, whereas those of unexposed well (red curve) remain same as before. Another reaction of the cell to the light is significant change in oscillation amplitudes. That in the stimulated well decreased largely soon after the light is turned on (indicated by horizontal dotted lines), but on the other hand, that in the non-stimulated well increased. This is because the cell is trying to escape from the exposed area by actively pumping the protoplasm from the stimulated to non-stimulated wells. When white light is removed, the oscillation at the stimulated well came back to the original period and oscillation amplitudes in both wells became almost equal (data not shown). Thus, white light does slow down the dumb-bell shaped *Physarum* plasmodium, and with this light stimulus, the cell can be steered from the outside. We are going to show a stimulated vehicle driven by the *Physarum* cell can be actually steered by light in the next section.

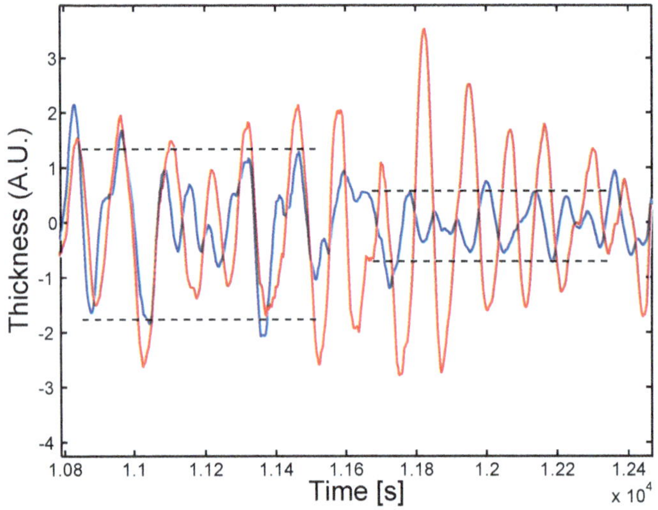

Fig. 5 A typical reaction of the *Physarum* plasmodium to light stimulus. Only one of two wells is exposed to white light (blue curve). A LED light is turned on at the point indicated by a downward arrow and kept on till the end of the plot. The oscillation period of the stimulated well (blue curve) become slightly longer when the light is on, whereas that of the unstimulated well (red curve) remained same. The oscillation amplitude significantly decreases in the stimulated well and increases in the other well (horizontal lines).

2.4 Vehicle Simulation Driven by Experimental Data

We assume that the power generated by a *Physarum* plasmodium is converted to drive wheels of a hypothetical microscale vehicle without any energy loss. Increases and decreases in thickness oscillations from two wells (cf. Fig. 4a) drive wheels of a Braitenberg-like vehicles [46] illustrated in Fig. 6a. The vehicle has two wheels driven by oscillations of a dumbbell-shaped *Physarum* plasmodium. It also has two lights which illuminate two wells of the cell integrated in the vehicle in order to steer the vehicle (described below). There are several outputs of the thickness oscillation that can be used to drive the wheels. For example, thickness amplitude, oscillation frequency, and phase difference between oscillations of two wells. Here we employ thickness amplitude to demonstrate vehicle control, i.e. the thickness oscillation pushes a hypothetical piston and rotates a crankshaft to change the speed of one of wheels. The wheels rotate at a constant speed and the thickness oscillations of the cell increase the speed depending on the amplitude. The motion of a *Physarum*-driven vehicle can be described as (Fig. 6b):

$$\begin{cases} x(t+1) = x(t) + |r_1|\sin(\theta+\alpha) + |r_2|\sin(\theta-\alpha) \\ y(t+1) = y(t) + |r_1|\cos(\theta+\alpha) + |r_2|\cos(\theta-\alpha) \end{cases} \quad (1)$$

where $x(t)$ and $y(t)$ are a position of the vehicle at time t, and r_1, r_2 are thickness displacement of the cell in left and right wells in a certain period (in this case, 21 s). α is a parameter that determines the contribution of the thickness displacement to the change of wheel running speed. The larger α is, the more wheels are accelerated. Hence, with larger α, vehicle's direction of movement becomes more sensitive to the difference of oscillation amplitudes from two wells (i.e. if $||r_1| - |r_2||$ is large, the vehicle turns either to the left or right).

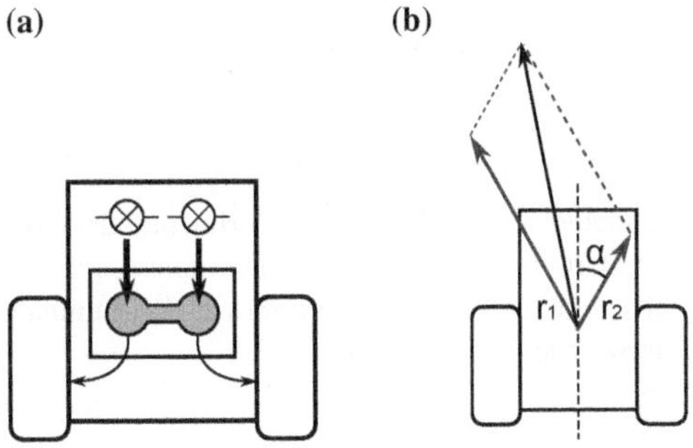

Fig. 6 (a) A schematic diagram of a vehicle driven by *Physarum* engine. The oscillation amplitudes of the cell determines the wheel rotation speed of the vehicle. Light illumination onto each well of a *Phyasrum* plasmodium is use to steer the vehicle. (b) The motion of *Physarum*-driven simulated vehicle. Vehicle's direction of movement can be represented as addition of vectors (Eq. 1).

Figure 7 shows a trajectory of the simulated vehicle driven by experimental data. The vehicle started from the origin (0,0) goes straight (black line) when no light inputs are on. It turns to the left (red line) as soon as the left well of the dumbbell-shaped cell is stimulated by light. When light is turned off, it drives straight ahead again. This is can be explained as follows in terms of *Physarum*'s behavior: When there is no stimulus to the cell, both wells oscillates with same amplitude (cf. Fig. 5). This drives the wheels with the same speed, and thus the vehicle goes straight. On the other hand, as described in the previous section, the oscillation amplitude in the stimulated well significantly decreases while the light is on, whereas that in the other well increases. This is because the cell moves out of the illuminated area and migrates towards the non-illuminated well. This slows down the left wheel driven by the stimulated well and accelerates the right wheel. As a result, the vehicle takes a left turn as long as the light is on. It drives in a straight line again after the light is turned off because oscillation amplitudes in two wells come back to nearly equal level when light stimulation is removed.

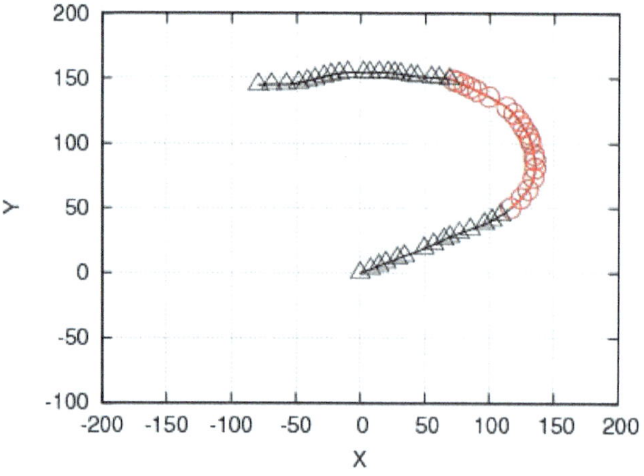

Fig. 7 Trajectory of the simulated vehicle driven by experimental data. $\alpha = \pi/720$ is used in this simulation. The vehicle started from (0,0). Black line indicates the period when there is no light illumination to wells, and red line is when a right well is illuminated by light. Periods Triangles and circles are drawn every 150 s.

3 The Emergence of Oscillatory Transport Phenomena in a Particle-Based Model

Due to the relatively slow time development of the *Physarum* plasmodium and natural variability of its 'performance' in terms of the chosen robotics tasks (the plasmodium is only concerned with survival and not, after all, with satisfying externally applied experimental tasks), it is helpful to develop computational modelling approaches which may facilitate the study of distributed robotic control. Such a modelling approach allows us to explore the emergence of complex oscillatory behaviour from simple components and assess its application to robotics tasks.

To investigate the use of emergent oscillatory phenomena for engine-like transport we employ an extension to the particle model in [47] which was shown to generate dynamical emergent transport networks. In this approach a plasmodium is composed of a population of mobile particles with very simple behaviours, residing within a 2D diffusive environment. A discrete 2D lattice (where the features of the environment arena are mapped to greyscale values in a 2D image) stores particle positions and also the concentration of a local factor which we refer to generically as chemoattractant. The 'chemoattractant' factor actually represents the hypothetical flux of sol within the plasmodium. Free particle movement represents the sol phase of the plasmodium. Particle positions represent the fixed gel structure (i.e. global pattern) of the plasmodium. Particles act independently and iteration of the particle population is performed randomly to avoid introducing any artifacts from sequential ordering. Particle behaviour is divided into two distinct stages, the sensory stage and the motor stage. In the sensory stage, the particles sample their local environment using three forward biased sensors whose angle from the forwards position (the sensor angle parameter, SA), and distance (sensor offset, SO) may be parametrically adjusted (Fig. 8a). The offset sensors represent the overlapping and inter-twining filaments within the transport networks and plasmodium, generating local coupling of sensory inputs and movement (Fig. 8c and d).

The SO distance is measured in pixels and a minimum distance of 3 pixels is required for strong local coupling to occur. During the sensory stage each particle changes its orientation to rotate (via the parameter rotation angle, RA) towards the strongest local source of chemoattractant (Fig. 8b). After the sensory stage, each particle executes the motor stage and attempts to move forwards in its current orientation (an angle from 0-360°) by a single pixel forwards. Each lattice site may only store a single particle and-critically-particles deposit chemoattractant into the lattice only in the event of a successful forwards movement (Fig. 9a). If the next chosen site is already occupied by another particle the default (non- oscillatory) behaviour is to abandon the move, remain in the current position, and select a new random direction (Fig. 9b). Diffusion of the collective chemoattractant signal is achieved via a simple 3x3 mean filter kernel with a damping parameter (set to 0.07) to limit the diffusion distance of the chemoattractant.

The low level particle interactions result in complex pattern formation. The population spontaneously forms dynamic transport networks showing complex evolution and quasi-physical emergent properties, including closure of network lacunae, apparent surface tension effects and network minimisation. An exploration of the possible patterning parameterisation was presented in [48]. Although the particle model is able to reproduce many of the network based behaviours seen in the *Physarum* plasmodium such as spontaneous network formation, shuttle streaming and network

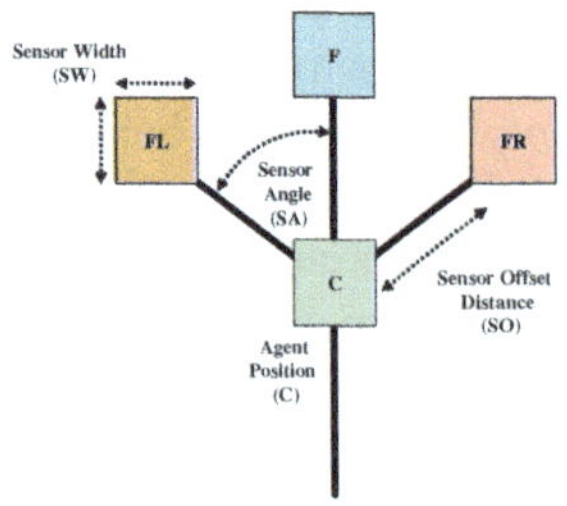

(a)

```
[Sensory stage]

- Sample chemoattractant map values
- if (F > FL) && (F > FR)
      - Continue facing same direction
- Else if (F < FL) && (F < FR)
      Rotate by RA towards larger of FL and FR
- Else if (FL < FR)
      Rotate right by RA
- Else if (FR < FL)
      Rotate left by RA
- Else
      Continue facing same direction
```

(b)

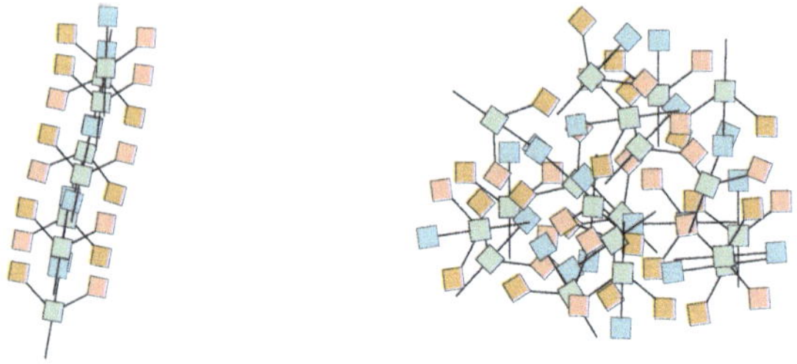

(c) (d)

Fig. 8 Agent particle morphology, algorithm, and aggregation types.

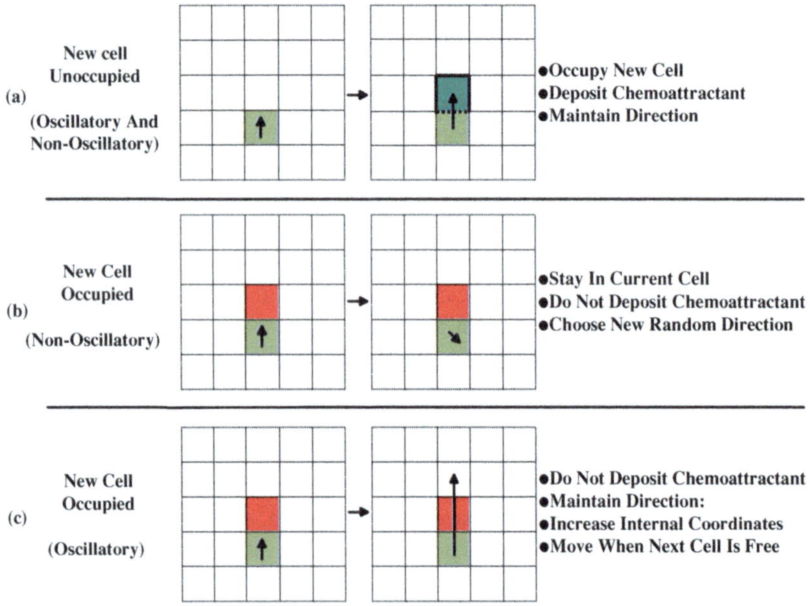

Fig. 9 Particle motor behaviour in non-oscillatory and oscillatory conditions

minimisation, the default behaviour does not exhibit oscillatory phenomena and inertial surging movement, as seen in the organism. This is because the default action when a particle is blocked (i.e. when the chosen site is already occupied) is to randomly select a new orientation-resulting in very fluid network evolution, resembling the relaxation evolution of soap films, and the lipid nanotube networks seen in [13].

The oscillatory phenomena seen in the plasmodium are thought to be linked to the spontaneous assembly / disassembly of actin- myosin and cytoskeletal filament structures within the plasmodium which generate contractile forces on the protoplasm within the plasmodium. The resulting shifts between gel and sol phases prevent (gel phase) and promote (sol phase) cytoplasmic streaming within the plasmodium. To mimic this behaviour in the particle model requires only a simple change to the motor stage. Instead of randomly selecting a new direction if a move forward is blocked, the particle increments separate internal co- ordinates until the nearest cell directly in front of the particle is free. When a cell becomes free, the particle occupies this new cell and deposits chemoattractant into the lattice (Fig. 9c). The effect of this behaviour is to remove the fluidity of the default movement of the population. The result is a surging, inertial pattern of movement, dependent on population density (the population density specifies the initial amount of free movement within the population). The strength of the inertial effect can be damped by a parameter (pID) which sets the probability of a particle resetting its internal position coordinates, lower values providing stronger inertial movement.

When this simple change in motor behaviour is initiated surging movements are seen and oscillatory domains of chemoattractant flux spontaneously appear within the virtual plasmodium showing characteristic behaviours: temporary blockages of particles (gel phase) collapse into sudden localised movement (solation) and vice versa. The oscillatory domains them- selves undergo complex evolution including competition, phase changes and entrainment. We utilise these dynamics below to investigate the possibility of generating useful patterns of regular oscillations which may be coupled to provide motive force.

3.1 Model Setup

The emergence of oscillatory behaviour in the model corresponds to differences in distribution of protoplasm within the plasmodium and subsequent changes in thickness of the plasmodium. In a real *Physarum* plasmodium the changes in thickness of the plasmodium membrane are used to provide impetus (pumping of material through the vein network, or bulk movement of the plasmodium). There is known to be a relationship between the spontaneous contraction of the plasmodium and the subsequent transport of protoplasm away from that region. Thus the region undergoing contraction becomes thinner (allowing more light to pass through when illuminated) and regions away from the contraction become thicker as more protoplasm in present (allowing less light to pass through). In the computational model the transport of particles represents the free flux of protoplasm within the material and the increase in flux (mass particle movement) is indicated in the supplementary video recordings by an increase in greyscale brightness (http://uncomp.uwe.ac.uk/jeff/robots.htm). A decrease in the bulk movement of particles represents congestion and a lack of transport and is indicated by a decrease in greyscale brightness (since deposition of chemoattractant factor only occurs in the event of successful forward movement). For clarity in the static images, the greyscale images are inverted (dark areas indicate greater flux).

The particle population's environment is a 2D lattice, represented by a digitised image configured to represent the habitat of the experimental plasmodium. We designed simple shapes in which the particle collective, which composes the virtual material, is confined. The particle population is free to move within unconfined areas. The shapes are composed of 'wall' regions where movement cannot occur, 'vacant' regions where movement was possible and (where relevant) 'stimulus' regions which provide attraction stimuli, or repulsion stimuli, to the particle population. At the start of each experiment the particle population is randomly distributed through the vacant space in the experimental arena and all particles have random initial orientations. A fixed population size was used (we do not discuss an adaptive population size in this report). The total amount of free possible particle movement is dependent on the population size as a fraction of vacant space. In these results we use a 90% occupancy rate unless otherwise specified, i.e. 90% of all vacant areas are occupied by particles. In all of the experiments there is an initial period where oscillatory behaviour is not initially activated and this results in self-organised regular domains. When oscillatory motor behaviour is induced these regular domains collapse and the emergence of small domains of regular oscillatory patterns begins. Over time these domains coalesce and compete, causing entrainment of the population into regular oscillation patterns, influenced by both the particle sensory parameters and also by the shape of the experimental arenas.

We use the arena shape to constrain the oscillatory behaviour to ascertain the possibility of utilising the oscillatory behaviours to provide useful 'engine-like' output. By this we mean that the oscillation patterns should be periodic, regular and reliable, in the same way that mechanical engines provide regularly and reliably timed patterns of impetus.

3.2 Data Analysis

We sampled regular frames from the emergent chemoattractant flux patterns and analysed the differences in particle flux by comparing the greyscale levels in different regions of the arena. We were particularly concerned with the reliable initiation of oscillatory behaviour and the characteristics of the behaviour (e.g. the period, the intensity and any coupling effects).

4 Results

We have previously shown that the particle population successfully reproduces the behaviour of the plasmodium when confined within a circular well, resulting in the spontaneous emergence of oscillation patterns (summarised in (Fig. 10). The type of pattern produced depends on the sensory parameters of each particle (SA and RA parameters) and the SO parameter which, when increased, causes a transition to a different pattern (see [49] for more information). In the circular well the patterns most frequently observed were rotational patterns. When the SO interaction distance was further increased lateral oscillations were observed, followed by an annular pattern. There appears to be a relationship between the circular confining shape of the arena and the type of pattern produced.

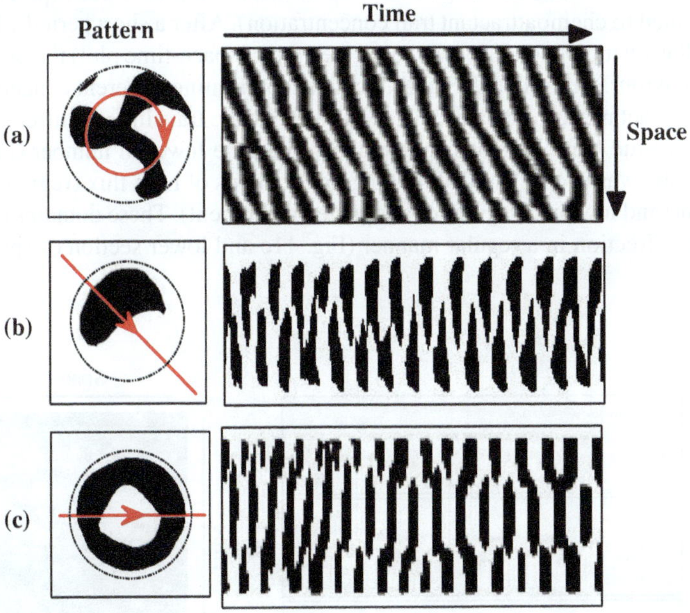

Fig. 10 Oscillation patterns observed in a particle model of *Physarum polycephalum* constrained within a circular well. Left: Snapshot of pattern with trajectory of space time plot overlaid, Right: Space time plot of oscillation pattern. a) Rotational pattern, b) Bilateral pattern, c) Annular pattern

We speculate that the reason for the change in pattern type (which, over time, is also observed in the real plasmodium) is that the scale of interactions between the oscillatory regions becomes too large for the circular well, the interactions are constrained and a new pattern is formed which can 'fit' within the confines of the well. This suggests that the intrinsic oscillations which spontaneously emerge can be shaped in some way by architectural changes to the environment. To further explore this relationship between oscillation pattern type and the environment shape we explored different methods of patterning the virtual arena in order to assess differences in oscillation pattern type and pattern evolution.

4.1 Transport Motion in Open Ended Patterns

We patterned the environment with a simple tube shape whose ends were looped around at the edge of the environment by invoking periodic boundary conditions for the particles. The wall boundaries confined the population to the tube. Snapshots and a space time plot are shown in Fig. 11. The space time plot is recorded by sampling every pixel along the width of the image at exactly halfway down the tube section and assembling an image based upon these sampled values (pixel brightness is related to chemoattractant trail concentration). After a short period where the non-oscillatory motor condition was used (see top of space-time plot) the oscillatory motor behaviour was initiated. Small domains representing different concentrations of chemoattractant flux appeared (Fig. 11a) which were travelling in different directions. Over time these domains competed and became fewer in number (Fig. 11b) until the tube was evenly divided into regular domains of high flux (regions of free movement) and low flux (regions of obstructed movement). These domains travelled in a single direction in a regular manner (Fig. 11c and lower section of space-time plot).

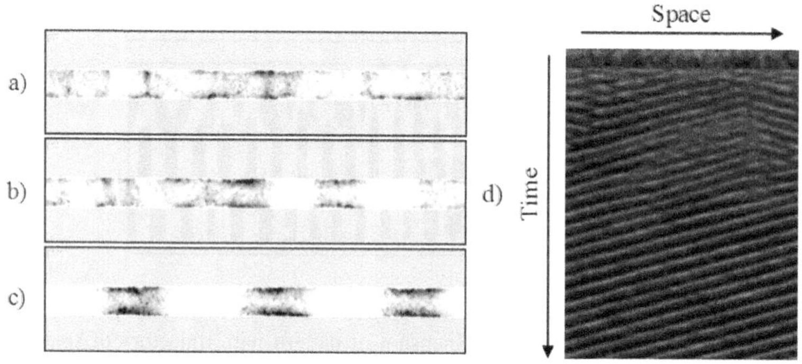

Fig. 11 Emergence of travelling wave transport in a tube shaped environment Left: Snapshots taken at a) 257, b) 868 and c) 3990 scheduler steps Right: Space time plot showing emergence of regular transport

The travelling waves arising from the collective particle behaviour actually travel in the opposite direction to the bulk particle motion. In the example shown in Fig. 12 the direction of the travelling wave (Fig. 12b, solid arrow) is right-to-left, whereas the actual movement of the particles (Fig. 12a, dashed arrow) tends to move from left-to-right. Since particles can only move when a space in the lattice becomes available the vacant spaces appear to move in the opposite direction of the particles and the particles themselves show a greater probability of moving when a region of relatively vacant space is near (Fig. 12c, also note that particle movement tends to occur in the lighter regions, which are regions with more vacant space and greater chemoattractant flux). The spaces themselves appear to move backwards because a particle moving from its current position to occupy a vacant space subsequently leaves a new vacant space at its previous position. Because chemoattractant is only deposited by the particles after a successful movement, the chemoattractant flux will be patterned by the distribution of vacant spaces and the wavefront thus moves in the opposite direction to the particles. The bulk movement of the particles is also much slower than the travelling waves, as indicated by Fig. 12, which shows that a single particle takes approximately 8000 scheduler steps to traverse the width of the arena, whereas the travelling wave crosses the arena in approximately 400 steps, a 20:1 difference (although the particle's progress is hindered somewhat by the resistance caused by the low numbers of vacant spaces and changes in direction on contact with the border regions). The opposite direction of the self-organised travelling wave with respect to particle movement is reminiscent of the characteristic backwards propagation seen, for example, in traffic jams [50].

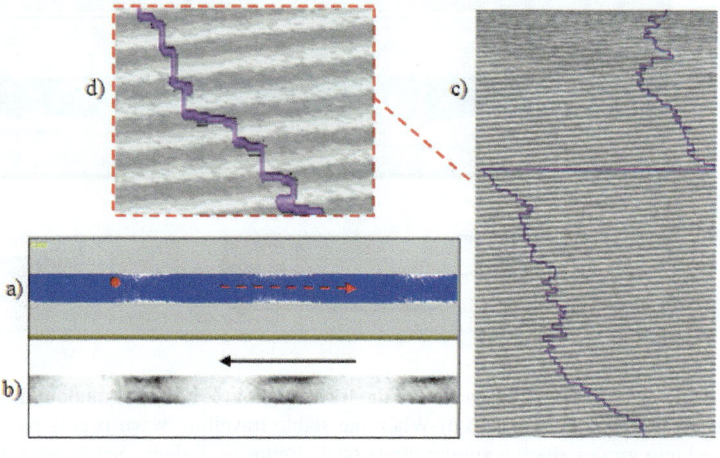

Fig. 12 Collective particle drift is opposite to the direction of wave propagation. a) Particle positions and tracer particle (circled), b) Chemoattractant wave propagation, c) Space-time plot overlaid with tracer particle position, d) Enlarged portion showing tracer particle movement (horizontal movement) tends to occur in vacant (high chemoattractant flux) areas.

Further analysis of the directional alignment of the particles revealed some unexpected properties of the particle population in relation to the travelling wave. At the start of an experiment the distribution and alignment of the particle population is randomly chosen (Fig. 13a) and the ratio of particles facing left and those facing right is typically equal at 50:50 (we consider particle alignment as being 'left facing' or 'right facing', depending on whether their actual orientation angle falls into either category). When the travelling wave has emerged and stabilised the distribution of the particles is such that they are grouped in regions (Fig. 13b) which are similar - but slightly offset - to the regions of high and low chemoattractant concentration seen in the travelling wave (Fig. 13c). The particles in each region share the same general alignment and the different regions alternate with respect to the alignment of the particles within them (i.e. regions are LEFT, RIGHT, LEFT, RIGHT and so on). The final alignment ratio at the end of an experiment was typically 53:47 (rounded average of ratios over ten runs), with the majority of particles facing in the direction of the travelling wave. The particles change their directional alignment at the same speed as the oncoming travelling wave. The actual movement of the particles, however, is much slower and in the opposite direction to the wave.

a)
b)
c)
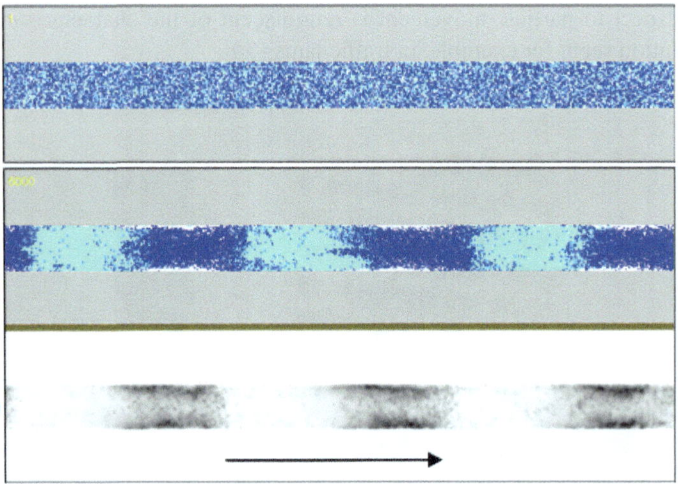

Fig. 13 Collective particle alignment and the travelling wave. a) Initial random distribution of particles and their alignments, b) When the stable travelling wave occurs particles are distributed into groups sharing similar directional alignment: lighter regions are oriented to face right and darker regions are oriented leftwards, c) Chemoattractant flux in travelling wave moving to the right.

By patterning the environment to remove all wall boundaries and ensure periodic boundary conditions the emergent oscillatory patterns self-organised into travelling waves (Fig. 14). However there was a much greater length of time needed for the competition between the wave patterns to complete and form synchronous travelling waves. The increase in time before synchronous waves emerge can be explained by the greater initial freedom of movement afforded by the lack of movement constraints from the environment.

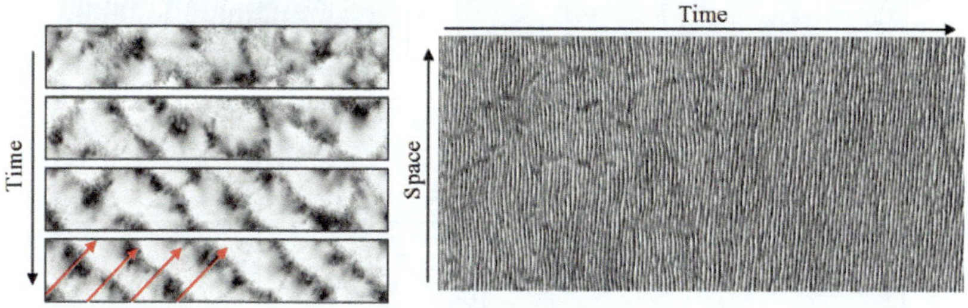

Fig. 14 Helical transport in non-patterned environment with periodic boundary conditions. Left: Snapshots of competition and entrainment of wavefronts with final helical-type movement arrowed, Right: Space-time plot showing long period before regular transport occurs.

By removing the movement at the boundaries and instead patterning the vacant space into looped structures rotary motion of the travelling waves was achieved (Fig. 15). The competition period before synchronisation was relatively brief (Fig. 15a, space-time plot), again because the environmental barriers reduce initial freedom of movement. The looped structures also enabled travelling wave motion even when the environment added non-circular elements, and more tortuous paths (Fig. 15 b and c). The successful initiation of rotary motion suggests the possibility of generating reliable conveyor type transport.

By combining two circular patterns and introducing a region which overlapped and exposed to two separate wavefronts it was possible to achieve entrainment, after 5000 scheduler steps, of the pulses from one rotary 'motor' to another, mimicking the transmission and synchronisation of movement to another ring by a fluidic coupling effect (Fig. 16). Like a conventional gear transmission, the rotation of two facing rings was in opposite directions, however, unlike conventional gear trains, the 'teeth' (wavefront peaks) did not overlap.

4.2 Transport in Closed Path Patterns

The transport motion in open ended looped patterns stabilises because the bulk particle drift eventually synchronises with the travelling waves. The distribution of the particle population becomes relatively evenly distributed within the path, punctuated by regularly spaced changes of particle occupation density. When closed path patterns were used, however, the uniform distribution cannot occur because separate ends of the path cannot communicate the transport of particles. Thus, over time, the drift of particles results in a tendency for the particle population density to become greater at one end of the chamber. Once there is an imbalance at one end the number of vacant spaces at that end falls and the particles are then attracted to the

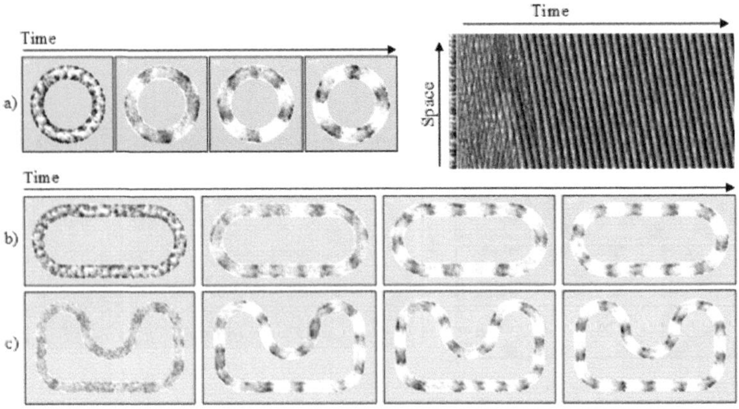

Fig. 15 Emergence of rotary motion oscillations. a) Emergence of rotary motion and visualisation of space-time plot showing regular motor pulses. Space time plot was created by recording 360 points inside the vacant track of the circle, b) Tracked rotary movement from a combination of circular and straight regions, c) Conveyor type motion from a more tortuous looped structure.

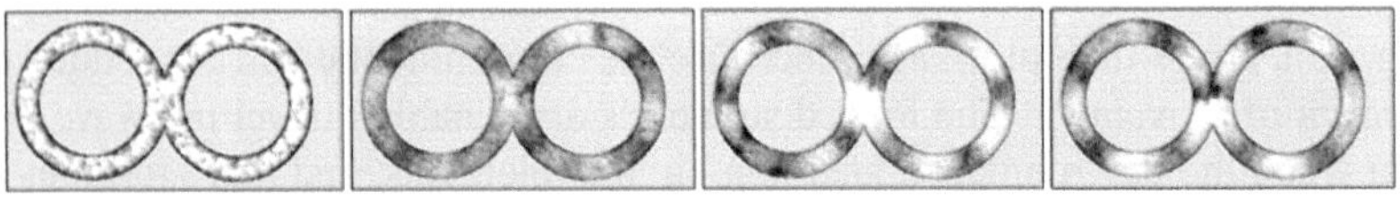

Fig. 16 Emergence of gear-like coupling in an overlapping two rotor pattern. Snapshots taken at 20, 121, 981 and 6997 scheduler steps.

opposite side of the chamber where more vacant areas exist. Short term oscillatory transport and competition within the chamber still occurs (Fig. 17a and b), as with looped patterns, but this is mediated by a second order of oscillations, which occurs over a much longer timescale since it is caused by the slower bulk drift of particles (Fig. 17c). The effect is regular changes in direction of transport direction and this suggests a possible mechanism of how changes in direction in open-ended looped systems could be achieved - by temporarily introducing a blockage in a looped path until a change in direction occurred.

Another method to confine the collective to a region is by using strong sources of attractant to effectively 'pin down' the collective in place instead of confining it physically within a region by its boundaries. Attractant sources are represented by projecting values at every scheduler step into the chemoattractant diffusion field. These sources diffuse and attract the individual particles of the collective. By spacing the placement of attractants the collective is retained in place in a sheet-like fashion by the attraction of the particles to the sources and the mutual attraction of particles to their own deposition of chemoattractant into the diffusion field (Fig. 18a).

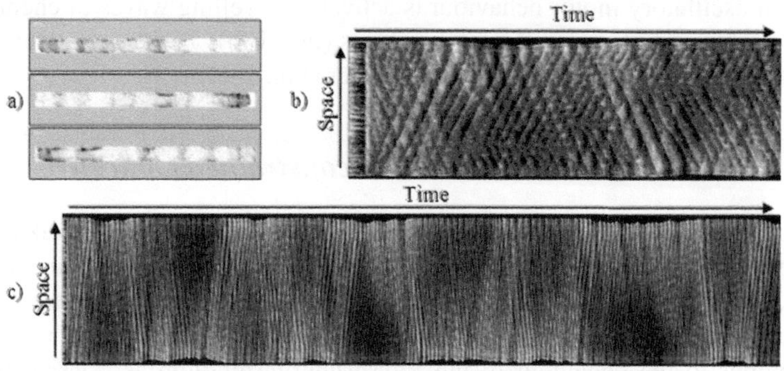

Fig. 17 Complex second-order oscillations caused by bulk drift in closed path environments. a) Pulsatile oscillations at different sides of a closed chamber, b) Initial phase of space-time plot showing initiation of oscillatory behaviour and competition between oscillatory domains, c) long term (20,000 steps) space-time plot showing second order oscillations as bulk particle positions oscillate from one side of the chamber to the other. Dark regions at either side of plot indicate periods of second-order oscillations.

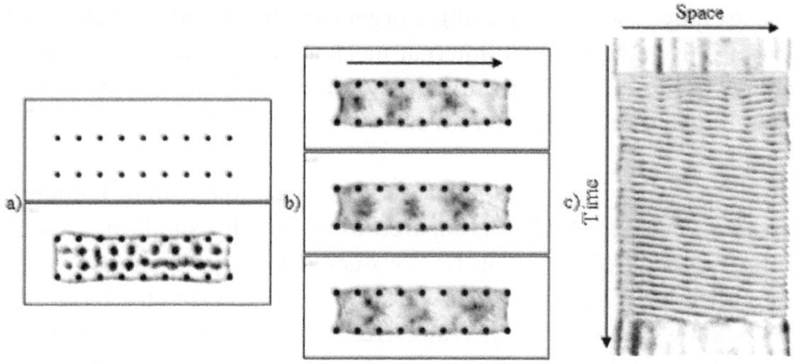

Fig. 18 Confining the collective by attractant projection and emergence of travelling wave. a) Regularly spaced projection of chemoattractant sources (top) confines position of collective (bottom), b) When oscillatory motor behaviour is activated a travelling wave emerges across the confined collective, c) Space-time plot of the emergence of travelling waves in a pinned collective.

When oscillatory motor behaviour is activated travelling waves of chemoattractant movement emerge in different directions (Fig. 18b). These waves compete for a short time before one direction predominates (Fig. 18c).

4.3 Amoeboid Movement in an Unconstrained Collective

The previous examples of collective transport consider the particle collective as a sheet of virtual material upon which oscillatory phenomena are spontaneously generated. The material is then patterned to control the movement of the travelling oscillation waves for transport purposes (pumps, motors, gears, conveyors etc.). This corresponds to the behaviour of the *Physarum* plasmodium as a large sheet-like structure which, via adhesion to its anchoring substrate, transports material within the plasmodium using hydrostatic pressure generated by the spontaneous oscillation rhythms of its actin-myosin network. In addition to these general transportation phenomena, however, *Physarum* also has the ability to migrate towards a nutrient source [51], away from a hazardous source [44], and adapt its gross body plan to a complex environment [52]. Small *Physarum* cultures can even migrate the entire plasmodium away from unfavourable conditions in certain circumstances such as bacterial contamination. Furthermore, the plasmodium is also famous for its ability to survive physical damage - pieces of plasmodium excised can survive independently and individual pieces may fuse to form a single organism. To extend the vehicle analogy, *Physarum* not only represents the internal mechanicals (motive force mechanism, transmission coupling), but also the moving vehicle itself - and a vehicle which can survive the removal of parts, the introduction of new foreign parts and the repair of damaged parts.

We set out to explore the behaviour of an unconstrained small particle collective to assess its behaviour when compared to an equally small and independent plasmodium fragment. Rather than be considered as a rigid sheet-like material, the smaller collective may be considered as a generic amorphous 'blob' of virtual material. It is known that a smaller collective, without using the oscillatory motor behaviour condition, will condense into a uniform circular shape as the initial transport network condenses (Fig. 19a). The non-oscillatory blob shows regular vacancy domains (dark areas) and the fluid particle motion afforded by the non-oscillatory motor condition ensures that the blob is cohesive and takes a minimal shape. The non-oscillatory blob is also resilient to external perturbation. When excited by an externally applied source of chemoattractant (Fig. 19a, mouse position in fourth image), the deformation of the collective induced by the stimulus as it is attracted to the stimuli is repaired when the stimulus is removed, the collective returning to its minimal shape.

When oscillatory motor behaviour is initialised the regular domains collapse as the particle motion becomes less fluid (Fig. 19b) and oscillations travel through the collective. Because the small collective is not constrained by any externally applied pattern the oscillations distort the shape of the collective. When the pID parameter is further reduced to 0.01 there is even greater restriction on the fluidity of individual particle movement and the oscillations become stronger and distort the collective's boundary significantly (Fig. 19c). The large shift of a mass of particles causes the collective to move across its environment. The cohesion of the collective is maintained but other SA/RA parameter settings, combined with lower sensor interaction (SO) distance, can result in the fragmentation of the collective (see supplementary material for examples of oscillation patterns using different SA/RA combinations). There is significant interplay between the pattern of oscillatory activity within the collective and the global shape of the collective, as is the case with the real plasmodium [32]. The oscillation patterns are determined - in part - by the size and shape of the collective. These oscillations typically travel in waves traversing the collective. The travelling waves (which are essentially differences of chemoattractant flux) cause changes in particle flow within the collective which, in turn, causes a change in shape of the collective. The changed shape of the collective thus has a repeated effect on subsequent oscillation patterns and the collective shape.

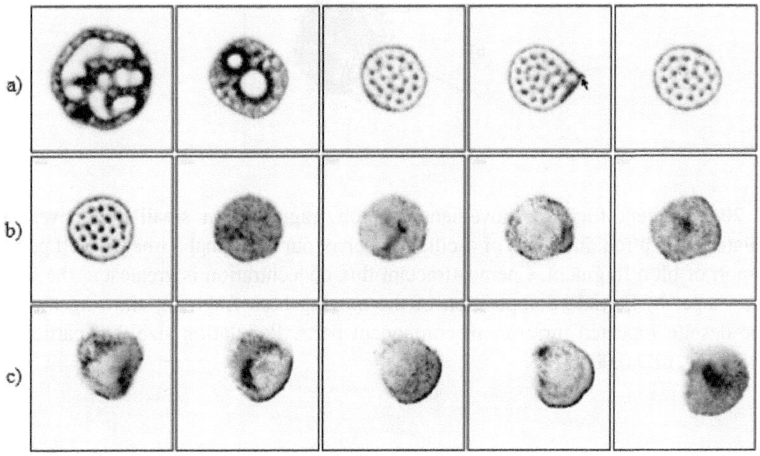

Fig. 19 Initiation of oscillatory behaviour and amoeboid movement in unconstrained collective. a) Condensation of blob material in non-oscillatory behaviour shows vacancy domains and resilience to deformation, b) Initial non-oscillatory collective with regular vacancy domains (left) and onset of oscillatory behaviour at 1950, 2039, 2057, 2086 and 2122 scheduler steps. 9380 particles, SA90, RA45, SO15, pID 0.05. All other parameters as in (a), c) Reduction of pID to 0.01 gives stronger oscillations and amoeboid movement at 4648, 4684. 4720, 4768 and 4836 scheduler steps. All other parameters as in (a).

4.4 Persistent Movement in a Small Blob Fragment

The amoeboid movement seen in Fig. 19 occurs because the oscillation waves distort the boundary of the collective whilst it is still able to maintain a cohesive whole. Because the population maintains its cohesion, any distortion of the boundary on one side must result in a shift in population distribution from the opposite side (since the collective is non-compressible and occupies a fixed area). The diameter of the collective (a function of the number of pixel sized particles comprising it) must be large enough to for oscillations to emerge and to confine an oscillation pattern within it.

a)

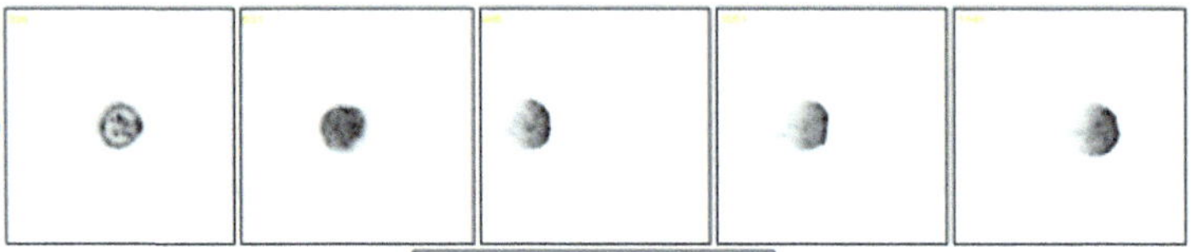

b)

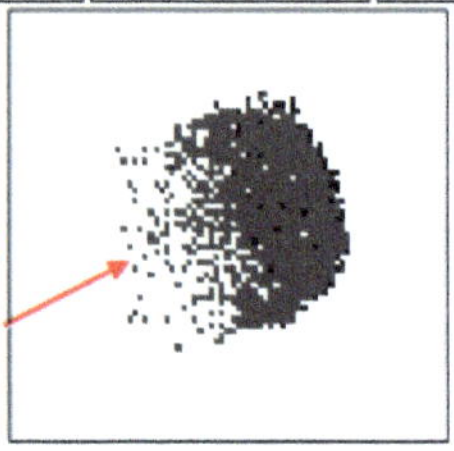

Fig. 20 Persistent forward movement of blob fragments in small collectives. a) Non-oscillatory condition, initiation of oscillatory behaviour and (final 3 images) self perpetuating transport of blob fragment. Chemoattractant flux concentration is greatest at the front of the dome shape, b) Particle composition of the moving blob fragment showing the persistent shape despite repeated turnover of component parts. Population size 900 particles, SA90, RA45, SO9, pID 0.001.

When the collective is comprised of only a relatively small number of particles, the distortion of the boundary forms an approximately semi-circular domed shape and the small number of particles ensures that the collective cannot maintain a fully circular shape. However the persistence of forward movement generated by the oscillatory motor behaviour at low pID values causes the dome shape itself to be maintained over time, and the small blob fragment is able to move forwards (Fig. 20a). Movement of the fragment is relatively smooth and different from the pulsatile motion observed in larger collectives. Particles move towards the front of the domed profile and then, over time, move to the side (Fig. 20b). Particles at the sides of the fragment ultimately fall behind only to re-enter the dome at the centre. Higher pID values result in more frequent changes of direction of the fragment as the dome shaped front profile cannot be maintained. If the population size is increased, the single sided dome shape cannot be maintained and the resultant motion becomes pulsatile and chaotic.

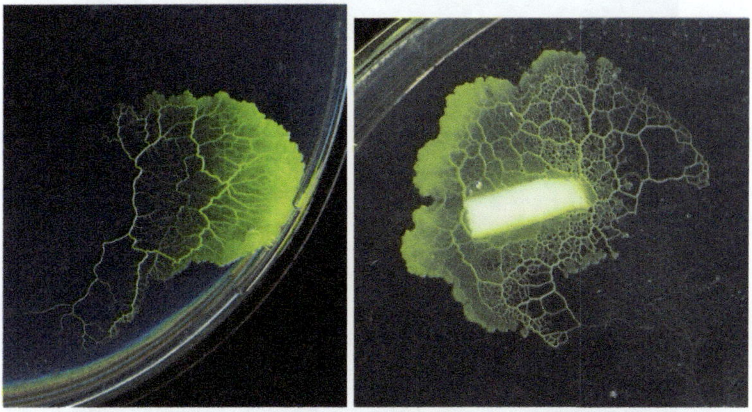

(a) (b)

Fig. 21 Examples of plasmodiual blobs: (a) self-propelled and (b) stimulated by food. See details in [53].

The model matches biological reality pretty well. Examples of blob fragments recorded in experiments with living plasmodium are provided in Fig. 21. A self-propelled, i.e. without attractive or repelling control stimuli, motion of blob is shown in Fig. 21a. A blob occupying source of food (egg white in this particular example) is shown in Fig. 21b.

4.5 External Influence of Collective Movement - Attraction and Repulsion

Movement of the *Physarum* plasmodium is strongly affected by local environmental conditions. Attractant sources (such as increasing temperature gradients and chemoattractant nutrients) cause the plasmodium to move and grow towards the attractants whereas repulsive sources (salts, dry regions) cause the plasmodium to try and avoid such regions. The plasmodium is able to integrate many separate localised inputs to compute its response to the environment. One method in which this is achieved is by the modulation of local oscillation patterns in response to attractants or hazards - attractants tend to increase localised oscillation strength and hazards decrease oscillation strength. We set out to see if a localised response to external influences could be used to govern the collective movement of the particle population.

Fig. 22 Guiding plasmodium by a series of food sources, oat flakes. The plasmodium is inoculated in north-most oat flake and guided southward. Experimental photo. See details in [53].

Attractant sources were previously as a method to confine the collective to a region by pinning it down. By externally presenting an attractant source (effectively a nutrient source) to a cohesive blob of virtual material (Fig. 23a) we found that the presentation of the source resulted in diffusion from the source (Fig. 23b). This is supported by experimental evidences, see e.g. Fig. 22: one can guide a plasmodium along almost any non-intersecting trajectory by arranging point-wise sources of attractants.

When the diffusion reached the sensors of the closest particles at the front of the collective it provoked movement towards the source. A pseudopod-like extension of the border region emerged and extended towards the source (Fig. 23c), ultimately engulfing it. Travelling waves spontaneously emerged within the collective which were directed at the source, caused the collective to shift its position towards the source (Fig. 23d). Consumption of the source was simulated by simply decrementing the value projected to the diffusion field when the source was covered by a particle. When the source was 'consumed' by the population, the collective regained its previous approximately circular shape.

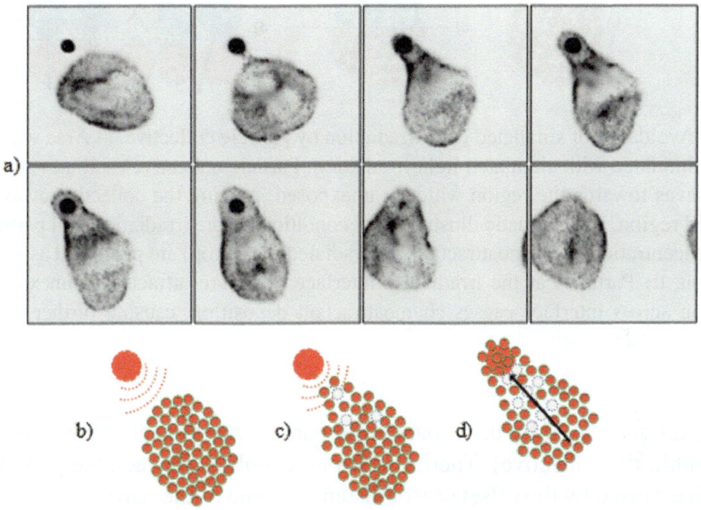

Fig. 23 External control of amoeboid movement by attraction. a) A persistent source of chemoattractant is placed into the diffusion field which provokes the extension of the collective. The collective engulfs the source by moving via travelling waves towards the source. When the source is exhausted the collective re-adopts it original approximately circular shape, b) Schematic illustration of attraction of the collective by the nutrient source, c) Migration of leading particles towards the source, d) Emergence of travelling waves pulling the collective to engulf the source.

To approximate the repulsion of the collective to hazardous sources such as the simulated response to irradiation by visible light we added a condition to the sensory stage of the algorithm to the effect that if any particles of the collective were in a region exposed to 'light' (a defined area within the arena), those particles would have their sensitivity to chemoattractant diminished whilst they remained in this region (this achieved by multiplying the sampled sensor values with a weighting factor less than 1, lower values generating a stronger response to irradiation). The effect of exposing regions of the collective to simulated light damage was that the collective immediately started to move away from the irradiated region (Fig. 24a). Specifically, oscillation waves moved from the irradiated region towards the unexposed regions. The shift of particles from the irradiated region eventually moved the collective away from the stimulus. The cause of movement away from the light can be found at the interface between irradiated and unexposed areas. Before irradiation (Fig. 24a) all regions of the collective are equally attractive to the particles (subject

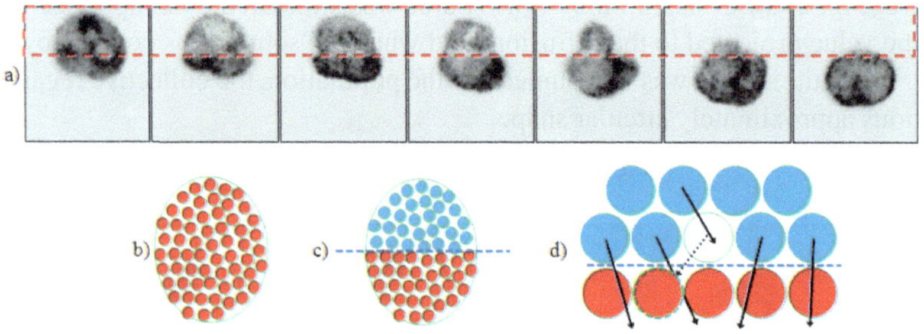

Fig. 24 Avoidance of simulated light irradiation by particle collective. a) Area within dashed box is stimulated with simulated light irradiation. Particle collective oscillates sending travelling waves towards the region which is unexposed, moving the collective away from the irradiated region, b) Schematic illustration of condition before irradiation - all particles sense equal concentration of chemoattractant, c) Irradiated areas (top) are perceived as weaker concentration, d) Particles at the irradiation interface are more attracted to unexposed areas. Migration across interface causes chemoattractant deposition, causing further attraction to region and vacant space.

to fluctuations caused by discrepancies in particle movement and intrinsic oscillations within the collective). There is a strong coupling between the particles in the collective caused by the offset sensor distance. Some of the particles at the interface of the irradiated region will receive input from the unexposed region and will be attracted to that area because the chemoattractant concentration in unexposed areas is perceived as greater due to the damping in irradiated regions (Fig. 24a). The movement of particles near the interface towards unexposed regions causes both new vacant spaces (Fig. 24a) and also an increase in chemoattractant concentration (because only mobile particles deposit chemoattractant). This results in a further increased attraction to the interface region, until eventually the entire collective has migrated from the irradiated region.

A range of repulsive fields, applicable to control of Physarum polycephalum, is discussed in [53]. Light and salt are the simplest physical meanings of implementing repulsive control. And example of splitting plasmodial blob with a grain of potassium chloride is shown in Fig. 25. Initially the plasmodium-blob travels north-east. When approaching salt crystal the blob experience repelling forces and self-divides into two blobs – one travels north-west-west, another south-east [53].

4.6 Morphological Adaptation of the Collective

The previous results demonstrated that the collective changes its shape during self-oscillatory behaviour and also in response to simulated attractants and hazards. The relatively amorphous – typically variations of circular shaped patterns – collective retains its shape due to the cohesion of the individual particles making up the population. When the morphology of the collective is disturbed by its movement towards,

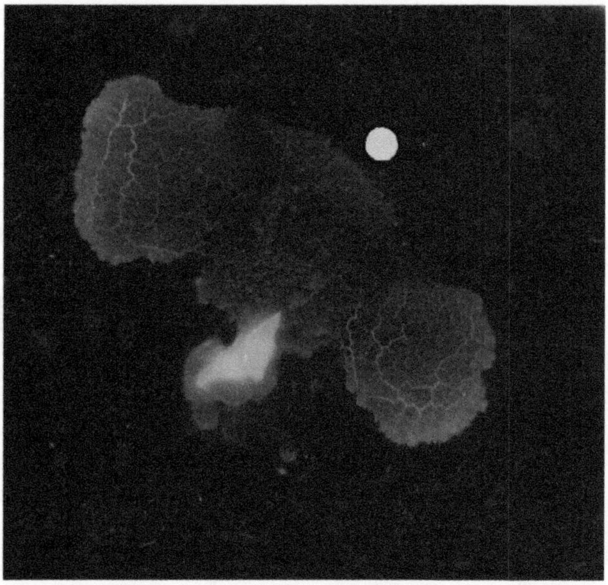

Fig. 25 Splitting plasmodial blob with a grain of potassium chloride. Location of salt crystal is shown by white disc. See details in [53].

or away from, externally applied stimuli it can reform the original shape when the stimulus is removed. An adaptive morphology is a very desirable property in robotic devices since it imbues the robot with great flexibility of size and movement, enabling it to traverse environments which traditionally are difficult to navigate (for example narrow spaces, gratings etc.). This feature is only possible because the properties of the movement and guidance of the collective are distributed throughout the collective and not located in fixed sized and inflexible units as is the case with conventional robotic systems. This is also the case with the *Physarum* plasmodium which adapts its shape and growth patterns in response to its environment. One of the most remarkable properties of the *Physarum* plasmodium is the ability to survive external damage beyond simple attraction and repulsion. A piece of plasmodium excised from the growing tip can survive, and indeed continue to move and grow as an independent entity. Furthermore two independent plasmodia can fuse to form a single plasmodium when placed in close proximity. These phenomena are not only desirable from a robotics perspective in terms of resilience and damage repair, but offer new and as yet little explored opportunities in robotic movement and control.

We set out to find if the particle collective could also replicate these highly desirable features as seen in the real plasmodium. We took a large single oscillating collective (5000 particles) and applied a narrow band of hazardous simulated light irradiation through its centre (Fig. 26a, dashed box represents irradiated area). Particles immediately began to surge away from the irradiated region on both sides and the collective narrowed in diameter and became further 'pinched' until the collective was cleaved into two independently controllable 'blobs'. The cleavage mechanism can be applied in different ways. For example both resulting blobs can be of equal size and have similar oscillatory properties. Alternately it is possible to cleave the collective in such a way to have one large blob and one smaller - recall that a blob which is small enough will be able to move spontaneously in a persistent direction. It is possible to guide each blob independently using either a 'pulling' mechanism (externally applied attractants) or a 'pushing' mechanism (simulated irradiation). In Fig. 26b we guide the lower right blob (arrowed) towards the larger blob by pushing it from its opposite side with simulated irradiated light. As the blobs become closer (specifically, to a separation distance which is sufficient for the border particles in each blob to sense the chemoattractant flux in the other blob) the closest border regions of each blob surge towards each other and a single larger collective is formed by the fusion.

a)
b)

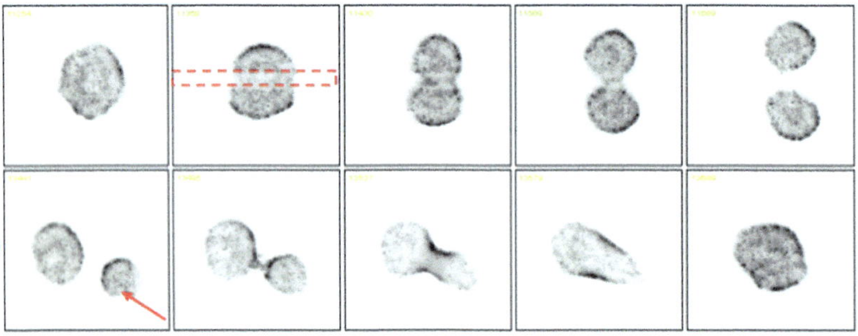

Fig. 26 Controlling blob morphology by splitting and fusion. a) A blob of aggregate particles may be split by applying simulated light irradiation (dashed box region), causing a disruption of flux in the dashed region. The single blob separates into two smaller blobs, each capable of individual oscillation and external control, b) Fusion of two independent particle aggregates. The blob on the lower right is guided diagonally upwards in the direction of the arrow towards the larger blob. The two independent blobs fuse forming a single aggregate.

As an example of the robotic flexibility endowed by the adaptive morphology of the collective, and the guidance mechanisms enabled by its external control, we show in Fig. 27 how a blob can be guided externally (in this case a repulsive 'pushing' by simulated irradiation) in order to traverse a path through a series of obstacles which are much narrower than the diameter of the collective itself. The blob automatically separates its structure in response to the different obstacles and reforms its shape by mutual attraction and cohesion of the particles. No fine control of the individual components, nor complex predeterminism of path choice, is necessary - the collective is guided by simple avoidance of the simulated irradiation (the irradiation location is not shown, but follows the previous examples of simply exposing the rearward part of the collective to push it forwards). When clear of the obstacles, the original circular shape is reformed. If the blob is returned backwards through the obstacle path again the path chosen can be somewhat different to the original path with the same result.

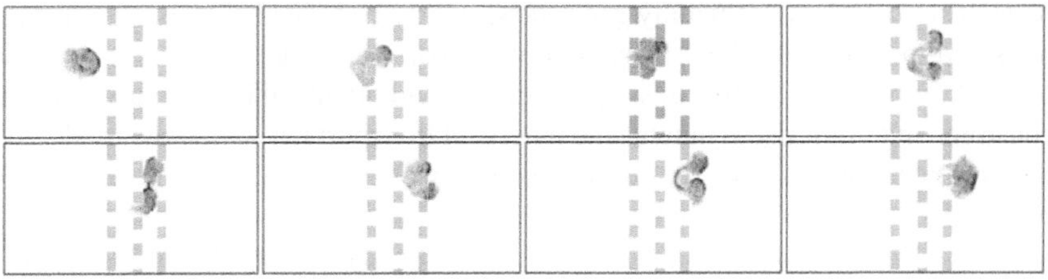

Fig. 27 Utilising the adaptive morphology of the collective. A blob is 'pushed' by application of external light irradiation (not shown) through the path of obstacles whose width is narrower than the blob itself (left to right).

5 Conclusion and Discussion

We have examined the problem of generating and controlling motive forces at very small scales using the true slime mould *Physarum polycephalum* as a prototype mechanism. *Physarum* is attractive because it satisfies many physical and computational properties which are desirable in robotics applications (self-oscillatory, simple components, distributed sensory and motor control, integration of multiple sensory stimuli, amorphous and adaptive shape, amenable to external influence, resilience to damage, self repair). Using the organism itself we have demonstrated and measured its ability to lift and pump material using its intrinsic oscillations and protoplasmic streaming. We also investigated mechanisms by which the pumping and transport behaviour may be externally influenced and inhibited, using irradiation stimuli, so that it may be used as a prototype steering mechanism. We used actual experimental data of the bilateral output from a dumbbell shaped plasmodium to drive a model mechanism, a Braitenberg-type vehicle and used the irradiation stimulus to act as a steering force.

Physarum can in many ways be regarded as a living example of a so-called 'smart material' because of the way in which it combines robotic movement and control functions. Perhaps the most interesting question about the organism is how can such complex behaviour is achieved by the interactions between such simple component parts? An answer to this question may provide insights into the development of non-living smart materials which have all of the advantages demonstrated by the *Physarum* plasmodium, but without some of its limitations. The limitations of the plasmodium itself include its relatively slow speed, its fragility as a living organism and its unpredictability, since the plasmodium is concerned only with survival and not the artificially applied goals of the experimenter.

We sought to investigate this question of the emergent complexity of *Physarum* by attempting to approximate the complex oscillatory behaviour in a particle-based computational model. Because of the simplicity of the plasmodium's components and structure it is clear that any search for a 'secret source' of its complexity would be fruitless. Instead we set about answering the question posed in reverse: rather than try to find out how the organism produces such complex behaviour from simple parts, is it possible to artificially *generate* similarly complex behaviour from simple parts and interactions? We utilised a previous particle model of emergent transport networks where a collective of simple particles with identical behaviour was used to construct and minimise synthetic and emergent transport networks, and modified its motor algorithm in a very simple way, so that instead of smooth network flow, we obtained a more resistant and interrupted flow of particles. The addition of resistance to particle flux was sufficient to generate complex oscillatory dynamics similar to those observed in the *Physarum* plasmodium.

By patterning a densely packed particle population into different shapes we were able to generate reliable and regular oscillatory movement as emergent travelling waves which were fashioned into rotary, reciprocal, helical, and coupled transport mechanisms. The waves, consisting of peaks and troughs of simulated chemoattractant flux, were found to travel at opposite directions, and with much greater velocity, than the underlying particle motion which generated the waves. Smaller unconstrained and isolated collectives resulted in cohesive 'blob-like' patches of virtual material which exhibited spontaneous oscillations, allowing an amoeboid movement of the collective as the particle population reorganised itself in response to the internal oscillations. External control of the blob-like patches was effected by stimulating the collective with attractants ('pulling' the collective) and repulsion ('pushing' by simulated light irradiation). Very small populations resulted in spontaneous and persistent non-amoeboid forward motion. We were also able to reproduce the resilience of the *Physarum* plasmodium to damage by cleaving the collective into two separate and independent blobs and fusing two independent blobs to form a single functional blob. Finally the adaptive morphology of the collective was demonstrated by guiding the collective through an obstacle field narrower than the diameter of the collective itself.

The results presented in this chapter demonstrate how very simple and local low-level interactions in simple materials (real and virtual) can generate complex and emergent behaviour which appear to transcend the capabilities of the simple matter of which they are composed. It has also been shown that the emergent behaviours are amenable to external influence by the presentation of attraction or repulsion stimuli (we hesitate to use the word *control* in living systems, but this may be applicable in synthetic systems). Of course there is nothing magical or special about the properties of these materials; the complexity emerges merely from their interactions. This complex behaviour is harnessed effortlessly by organisms such as *Physarum* as part of a parsimonious survival strategy, enabling their persistence in unpredictable, changeable and hazardous environmental conditions. By understanding the generative mechanisms underlying the complex behaviour it will be possible, we believe, to incorporate these features within physical materials for small scale robotic devices. By utilising the robotic material itself for distributed computation, transport and movement it will be possible to reduce the total number of component parts and also reduce the number of different *types* of components, thus further simplifying the production of the devices.

Acknowledgements. The work was partially supported by the Leverhulme Trust research grant F/00577/1 "Mould intelligence: biological amorphous robots". Experiments in Section 2.2 was conducted in Prof Hywel Morgan's lab at University of Southampton as a part of the project supported by the Life Sciences Interfaces Forum (Project manager: Dr Klaus-Peter Zauner).

References

[1] Fearing, R.S.: Survey of sticking effects for micro parts handling. In: Proceedings IEEE/RSJ International Conference on Intelligent Robots and Systems 1995 Human Robot Interaction and Cooperative Robots, vol. 2, pp. 212–217. IEEE, Los Alamitos (2002)

[2] Ai, B.Q., Xie, H.Z., Wen, D.H., Liu, X.M., Liu, L.G.: Heat flow and efficiency in a microscopic engine. The European Physical Journal B 48(1), 101–106 (2005)

[3] Sher, I., Levinzon-Sher, D., Sher, E.: Miniaturization limitations of HCCI internal combustion engines. Applied Thermal Engineering 29(2-3), 400–411 (2009)

[4] Zhang, L., Abbott, J.J., Dong, L., Kratochvil, B.E., Bell, D., Nelson, B.J.: Artificial bacterial flagella: Fabrication and magnetic control. Applied Physics Letters 94(6), 064107 (2009)

[5] Dreyfus, R., Baudry, J., Roper, M.L., Fermigier, M., Stone, H.A., Bibette, J.: Microscopic artificial swimmers. Nature 437(7060), 862–865 (2005)

[6] den Toonder, J., Bos, F., Broer, D., Filippini, L., Gillies, M., de Goede, J., Mol, T., Reijme, M., Talen, W., Wilderbeek, H., et al.: Artificial cilia for active micro-fluidic mixing. Lab Chip 8(4), 533–541 (2008)

[7] Suh, J.W., Darling, R.B., Bohringer, K.F., Donald, B.R., Baltes, H., Kovacs, G.T.: Fully programmable MEMS ciliary actuator arrays for micromanipulation tasks. In: Proceedings IEEE International Conference on Robotics and Automation, ICRA 2000, vol. 2, pp. 1101–1108. IEEE, Los Alamitos (2002)

[8] Trimmer, B.A., Takesian, A.E., Sweet, B.M., Rogers, C.B., Hake, D.C., Rogers, D.J.: Caterpillar locomotion: A new model for soft-bodied climbing and burrowing robots. In: 7th International Symposium on Technology and the Mine Problem. Citeseer (2006)

[9] Saga, N., Nakamura, T.: Development of a peristaltic crawling robot using magnetic fluid on the basis of the locomotion mechanism of the earthworm. Smart Materials and Structures 13, 566 (2004)

[10] Umedachi, T., Kitamura, T., Takeda, K., Nakagaki, T., Kobayashi, R., Ishiguro, A.: A Modular Robot Driven by Protoplasmic Streaming. Distributed Autonomous Robotic Systems 8, 193–202 (2009)

[11] Kubea, C.R., Bonabeau, E.: Cooperative transport by ants and robots. Robotics and autonomous systems 30, 85–101 (2000)

[12] Zhang, S.: Fabrication of novel biomaterials through molecular self-assembly. Nature biotechnology 21(10), 1171–1178 (2003)

[13] Lobovkina, T., Dommersnes, P.G., Tiourine, S., Joanny, J.F., Orwar, O.: Shape optimization in lipid nanotube networks. The European Physical Journal E: Soft Matter and Biological Physics 26(3), 295–300 (2008)

[14] Lobovkina, T., Gozen, I., Erkan, Y., Olofsson, J., Weber, S.G., Orwar, O.: Protrusive growth and periodic contractile motion in surface-adhered vesicles induced by ca2+-gradients. Soft Matter 6(2), 268–272 (2010)

[15] Lagzi, I., Soh, S., Wesson, P.J., Browne, K.P., Grzybowski, B.A.: Maze solving by chemotactic droplets. Journal of the American Chemical Society (2010), doi:10.1021/ja9076793
[16] Dillon, R.H., Fauci, L.J., Omoto, C., Yang, X.: Fluid dynamic models of flagellar and ciliary beating. Annals of the New York Academy of Sciences (Reproductive Biomechanics) 1101, 494–505 (2007)
[17] Ijspeert, A.J.: Central pattern generators for locomotion control in animals and robots: a review. Neural Networks 21(4), 642–653 (2008)
[18] Adamatzky, A., Arena, P., Basile, A., Carmona-Galan, R., de Lacy Costello, B., Fortuna, L., Frasca, M., Rodriguez-Vazquez, A.: Reaction-diffusion navigation robot control: from chemical to vlsi analogic processors. IEEE Transactions on Circuits and Systems I: Regular Papers [see also IEEE Transactions on Circuits and Systems I: Fundamental Theory and Applications] 51(5), 926–938 (2004)
[19] Adamatzky, A., de Lacy Costello, B., Skachek, S., Melhuish, C.: Manipulating objects with chemical waves: Open loop case of experimental belousov-zhabotinsky medium coupled with simulated actuator array. Physics Letters A 350(3-4), 201–209 (2006)
[20] Maeda, S., Hara, Y., Sakai, T., Yoshida, R., Hashimoto, S.: Self-Walking Gel. Advanced Materials 19(21), 3480–3484 (2007)
[21] Adamatzky, A., De Lacy Costello, B., Shirakawa, T.: Universal computation with limited resources: Belousov-zhabotinsky and physarum computers. International Journal of Bifurcation and Chaos (2008) (in press)
[22] Nakagaki, T. (ed.): Int. Journal of Unconventional Comput. Special Issue: The Birth of Physarum Computing. Old City Publishing (2010)
[23] Adamatzky, A.: Developing proximity graphs by physarum polycephalum: Does the plasmodium follow toussaint hierarchy? Parallel Process. Lett. 19, 105–127 (2008)
[24] Shirakawa, T., Gunji, Y.P.: Computation of Voronoi diagram and collision-free path using the Plasmodium of Physarum polycephalum. Int. J. Unconventional Computing 6(2), 79–88 (2010)
[25] Shirakawa, T., Adamatzky, A., Gunji, Y.P., Miyake, Y.: On simultaneous construction of voronoi diagram and delaunay triangulation by physarum polycephalum. International Journal of Bifurcation and Chaos 19(9), 3109–3117 (2009)
[26] Tero, A., Kobayashi, R., Nakagaki, T.: Physarum solver: A biologically inspired method of road-network navigation. Physica A: Statistical Mechanics and its Applications 363(1), 115–119 (2006)
[27] Tero, A., Takagi, S., Saigusa, T., Ito, K., Bebber, D.P., Fricker, M.D., Yumiki, K., Kobayashi, R., Nakagaki, T.: Rules for biologically inspired adaptive network design. Science 327(5964), 439–442 (2010)
[28] Adamatzky, A., Jones, J.: Road planning with slime mould: If physarum built motorways it would route m6/m74 through newcastle. International Journal of Bifurcation and Chaos (2010) (in press)
[29] Adamatzky, A.: Physarum machine: Implementation of a kolmogorov-uspensky machine on a biological substrate. Parallel Processing Letters 17(4), 455–467 (2007)
[30] Adamatzky, A., Jones, J.: Towards Physarum robots: computing and manipulating on water surface. Journal of Bionic Engineering 5(4), 348–357 (2008)
[31] Tsuda, S., Zauner, K.-P., Gunji, Y.-P.: Robot control with biological cells. In: Proceedings of the Sixth Int. Workshop on Information Processing in Cells and Tissues, St. William's College, York, August 30–September 1, pp. 202–216 (2005)
[32] Nakagaki, T., Yamada, H., Ueda, T.: Interaction between cell shape and contraction pattern in the Physarum plasmodium. Biophysical Chemistry 84(3), 195–204 (2000)

[33] Allen, P.J., Price, W.H.: The relation between respiration and protoplasmic flow in the slime mold, Physarum polycephalum. American Journal of Botany 37(5), 393–402 (1950)

[34] Kamiya, N.: The protoplasmic flow in the myxomycete plasmodium as revealed by a volumetric analysis. Protoplasma 39(3), 344–357 (1950)

[35] Gotoh, K., Kuroda, K.: Motive force of cytoplasmic streaming during plasmodial mitosis of physarum polycephalum. Cell Motility and the Cytoskeleton 2(2), 173–181 (1982)

[36] Smith, D.A., Saldana, R.: Model of the Ca2+ oscillator for shuttle streaming in Physarum polycephalum. Biophysical journal 61(2), 368–380 (1992)

[37] Kamiya, N., Kuroda, K.: Studies on the velocity distribution of the protoplasmic streaming in the myxomycete plasmodium. Protoplasma 49(1), 1–4 (1958)

[38] Tsuda, S., Jones, J.: The emergence of synchronization in *physarum polycephalum* and its particle approximation. Biosystems (2010) (in press)

[39] Takamatsu, A., Fujii, T.: Construction of a living coupled oscillator system of plasmodial slime mold by a microfabricated structure. Sensors Update 10(1), 33–46 (2002)

[40] Takamatsu, A., Fujii, T., Endo, I.: Control of interaction strength in a network of the true slime mold by a microfabricated structure. BioSystems 55, 33–38 (2000)

[41] Häder, D.-P., Schreckenbach, T.: Phototactic Orientation in Plasmodia of the Acellular Slime Mold, Physarum polycephalum. Cell 25(1), 55–61 (1984)

[42] Wolf, R., Niemuth, J., Sauer, H.: Thermotaxis and protoplasmic oscillations in Physarum plasmodia analysed in a novel device generating stable linear temperature gradients. Protoplasma 197(1-2), 121–131 (1997)

[43] Anderson, J.D.: Galvanotaxis of slime mold. J. Gen. Physiol. 35(5), 1–1 (1951)

[44] Nakagaki, T., Yamada, H., Ueda, T.: Modulation of cellular rhythm and photoavoidance by oscillatory irradiation in the Physarum plasmodium. Biophysical Chemistry 82, 23–28 (1999)

[45] Nakagaki, T., Uemura, S., Kakiuchi, Y., Ueda, T.: Action spectrum for sporulation and photoavoidance in the plasmodium of Physarum polycephalum, as modified differentially by temperature and starvation. Photochem. Photobiol. 64(5), 859–862 (1996)

[46] Braitenberg, V.: Vehicles: Experiments in Synthetic Psychology. MIT Press, Cambridge (1984)

[47] Jones, J.: The emergence and dynamical evolution of complex transport networks from simple low-level behaviours. International Journal of Unconventional Computing 6(2), 125–144 (2010)

[48] Jones, J.: Characteristics of pattern formation and evolution in approximations of physarum transport networks. Artificial Life 16(2), 127–153 (2010)

[49] Tsuda, S., Jones, J.: The emergence of complex oscillatory behaviour in *physarum polycephalum* and its particle approximation. In: Artificial Life XII, pp. 698–705 (2010)

[50] Flynn, M.R., Kasimov, A.R., Nave, J.C., Rosales, R.R., Seibold, B.: Self-sustained nonlinear waves in traffic flow. Physical Review E 79(5), 56113 (2009)

[51] Durham, A.C.H., Ridgway, E.B.: Control of chemotaxis in Physarum polycephalum. The Journal of Cell Biology 69, 218–223 (1976)

[52] Nakagaki, T., Kobayashi, R., Nishiura, Y., Ueda, T.: Obtaining multiple separate food sources: behavioural intelligence in the Physarum plasmodium. R. Soc. Proc.: Biol. Sci. 271(1554), 2305–2310 (2004)

[53] Adamatzky, A.: Physarum Machines. World Scientific, Singapore (2010)

Developing Self-Organizing Robotic Cells Using Organic Computing Principles

Alwin Hoffmann, Florian Nafz, Andreas Schierl,
Hella Seebach, and Wolfgang Reif

Abstract. Nowadays industrial robotics applications, which are often designed and planned with a huge amount of effort, have a fixed behavior during runtime and cannot react to changes in their environment. Failures can hardly be compensated and often can only be repaired by human involvement. The idea of Organic Computing is to enable systems to possess life-like properties, such as self-organizing or self-healing. In this chapter we present a layered architecture to bring these two worlds together. Further it is discussed what are the requirements of the respective layers to allow to engineer self-x properties into such systems. The presented approach allows for developing self-organizing robotic applications that are able to take advantage of Organic Computing principles and therefore are more robust and flexible during runtime.

1 Introduction

With respect to their structure, traditional automation systems are very static. The material flow is fixed and every component is optimized according to the planned system structure to reach maximum throughput. This approach is very suitable for mass production as in the automotive industries, where one product is manufactured for a considerable time. Even the use of industrial robots does not change this situation. In fact, industrial robots are very flexible and, given an appropriate tool, are able to perform a large variety of tasks [10]. However, the complex and tedious programming of today's industrial robots, the fixed wiring and difficult integration of additional devices, as well as the very static layout of shop floors do not exploit the possible flexibility of robotic solutions. As a consequence, high effort is needed to customize and adjust automation systems, making them hardly applicable for small-series production with varying products where regular adaptation is

Alwin Hoffmann · Florian Nafz · Andreas Schierl · Hella Seebach · Wolfgang Reif
Institute for Software & Systems Engineering, University of Augsburg,
86135 Augsburg, Germany

Y. Meng and Y. Jin (Eds.): Bio-Inspired Self-Organizing Robotic Systems, SCI 355, pp. 253–273.
springerlink.com

required. Moreover, failure tolerance and flexible optimization is hard to achieve with traditional automation systems.

On the other hand, the idea of Organic Computing [3, 20, 39] and Autonomic Computing [6, 15] is to develop systems which possess self-x properties, like self-healing (i.e. the compensation of failures), self-optimization (i.e. the autonomous optimization according to a given fitness function), or self-adaptation (i.e. the adaptation to new or changing tasks or a different system structure). In the context of production automation, the goal is to provide architectures and techniques to build *organic* automation systems, where organic means life-like behavior or more concise that the systems are capable of autonomously adapting to changes in their environment. This behavior is often realized by the use of bio-inspired paradigms and algorithms, e.g. genetic algorithms or pheromone-based approaches.

Hence, the well-designed combination of Organic Computing principles and robot technology can lead to hyper flexible and robust automation systems. While robots – mobile platforms as well as industrial manipulators – provide mechanical flexibility and the ability to perform a large variety of different tasks, Organic Computing introduces self-organization to the systems which enables them for self-healing, self-optimization or self-adaptation. For example, an Organic automation system can compensate failures due to its self-healing capabilities and continues to operate in *graceful degradation* – a requirement for robotic systems which is getting more and more important [11]. Especially in small and medium enterprises, where there are phantom shifts at night, systems of this kind are welcome. Using self-optimization allows for continuously and autonomously optimizing an automation system during its runtime and without external interaction. Finally, due to self-organization, organic systems are reconfigurable by design and are easily able to adapt to new tasks or products – a requirement in today's globalized economy with its turbulent markets and fast changing demands [17]. However, evolutionary approaches and bio-inspired principle which solely rely on the idea of emergence cannot directly be applied to production systems. Emergence rather has to be controlled and directed to accomplish a defined goal, i.e. manufacturing a product.

In this chapter, we want to outline how self-x properties and Organic Computing principles can be applied to industrial robotic manufacturing cells [2]. In previous work, we have introduced an approach for modeling and designing self-organizing resource-flow systems based on Organic Computing principles [31] and showed how to specify a behavioral corridor for this system class [8]. Moreover, we have developed a guideline for systematically engineering self-organizing resource-flow systems [30] and verified their functional correctness using formal methods [23].

Because robotic manufacturing cells usually constitute a resource-flow system, where a product (i.e. the resource) is manufactured step-by-step, our approach can be applied to the domain of industrial robotics. However, we identified three robotic-specific challenges one is facing in order to build self-organizing robotic cells. These challenges are ranging from uncontrolled emergence over the problems in the layout of robotic cells to limitations in current software architectures and will be described in detail in Section 2. Based on these challenges, we present a multi-layer architecture (Section 3) which allows for developing self-organizing robotic manufacturing cells using Organic Computing. Every layer addresses the system at a different level of abstraction and has distinct responsibilities in order to comply with the architectural requirements. In Section 4, we illustrate our approach with a simple but evident case study. Finally, in Section 5, a conclusion is drawn and future research steps are highlighted.

2 Challenges

From our point of view, the development of self-organizing robotic cells using Organic Computing principles poses three major challenges:

1. To render organic systems acceptable for industry, *emergence must be controlled* to accomplish a defined goal.
2. To apply self-x properties, the layout of robotic cells must provide *additional degrees of freedom.*
3. To utilize these additional degrees of freedom, robotic software architectures must provide *flexibility with regard to programming techniques*, coping with geometric uncertainty and device integration.

These challenges and their influence on the development of organic automation systems are described in detail in the following sections.

2.1 Controlling Emergence

A main concept of self-organizing systems is emergence. Emergence describes the appearance of complex system behavior caused by relatively simple and local interactions of individuals without the control of a central instance. Hence, the system behavior is not explicitly programmed, but a result of these local interactions. An example of emergent behavior is an ant colony where no central control is present. Instead, each ant is an autonomous unit that reacts depending on local information, i.e. pheromones, and genetically encoded rules.

Thus, the behavior of the complete system cannot be exactly predicted. Müller-Schloer [18] calls this kind of behavior *bottom-up constraint propagation* which stands in contrast to the classical top-down design of technical systems. In the latter approach, the developer tries to model and implement all possible system states. This usually starts with a high-level specification, until after a number of transformations and refinements, executable code is generated.

However, having an exhaustive model of a complex system is often not feasible and even contradictory to the idea of emergence. In order to solve this contradiction,

Fig. 1 Organic production systems require a corridor of expected behavior. Inside this corridor, emergent behavior is approved and even desired.

we suggest defining a corridor of *good* expected behavior [8] for every organic production system. Inside this corridor, emergent behavior is approved and even desired, whereas the system is in an exceptional state when this corridor is left (cf. Fig. 1).

This corridor is defined through constraints by the system developer and allows *controlled emergence*. Usually, these constraints can be observed locally by each autonomous component of the system. If one or more constraints are violated, the component tries to restore the constraints locally. If this is not possible, it starts to involve surrounding components until a valid solution is found that satisfies the constraints. In Organic Computing this kind of architecture is called an Observer/Controller architecture [18, 31]. By doing so, organic automation systems are self-organizing and can be directed to accomplish a defined goal, e.g. to manufacture products. Besides, behavioral guarantees in terms of functional correctness can be given [21].

2.2 Adding Degrees of Freedom

Usually, automation systems are designed and tuned to accomplish pre-defined tasks for a long period. In single-station automated cells, a production machine is typically equipped with a material handling system (e.g. a robot for loading and unloading the machine) and a storage system. Due to this setting, the cell is able to operate unattended but the system fails if any of the components breaks. An automated production line consists of multiple workstations that are automated and linked by a transport system which transfers parts from one station to the next. Again, if one components breaks, the whole system fails. According to [40], flexible manufacturing systems still have limited capabilities regarding customized products and failure compensation. They even state that the need for flexible manufacturing and adaptive systems cannot be supplied with traditional approaches.

In order to become self-organizing, automation systems require additional degrees of freedom and redundancy in the available hardware. Without these prerequisites, the system is not able to adapt to new environmental conditions or to compensate failures:

- For *self-healing*, an organic automation system needs redundant hardware components. Otherwise, it cannot compensate for the failure of one component and continue operation in graceful degradation.
- Regarding *self-adaptation*, an organic production system needs degrees of freedom, i.e. flexible tools or transport systems, in order to adapt to changing or new tasks as well as to a modified system structure.
- Finally, *self-optimization* is only possible if there are several degrees of freedom which can be optimized with respect to a given fitness function.

Due to these reasons, we believe that robotic cells are well-suited for self-organization by using Organic Computing. In robotic-based systems, additional degrees of freedom can be achieved by adding robots, redundant tools, or tool-changing systems. Concerning transportation, robots can be connected using carousels, two-way conveyors, or even mobile platforms. Further details are given in Sect. 3, but here it is worth mentioning that the concrete choice of how redundancy is added can impact the system's robustness and its mean time to failure, as the example in Sect. 4 shows. Giving one component all redundancy is in general a bad choice, as a component failure will lead to a complete loss of the available redundancy. To find good distribution strategies for redundancy, the ADCCA[1] technique [9] can be used, which calculates minimal combinations of failures which lead to a standstill of the whole production system. Also similar safety analysis techniques like Fault Tree Analysis [37] can be used to identify single-point or n-point failures and optimize the redundancy distribution accordingly.

2.3 Requiring Software Flexibility

By adding degrees of freedom and redundancy to the available devices and to the shop floor layout, self-organization becomes feasible. However, to completely utilize self-x properties, additional requirements to the architectures of robotic systems with regard to software flexibility are necessary.

Flexible and reconfigurable automation systems require the introduction of *smart products* carrying information about how to be processed by the system. This can be e.g. realized by using RFID [40]. As a consequence, a *product-centric* approach of configuring and commanding industrial robots and their tools is required. Predefined motion sequences have to be replaced by more dynamic motion planning considering the environment and avoiding obstacles. Due to the dynamic system behavior, the use of previously taught motions cannot be sufficient anymore. Instead, the use of sensor feedback (e.g. vision) or compliant devices should be considered.

[1] Adaptive Deductive Cause-Consequence Analysis.

With sensor-based or compliant motions [16], an error-tolerant execution of complex robot tasks in uncertain and unknown environments is possible [34].

In contrast to these demanding requirements, industrial robots are still programmed with special robot programming languages which are derived from early imperative languages and have not evolved much since then. Due to these low-level programming techniques, developing software for an industrial robot is a complex and tedious task requiring considerable technical expertise [26]. Hence, industrial robots are usually equipped and programmed to perform only a set of pre-defined tasks. This contradiction between low-level programming and high-demanding requirements must be solved in the future to realize self-organizing robotic systems.

Furthermore, the integration of external devices must be facilitated. Today, tools are usually connected by a fixed wiring to a robot controller and communicates over digital and analog I/O ports. But when using tool changing systems, no human interaction should be required. The software of the robot controller must be able to independently cope with different tools mounted to the robot. Moreover, it must be possible to integrate arbitrary sensors for intelligent perception and sophisticated tools that allow e.g. complex grasping strategies and dexterous manipulation [24]. The introduction of plug-and-play mechanisms as proposed in [11] would cover this requirement for flexible device integration.

3 Architecture

Our approach uses a layered software architecture which addresses the system at different levels of abstraction. The proposed architecture is depicted in Fig. 2.

On top of the hardware layer, two software layers are located for controlling the robot. The lower one, the *Robot Control Layer*, is responsible for the real-time critical, low-level hardware control, whereas the upper layer, the *Robot Programming Layer*, is used for defining the control flow and specifying required motions and tools actions. For traditional production systems, these three layers are sufficient, as they allow the robot to execute arbitrary, pre-programmed tasks in a reliable, repeatable fashion. However, to extend the system towards self-organization, additional communication and control software is required.

Therefore, our approach adds two more layers on top, which control the robotic system according to Organic Computing principles. The first layer, the *Organic Control Layer*, wraps the components of the robotic cell and turns them into software agents coordinating with other agents through communication. Furthermore, it is responsible for the execution of *capabilities* which are to be applied to the workpiece. When this layer detects a locally unrecoverable error, the *Organic Planning Layer* takes control and searches for a new configuration to achieve the task. Once a solution has been found, control is returned to the Organic Control Layer for further execution. The layers of the architecture are explained below from bottom up.

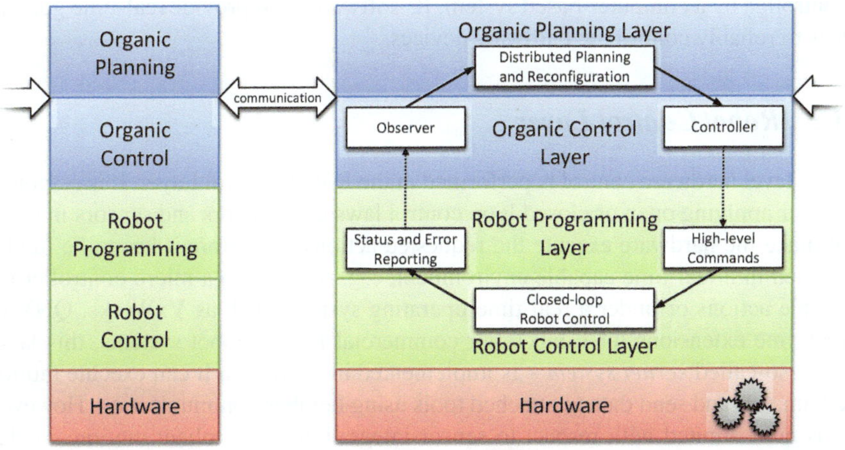

Fig. 2 The proposed architecture for self-organizing robotic cells showing two individual components.

3.1 Hardware

The foundation of each robotic cell is a set of robots with tools that are interlinked with a transportation system. In order to become a self-organizing production system, additional degrees of freedom are required as stated in Sect. 2.2. This means that a robot cannot only be equipped with one static tool corresponding to its preassigned task. For simple cases, it might e. g. be enough to equip the robot with a set of equal drills so that it can replace them when they fail during production, but for exploiting all advantages of self-organizing systems, different tools are needed that can perform a variety of diverse tasks, and a way to interchange them without human interaction. This can be achieved by the use of external tools, by an automatic tool exchange system, or by using advanced tools like anthropomorphic hands which allow dexterous manipulation.

If different tasks have to be executed, or the different task steps should be assigned to different robots, the transportation system also has to become flexible. Instead of a single conveyor connecting the robots in a given order, this set-up requires a way to change the order a workpiece passes the different robots. Similar to existing systems, robots can be connected using a carousel or two conveyors, one moving forward and one moving backwards. Thus, each robot can forward the workpiece to any other robot by placing it onto the right conveyor. Corresponding to the idea of hyper flexible manufacturing systems [11], another solution is to replace the conveyor by a set of mobile platforms navigating between the robots, transporting partly processed workpieces.

This allows a system to show a dynamic behavior. However, as the hardware devices are expected to perform different tasks over time, all of them have to be controlled by a computer-based system. Its software must provide real-time guarantees to reliably control the hardware devices.

3.2 Robot Control Layer

Low-level hardware control is performed in the Robot Control Layer. It is responsible for applying open or closed loop control laws on actuators and sensors in order to make the hardware execute the requested actions. Therefore it has to be implemented in a real-time capable environment, e. g. running on a micro controller for simple actions or under a real-time operating system (such as VxWorks, QNX or real-time extensions for Linux). For commercial KUKA robot systems, this layer – the so-called *kernel system* – is implemented in VxWorks. It can execute motion commands and send data to attached tools using fieldbus communication. However, it is quite limited with respect to sensor integration or compliant motion – fields where research robot controllers like OROCOS [32] are more advanced. Furthermore the control layer has to monitor the attached hardware for errors, and report them to the above layer to allow reasonable failure strategies.

As a typical robot action consists of more than the application of one single control law with given parameters (e. g. one motion to a point), the robot control layer has to provide an interface allowing to specify multiple control laws or commands that are to be applied sequentially or in parallel. This interface can be used by the programming layer. Examples for action task descriptions specified over such an interface are manipulation primitives [5], constraint-based task specifications [4] or realtime primitive nets [38].

Realtime primitive nets describe actions executed by one or multiple cooperating robots. These actions are composed of calculation primitives (blocks) and data flows between them, which are evaluated in a real-time loop and form the corresponding control loop for the hardware devices. All actions that have to be executed with exactly given timing constrains or depend on each other's progress can thus be combined into one realtime primitive net that will be executed atomically. This allows to specify complex or composed tasks with realtime requirements as a single transaction and execute them in the control layer, removing the need for real-time capability in the higher layers.

A limited example for a realtime primitive net is given in Figure 3. It shows a motion of a robot along a given trajectory, followed by a control action enabling the gas flow of a welding torch. The dotted boxes represent the high-level constructs used in the programming layer to define the task, whereas the solid boxes show realtime primitives and their dataflow links. In this example, position values from the trajectory generator are sent to the robot block as set points (a typical example of open loop control), and the digital output representing the welding torch is enabled immediately once the trajectory has finished. Of course, a real world welding task consists of more actions to be included in the realtime primitive net, e. g. ignition of the welding torch, going along the welding seam and disabling the torch once the

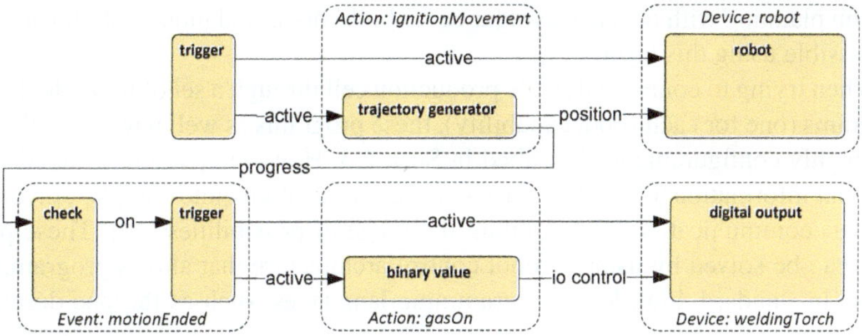

Fig. 3 An example realtime primitive net describing the motion of a robot followed by the execution of a tool action.

destination has been reached. Further details about the realtime primitives interface can be found in [38].

Commercial robot controllers usually omit a clear separation between control and programming layer and execute complete robot programs on the control layer. However, by separating these layers and thereby encapsulating the realtime requirements on the control layer, a standard programming language can be used for the programming layer. This allows making the programming layer extensible and simplifies the integration of robot programs into the surrounding software system.

3.3 Robot Programming Layer

The Robot Programming Layer offers an interface which accepts high-level commands to be executed by the robot. It is responsible for transforming them into control laws or task descriptions that can be executed with real-time guarantees in the robot control layer. Furthermore, it transfers them to the robot control layer and monitors execution progress, errors and sensor events. For KUKA robots, this layer can be seen in the robot programming language KRL which allows writing robot programs including extended control flow (e.g. conditional statements and loops), motions and tool commands. Similar features are available in the languages RAPID for ABB and Karel for FANUC robots.

However, the self-organizing robot cell – opposed to traditional production cells – does not have a fixed processing or material flow order, thus it is not possible to write one program for each robot that can be executed repeatedly to perform the unchanging robot task. Each robot needs a set of robot programs (one for each robot capability) that can be started and controlled from a higher architecture layer.

As the dynamic nature of a flexible production system makes it hard to guarantee exact positioning of the workpieces during transportation, these systems also have to cope with greater uncertainty about object locations. Thus the integration of sensor feedback for object localization becomes more important here, as well as the possibility to program tolerant or compliant manipulators or tools. Also dynamic motion planning with obstacle avoidance for both robots and mobile platforms must be possible using this layer.

When trying to control a flexible production cell through a set of individual robot programs (one for each robot capability), these programs as well have to be flexible and highly configurable as described in Sect. 2.3. However, passing detailed environment information to traditional robot programs is often quite complex, involving fieldbus communication, thus limiting the range of possibilities [13]. These problems can be solved by using a robot control architecture that allows programming robots in standard, high-level programming languages, such as the one described in [1].

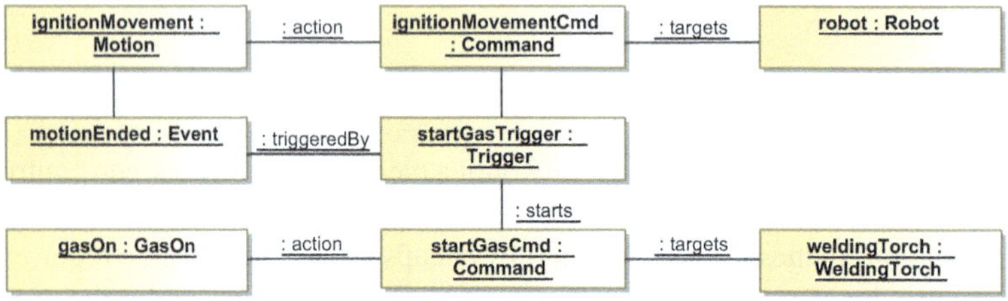

Fig. 4 The robot task from Fig. 3 represented as an object structure with actions, devices and events.

It provides a high level, object-oriented API for programming robots which can be directly used from the higher layers or encapsulated into a service that can be e.g. accessed via standard service-oriented methods.

Figure 4 shows an example of a robot task created using the object-oriented API. It contains two robot commands, one targeted at a robot and containing a motion, and the second enabling the gas flow of a welding torch. These commands are connected using a trigger that starts the second command once the motion of the first command has ended. This (and many more, when dealing with a real welding scenario) has to be executed with real-time guarantees to ensure that the welding seam is created with repeatable quality, so it is converted into a realtime primitive net (like the one shown in Fig. 3) and executed using the robot contorl layer.

As an overview, the two robot software layers are shown in figure 5. The upper part (programming environment and Robotics API) of the robot software architecture represents the programming layer, and the lower part (realtime primitives interface and robot control core) is an example for a robot control layer. Using this architecture, all layers above can communicate with the robot system using the standard means of object-oriented software development. This simplifies the development of the two organic layers located on top of the robotic layers in the proposed software architecture.

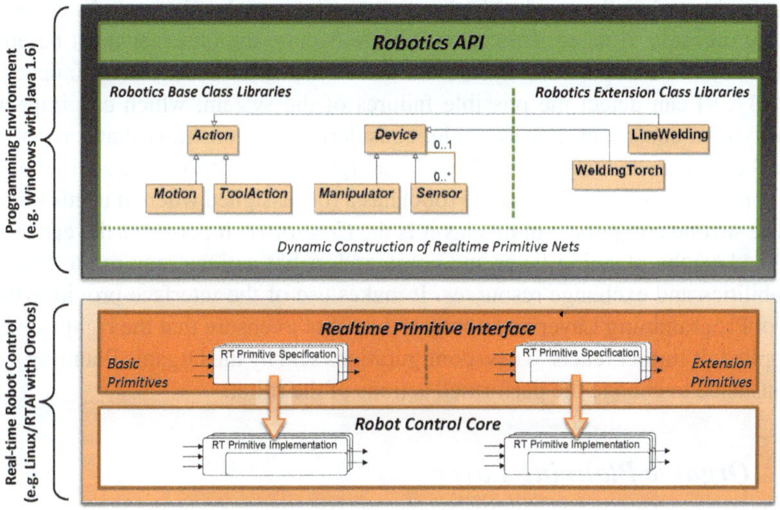

Fig. 5 Overview of the robot software layers.

3.4 Organic Control Layer

The presented architecture for the top two layers is similar to observer/controller architectures often used in the field of Organic Computing to realize the self-x features of a system [19, 28]. The main task of these layers is to maintain the behavioral corridor of the system (see Sect. 2.1). The corridor is specified by OCL Constraints [25, 31], which are annotated to the particular models during the design process and describing "good" system configurations leading to functionally correct behavior. By not explicitly forcing the system into a fixed set of configurations an additional degree of freedom is gained, in which the system can pick the configuration it assesses as good. Further the constraints ensure that only configurations are chosen that lead to a functional correct system. These constraints define a kind of invariant over the system state and distinguish good from erroneous states. They specify how correct configurations of the robots must look like. The system then tries to preserve these constraints as long as possible. In case of a violation information is forwarded to the planning layer which tries to restore them, by calculating a new reconfiguration for the system. This approach is called the *restore invariant approach* and described in detail in [8].

The Organic Control Layer therefore consists of two main components. An *observer* component which constantly evaluates the constraints, based on the status information of the system it receives from the lower layers. Here interfaces which allow to receive feedback from the Robot Programming Layer (e.g. example error-messages or sensor data) are needed. Whenever the observer detects a violation, it activates the planning layer and forwards all gathered information. The main challenge here is to formulate the constraints in such a way and granularity that the robot is able to locally decide whether a constraint is violated or not. As constraints usually are only violated if a system failure occurs, the observer must be able to reason about the impact of a system failure on the constraints. Here a failure analysis [37, 9] can detect the possible failures of the system, which can impact the validity of a constraint. For example, considering the constraint that a robot must have the tool needed to perform the roles assigned to it. A tool failure will than lead to a constraint violation in case the robot has a role assigned where it needs this tool.

The second component in this layer is the *Controller*. It performs the capabilities assigned by the planning layer and commands robot actions required to apply the capabilities and exchange resources. It makes use of the interface provided by the Robot Programming Layer and controls the robot to ensure that the right capability is applied. It further reacts to new configurations sent by the Organic Planning Layer, for instance to change the performed actions of the robot.

3.5 Organic Planning Layer

On top of the control layer is the Organic Planning Layer. It is triggered by the observer of the control layer and responsible for calculating new configurations if an error occurs. It analyzes the current situation and, as most of the failures cannot be compensated by one robot alone, it has to communicate with the planning layers of other robots to gather information about available robots and their capabilities. Then, the planning component tries to find a common solution to reach the objectives. After a consensus is found, the planning layer forwards the new configuration for its responsible robot to the Organic Control Layer, which then commands the robot accordingly.

The advantage of moving all the self-organization into this high-level layer is to be able use the full bandwidth of planning approaches, like bio-inspired or genetic algorithms as well as simple planners. Therefore this layer provides a plug-in interface to allow the use of several methods and algorithms for coordination and planning, implemented as centralized or decentralized variants. System architects can choose what is best suited for their kind of system and problem to solve. Another reason is that on this layer real time must not be considered, as all real time critical commands are dealt with on a lower layer.

The planning task is basically a constraint satisfaction problem on the systems configurations [36, 22]. Depending on the application and the parameters, this can be rather complex, especially as the robots do not have global knowledge. Here the challenge is to find proper communication protocols and algorithms which can deal with specific requirements of the application. One may think of a simple gathering of the global knowledge at one robot and then calculating a new configuration on this robot. The solution is then spread to the other robots. While this may applicable for smaller manufacturing cells it is not for larger systems. Further one does not want to always stop the complete cell, instead a local reconfiguration is preferred, where only a few robots are participating in the reconfiguration process.

Usually not just any solution for the problem is wanted, but an optimal solution for the actual situation. Here the planner's task is extended to find a best or nearly optimal configuration for the system according to the given optimization criteria, load balancing criteria or minimum number of reconfigured robots, for example.

4 An Adaptive Production Cell Example

In this section we want to illustrate the presented approach on a vision of a future adaptive production cell. It shows the benefits of applying organic principles to traditional robot systems. Traditional engineering would handle and design such a production cell in a rather static way, consisting of individual machines that process workpieces with their tools and linked to each other in a strict sequential order using conveyors or similar mechanisms. The layout of the cell is therefore predefined, very inflexible, and rigid. Additionally, and maybe more important, such a system is extremely prone to system errors as the failure of one component will stop the whole system. However, the adaptive production cell is self-organizing which means that it is adaptive according to user-defined tasks (work plans) and compensates for component failures. Furthermore, it tries to optimize the throughput by finding a configuration which is best suited for the actual work plan.

4.1 System Description

The adaptive production cell consists of KUKA Light-weight Robots (LWR), which are capable of using different tools. The traditional conveyor belt has been replaced by flexible and autonomous transportation units, which can carry workpieces. Some interesting concepts and ideas for flexible transportation units or conveyor belts are described by Bussmann in [29]. The goal of the example cell is to process workpieces in a user-defined sequence of tool applications (task).

Sect. 2 expounds that redundancy and software flexibility are needed to enhance traditional systems with self-organization. To achieve the maximum benefit from redundancy, it is important how the redundancy is distributed within the system. For example, it would be possible that a robot has the same tool three times and is the only one with this tool. Then, the robot is capable of reacting two times on tool breaks, but the breakdown of the whole robot stays a single-point of failure. Therefore a more failure tolerant distribution is to give one tool of each type to each robot, here a system breakdown needs at least a three-point of failure.

According to these results, the case study is arranged as follows (Fig. 6). We have three LWRs for processing, four carts for transportation and two storages, which provide unprocessed workpieces and store finished ones. Each LWR is able to perform all three capabilities: drill a hole into a workpiece, insert a screw into the drilled hole and tighten the inserted screw. The user-defined standard work plan is to process all workpieces with all three tools. The given order is first to drill the hole, then insert the screw and at last to tighten it. In principle, an easy but not very high-performance solution is to let each robot perform all capabilities and change tools after each step. As switching tools is very time consuming compared to the time for applying the capability, the standard configuration is to let every robot perform a different task. The distribution of processing steps among different robots requires flexible routing of carts so that the correct order is maintained. One such configuration is sketched in Fig. 6.

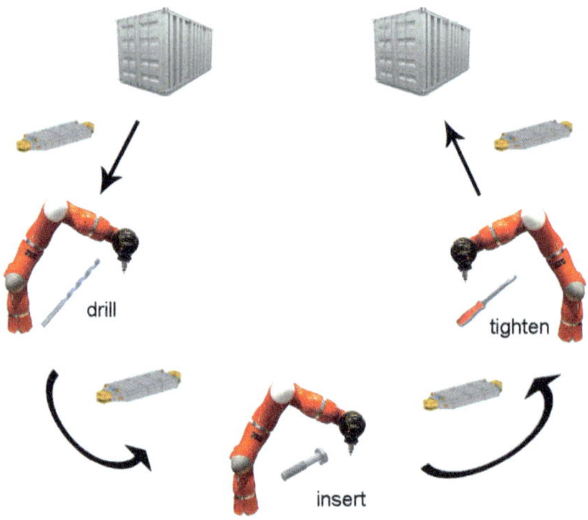

Fig. 6 Adaptive production cell

4.2 *Design of Self-organizing Resource-Flow Systems*

So far only the hardware of the adaptive production cell is described. But at least as interesting is the software for this example. For the two organic layers a software engineering guideline exists, which guides the engineer through several steps for developing self-organizing resource-flow systems [30]. The presented robotic cell is one simple instance of the class of resource-flow-systems. Other instances are all kinds of production automation, where you have a product running through a manufacturing process. One core concept of the guideline is the Organic Design Pattern (ODP, see Fig. 7), which determines the architecture and behavior of the system. It identifies the different components and artifacts of this domain and their relations.

The central components in the system are the *agents*, representing the robots and carts. They are processing the *resources* according to a given *task*. In case of the production cell every agent has several *capabilities*, divided into producing, processing,

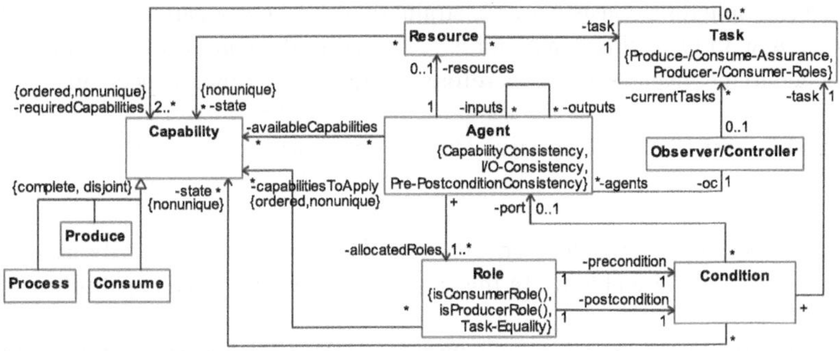

Fig. 7 Components of a Resource-Flow System

and consuming capabilities (*produce*, *process*, and *consume*). Consequently, the task is a sequence of capabilities beginning with a producing capability and ending with a consuming capability. Furthermore, the agent knows a couple of agents he can interact with and hand over resources (in case of the production cell, the workpieces). This is encapsulated in the *inputs* and *outputs* relation. The *role* concept is introduced to define correct resource-flows through the system. This means an agent has roles allocated telling him from which agent he receives the resource (*precondition/port*), which capabilities to apply, and then to which agent to hand over the resource (*postcondition/port*). Thus, the roles establish the connections between the agents and the combination of all roles forms the resource-flow. A system configuration is then a specific set of roles allocated to the agents (in this case robots and carts). For more details on the SE process and modeling of self-organizing resource-flow systems see [30]. In this case study, self-organization is done by role allocation. In case of a failure the system calculates a new valid set of roles, which is sufficient to fulfill the task again.

4.3 Specifying Self-x through Behavioral Corridors

To receive correct behavior the allocation of roles to the agents is curbed as already mentioned in 3.4 by the specification of behavioral corridors. This is realized with OCL constraints, which are annotated to the ODP (Fig. 7). This means the configuration of the system which is planned by the Organic Planning Layer is restricted to configurations within the specified corridor. This is sufficient for a correct behavior, as the execution semantics of roles is predefined. In other words it is specified how roles are executed. Therefore the challenge is to restrict the roles which are assigned to the robots or carts in such a way that the resulting behavior leads to the desired system goal – in our example the correct production of the workpiece and the completion of the defined task.

One example for a consistency constraint for the robots or carts is the *Capability-Consistency*. In OCL this constraint is evaluated in the context of an agent as it is annotated to the *agent* concept, therefore *self* refers to a robot respectively cart.

```
(self.availableCapabilities -> includesAll(
   self.allocatedRoles.capabilitiesToApply))
```

The *Capability-Consistency* ensures that the robots and carts only accept and perform roles they are able to. In this case, only roles that need capabilities which are available.

Another interesting consistency constraint is the *I/O-Consistency* for a robot or cart (referred as `self`):

```
    (self.inputs -> includesAll(
       self.allocatedRoles.precondition.port))
and (self.outputs -> includesAll(
       self.allocatedRoles.postcondition.port))
```

The agents know a couple of neighboring agents. They are documented in their "input" and "output" relation. These indicate with whom robots and carts can exchange workpieces. The pre- and postconditions of the particular roles determine from which robot/cart the workpiece comes from and to which the workpiece should be given. The *I/O-Consistency* indicates that the robots/carts recognize the breakdown of their required partners in the actual resource-flow. The required partners are the ports in the pre- and postconditions of the roles and must be part of the input respectively output relation of the agents. During runtime the robots/carts ping these neighbors to ensure that they are still available for receiving workpieces.

The carts are constrained in the way that they only take transport assignments between robots that are reachable for them. All these constraints can be monitored during runtime by the agents themselves as they can be evaluated locally. In general, also quantitative constraints for a configuration can be of interest, such as the assigned capabilities will not exceed a defined load or that the throughput has a certain threshold. These constraints usually are not monitored as they are violated if a failure occurs, which implies a previous violation of another monitored constraint.

More details about specification of behavioral corridors by constraints can be found in [8].

4.4 System Behavior at Runtime

The system starts with an initially calculated role allocation as depicted in Fig. 6. The needed tool applications are spread to the robots and the carts are assigned different routes to move the workpieces around. If a failure occurs, e.g. the drill tool of the drilling robot breaks, the robot monitors this violation of its *Capability-Consistency* constraint and starts a reconfiguration. It collects information about the neighboring robots and carts, calculates a new distribution of tool assignments and re-routes the carts in a way that production can continue. A traditional system would stop and a human interaction would be needed here. The reconfigured situation is depicted in Fig. 8.

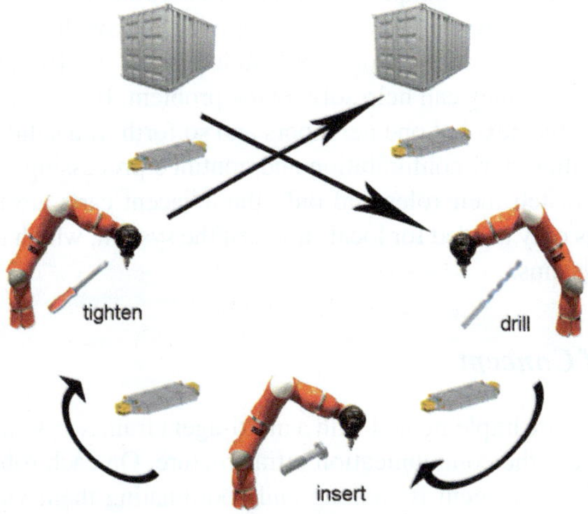

Fig. 8 Adaptive production cell after reconfiguration

In this case study, the robots and carts have only local rules and interaction possibilities. The resulting system is a self-organizing production cell which is capable of reacting to changes in the environment and new work plans. The configurations which are calculated by the robots or carts in case of a local constraint violation (e.g. capability or input, output loss) fulfills the constraints specified for the system. This means that the occurring emergence is restricted to positive emergence as claimed in Sect. 2.1.

4.5 Realizing Self-reconfiguration

There are several possibilities to implement self-reconfiguration. Currently, reconfiguration is done using a constraint solver, here Kodkod [35], to receive valid configuration for each robot and cart. Therefore the actual system state and the OCL constraints are converted into a formal model representing a constraint satisfaction problem (CSP). The model can directly be derived from the design pattern and the annotated constraints (see Sect. 4.2). This solver then tries to find a solution fulfilling all constraints, which is then spread to the agents. More details about the transformation and the use of constraint solvers for this class of systems can be found in [21, 22].

The advantage of integrating common techniques in contrast to stochastic algorithms is, that it is easier to give behavioral guarantees and ensure correct reconfiguration.

But also heuristic and stochastic algorithms, like genetic algorithms [7], can be used to realize the self-reconfiguration. For large problems they are often faster and allow to integrate self-optimization by defining adequate fitness functions, which also takes the quality (e.g. load balancing or throughput) of a solution into account.

For the production cell example a distributed coordination mechanism was developed, which realizes reconfiguration by applying a wave-like self-organization strategy [33]. The agent which recognizes a failure starts a self-reconfiguration by asking its neighbors if they can help solving the problem. If not, the search is propagated forward to the next but one neighbors and so forth. If a solution is found the agents switch to their new configuration and continue processing. In the best case two agents just switch their roles and only the adjacent carts are re-routed. Here reconfiguration is only needed for local subset of the system, which is advantageous in larger scale systems.

4.6 *Proof of Concept*

The organic layers are implemented with a multi-agent framework called Jadex [27], which also provides the communication infrastructure. On each robot and cart one Organic Control Layer agent is running and coordinating them via the interfaces provided by the Robot Programming Layer. Whenever a failure occurs or a reconfiguration request of another agent is received it spawns an Organic Planning Layer agent which then is handling the reconfiguration for this agent. There are different implementations (see 4.5) which are integrated into these planning layer agent and can be used for reconfiguration. The reconfiguration is based only on local knowledge and after successful reconfiguration the Organic Planning Layer terminates itself. Thus, there is no global knowledge base generated during runtime.

For the production cell scenario we implemented a prototypical implementation using Microsoft Robotic Studio. It provides a physical simulation environment for robotic applications and allows for prototypical testing of the developed concepts. A first version is described in [14].

5 Conclusion

In the field of Organic Computing, we were looking at the domain of production automation, in particular the field of adaptive production cells. Further in robotics research, we looked at facilitating the software development for industrial robots and improving software quality. In this chapter we presented how both worlds can fit together and how Organic Computing principles can be used to realize a flexible automation system. To be more concise, how the architecture of a self-organizing robotic cell can look like and how it can be implemented.

The lower layers were prototypically substituted by a simulation and coupled to the implementation of the organic layers within a multi-agent system, as described in Sect. 4.6. Nevertheless, these systems can benefit from the application of Organic Computing principles, especially in terms of failure tolerance and flexibility.

One major advantage of the proposed architecture and its implementation is that it is formally grounded and, therefore, allows to give behavioral guarantees with respect to the assigned configurations, which always leads to correct processing of the resources. The definition of a behavioral corridor and the assurance of remaining inside this corridor allows flexibility and gives firm guarantees about the system, which is very important for the acceptance in industrial applications. However, the drawback of moving self-organization into high-level layers is that no real-time critical behavior can be considered. Hence, only non real-time critical reconfigurations are possible.

Different production scenarios and factory settings need diverse reconfiguration mechanisms, e.g. completely decentralized coalition formation or wave propagations. We are currently working on different plug-ins for the Organic Planning Layer to enhance it by several reconfiguration algorithm implementations. In order to meet the flexibility requirements as proposed in Sect. 2, robotic software architecture (see [12]) which corresponds to both the Robotic Programming Layer and the Robotic Control Layer is currently extended.

Acknowledgements. This work has been partly sponsored by the priority program *Organic Computing* (SPP OC 1183) of the German research foundation (DFG).

References

1. Angerer, A., Hoffmann, A., Schierl, A., Vistein, M., Reif, W.: The Robotics API: An object-oriented framework for modeling industrial robotics applications. In: Proceedings of the IEEE/RSJ International Conference on Intelligent Robots and Systems (IROS 2010), Taipeh, Taiwan. IEEE Computer Society Press, Los Alamitos (2010)
2. Black, J.T., Musunur, L.P.: Robotic manufacturing cells. In: Nof, S. (ed.) Handbook of Industrial Robotics, ch.35, pp. 697–716. John Wiley & Sons, Hoboken (1999)
3. Branke, J., Mnif, M., Müller-Schloer, C., Prothmann, H., Richter, U., Rochner, F., Schmeck, H.: Organic Computing – Addressing complexity by controlled self-organization. In: Proceedings of the 2nd International Symposium on Leveraging Applications of Formal Methods, Verification and Validation (ISoLA 2006), Paphos, Cyprus, pp. 185–191. IEEE Computer Society Press, Los Alamitos (2006)
4. De Schutter, J., De Laet, T., Rutgeerts, J., Decré, W., Smits, R., Aertbeliën, E., Claes, K., Bruyninckx, H.: Constraint-based task specification and estimation for sensor-based robot systems in the presence of geometric uncertainty. Int. J. Rob. Res. 26(5), 433–455 (2007), doi:http://dx.doi.org/10.1177/027836490707809107
5. Finkemeyer, B., Kröger, T., Wahl, F.M.: Executing assembly tasks specified by manipulation primitive nets. Advanced Robotics 19(5), 591–611 (2005)
6. Ganek, A.G., Corbi, T.A.: The dawning of the autonomic computing era. IBM Systems Journal 42(1), 5–18 (2003)
7. Goldberg, D.E.: Genetic Algorithms in Search, Optimization, and Machine Learning, 1st edn. Addison-Wesley Professional, Reading (1989)
8. Güdemann, M., Nafz, F., Ortmeier, F., Seebach, H., Reif, W.: A specification and construction paradigm for organic computing systems. In: Proceedings of the Second IEEE International Conference on Self-Adaptive and Self-Organizing Systems (SASO 2008), Venice, Italy, pp. 233–242. IEEE Computer Society Press, Los Alamitos (2008)

9. Güdemann, M., Ortmeier, F., Reif, W.: Safety and dependability analysis of self-adaptive systems. In: Proceedings of ISoLA 2006. IEEE CS Press, Los Alamitos (2006)
10. Hägele, M., Nilsson, K., Pires, J.N.: Industrial robotics. In: Siciliano, B., Khatib, O. (eds.) Springer Handbook of Robotics, ch.42, pp. 963–986. Springer, Heidelberg (2008)
11. Hägele, M., Skordas, T., Sagert, S., Bischoff, R., Brogårdh, T., Dresselhaus, M.: Industrial robot automation. White paper, European Robotics Network (2005)
12. Hoffmann, A., Angerer, A., Ortmeier, F., Vistein, M., Reif, W.: Hiding real-time: A new approach for the software development of industrial robots. In: Proceedings of the 2009 IEEE/RSJ International Conference on Intelligent Robots and Systems (IROS 2009), pp. 2108–2113. IEEE Computer Society Press, St. Louis (2009)
13. Hoffmann, A., Angerer, A., Schierl, A., Vistein, M., Reif, W.: Towards object-oriented software development for industrial robots. In: Proceedings of the 7th International Conference on Informatics in Control, Automation and Robotics (ICINCO 2010). INSTICC Press, Funchal (2010)
14. Hoffmann, A., Nafz, F., Ortmeier, F., Schierl, A., Reif, W.: Prototyping plant control software with microsoft robotics studio. In: Proceedings of the Third International Workshop on "Software Development and Integration in Robotics" (SDIR-III). IEEE Computer Society Press, Los Alamitos (2008)
15. Kephart, J., Chess, D.: The vision of autonomic computing. Computer 36(1), 41–50 (2003)
16. Mason, M.: Compliance and force control for computer-controlled manipulators. IEEE Transactions on Systems, Man, and Cybernetics 11(6), 418–432 (1981)
17. Mehrabi, M., Ulsoy, A., Koren, Y., Heytler, P.: Trends and perspectives in flexible and reconfigurable manufacturing systems. Journal of Intelligent Manufacturing 13(2), 135–146 (2002)
18. Müller-Schloer, C.: Organic computing: on the feasibility of controlled emergence. In: Proceedings of the 2nd IEEE/ACM/IFIP International Conference on Hardware/Software Codesign and System Synthesis, Stockholm, Sweden, pp. 2–5. ACM, New York (2004)
19. Müller-Schloer, C., Sick, B.: Controlled emergence and self-organization. In: Würtz (ed.) [39], pp. 81–104.
20. Müller-Schloer, C., von der Malsburg, C., Würtz, R.P.: Organic computing. Informatik Spektrum 27(4), 332–336 (2004)
21. Nafz, F., Ortmeier, F., Seebach, H., Steghöfer, J.P., Reif, W.: A generic software framework for role-based organic computing systems. In: Proc. Intl. Workshop on Software Engineering for Adaptive and Self-Managing Systems, pp. 96–105 (2009)
22. Nafz, F., Ortmeier, F., Seebach, H., Steghöfer, J.-P., Reif, W.: A universal self-organization mechanism for role-based organic computing systems. In: González Nieto, J., Reif, W., Wang, G., Indulska, J. (eds.) ATC 2009. LNCS, vol. 5586, pp. 17–31. Springer, Heidelberg (2009), http://dx.doi.org/10.1007/978-3-642-02704-8-3
23. Nafz, F., Seebach, H., Steghöfer, J.P., Bäumler, S., Reif, W.: A Formal Framework for Compositional Verification of Organic Computing Systems. In: Proceedings of the seventh International Conference on Autonomic and Trusted Computing, ATC-2010 (2010)
24. Okamura, A., Smaby, N., Cutkosky, M.: An overview of dexterous manipulation. In: Proceedings of the 2000 IEEE International Conference on Robotics and Automation (ICRA 2000), pp. 255–262. IEEE Computer Society Press, San Francisco (2000)
25. OMG. Object Constraint Language, OMG Available Specification (2006)
26. Pires, J.N.: New challenges for industrial robotic cell programming. Industrial Robot 36(1) (2009)

27. Pokahr, A., Braubach, L., Lamersdorf, W.: Jadex: A bdi reasoning engine. In: Bordini, R., Dastani, M., Dix, G., Seghrouchni, A.E.F. (eds.) Multi-Agent Programming, pp. 149–174. Springer, Heidelberg (2005); Book chapter
28. Richter, U., Mnif, M., Branke, J., Müller-Schloer, C., Schmeck, H.: Towards a generic observer/controller architecture for organic computing. In: GI Jahrestagung, vol. (1), pp. 112–119 (2006)
29. Schild, K., Bussmann, S.: Self-organization in manufacturing operations. Commun. ACM 50(12), 74–79 (2007), `http://doi.acm.org/10.1145/1323688.1323698`
30. Seebach, H., Nafz, F., Steghöfer, J.P., Reif, W.: A software engineering guideline for self-organizing resource-flow systems. In: Proceedings of the Fourth IEEE International Conference on Self-Adaptive and Self-Organizing Systems (SASO 2010), Budapest, Hungary. IEEE Computer Society Press, Los Alamitos (2010)
31. Seebach, H., Ortmeier, F., Reif, W.: Design and construction of organic computing systems. In: Proceedings of the IEEE Congress on Evolutionary Computation (CEC 2007), Singapore, pp. 4215–4221. IEEE Computer Society Press, Los Alamitos (2007)
32. Smits, R., De Laet, T., Claes, K., Bruyninckx, H., De Schutter, J.: iTASC: A tool for multi-sensor integration in robot manipulation. In: Proceedings of the IEEE International Conference on Multisensor Fusion and Integration for Intelligent Systems (MFI 2008), Seoul, Korea, pp. 426–433. IEEE Computer Society Press, Los Alamitos (2008)
33. Sudeikat, J., Steghöfer, J.P., Seebach, H., Renz, W., Preisler, T., Salchow, P., Reif, W.: Design and simulation of a wave-like self-organization strategy for resource-flow systems. In: 4th International Workshop on Multi-Agent Systems and Simulation (MAS&S) (2010) (accepted)
34. Thomas, U., Finkemeyer, B., Kröger, T., Wahl, F.M.: Error-tolerant execution of complex robot tasks based on skill primitives. In: Proceedings of the 2003 IEEE International Conference on Robotics and Automation (ICRA 2003), Taipei, Taiwan, pp. 3069–3075. IEEE Computer Society Press, Los Alamitos (2003)
35. Torlak, E., Jackson, D.: Kodkod: A relational model finder. In: Grumberg, O., Huth, M. (eds.) TACAS 2007. LNCS, vol. 4424, pp. 632–647. Springer, Heidelberg (2007), `http://dx.doi.org/10.1007/978-3-540-71209-1_49`
36. Tsang, E.: Foundations of constraint satisfaction (1993)
37. Vesely, W.E., Goldberg, F.F., Roberts, N.H., Haasl, D.F.: Fault Tree Handbook. U.S. Nuclear Regulatory Commission, Washington, DC (1981)
38. Vistein, M., Angerer, A., Hoffmann, A., Schierl, A., Reif, W.: Interfacing industrial robots using realtime primitives. In: Proceedings of the 2010 International Conference on Automation and Logistics (ICAL 2010), Hong Kong, China. IEEE Computer Society Press, Los Alamitos (2010)
39. Würtz, R.P. (ed.): Organic Computing (Understanding Complex Systems). Springer, Heidelberg (2008)
40. Zaeh, M., Ostgathe, M.: A multi-agent-supported, product-based production control. In: Proceedings of the 7th IEEE International Conference on Control and Automation, Christchurch, New Zealand, pp. 2376–2383. IEEE Computer Society Press, Los Alamitos (2009)

Author Index

GPSR Compliance

The European Union's (EU) General Product Safety Regulation (GPSR) is a set of rules that requires consumer products to be safe and our obligations to ensure this.

If you have any concerns about our products, you can contact us on ProductSafety@springernature.com

In case Publisher is established outside the EU, the EU authorized representative is:

Springer Nature Customer Service Center GmbH
Europaplatz 3
69115 Heidelberg, Germany

Batch number: 10755837

Printed by Printforce, the Netherlands